Algebra

Volume II

Springer
New York
Berlin
Heidelberg
Hong Kong
London
Milan
Paris
Tokyo

B.L. van der Waerden

Algebra

Volume II

Based in part on lectures by E. Artin and E. Noether

Translated by John R. Schulenberger

Springer

B.L. van der Waerden
University of Zürich (*retired*)

Present address:
Wiesliacher 5
(8053) Zürich, Switzerland

Originally published in 1970 by Frederick Ungar Publishing Co., Inc., New York

Volume II is translated from the German *Algebra II*, fifth edition, Springer-Verlag Berlin, 1967.
The work was first published with the title *Moderne Algebra* in 1930-1931.

Mathematics Subject Classification (2000): 00A05 01A75 12-01 13-01 15-01 16-01

ISBN 0-387-40625-5 Printed on acid-free paper.

First softcover printing, 2003.

Printed in the United States of America.

9 8 7 6 5 4 3 2 1 SPIN 10947678

www.springer-ny.com

Springer-Verlag New York Berlin Heidelberg
A member of BertelsmannSpringer Science+Business Media GmbH

PREFACE TO THE FIFTH EDITION

P. Roquette has been kind enough to provide me with a nice proof of the residue theorem for algebraic differentials *udz*. The chapter "Algebraic Functions" has thereby been brought to a satisfactory conclusion.

In the chapter "Topological Algebra," following Bourbaki, the completion of groups, rings, and fields has been carried out by means of filters without using the second countability axiom.

The chapter "Linear Algebra," which is important for many applications, now appears at the beginning of the volume, and topological algebra is treated in the last chapter. The book now consists of three independent groups of three chapters each:

Chapters 12–14:	Linear Algebra, Algebra, Representation Theory
Chapters 15–17:	Ideal Theory
Chapters 18–20:	Fields with Valuations, Algebraic Functions, Topological Algebra

This subdivision of the material is now expressed more clearly in the schematic guide on page xv.

Zurich, March 1967 B. L. VAN DER WAERDEN

FROM THE PREFACE TO THE FOURTH EDITION

Two new chapters have been added at the beginning of the second volume: a chapter on algebraic functions of one variable, which goes as far as the Riemann-Roch theorem for arbitrary fields of constants, and a chapter on topological algebra, which is mainly concerned with the completion of topological groups, rings, and skew fields. I should like to thank Dr. H. R. Fischer, who read these two chapters in manuscript form, for many useful remarks.

The chapter "General Ideal Theory" has been extended to include the important theorems of Krull on symbolic powers of prime ideals and chains of prime ideals. The relation of the ideal theory of integrally closed rings with valuation theory has been brought out more clearly. A section on antisymmetric bilinear forms has been added to the chapter "Linear Algebra."

In the chapter "Algebras," more examples are given, the theory of the radical has been developed, following Jacobson, without a finiteness condition, and the fundamental ideas of Emmy Noether on direct sums and intersections of modules have been more strongly emphasized. It was possible to considerably simplify the proofs of the principal theorems by combining the methods of Jacobson with those of Emmy Noether.

By omitting some material I have tried to keep the size of the book within reasonable bounds. Thus, the chapter "Elimination Theory" has been omitted. The theorem on the existence of the resultant system for homogeneous equations, which was formerly proved by means of elimination theory, now appears in Section 121 as a corollary of Hilbert's Nullstellensatz.

Zurich, June 1959 B. L. van der Waerden

GUIDE

A survey of the chapters of Volumes 1 and 2 and their logical dependence.

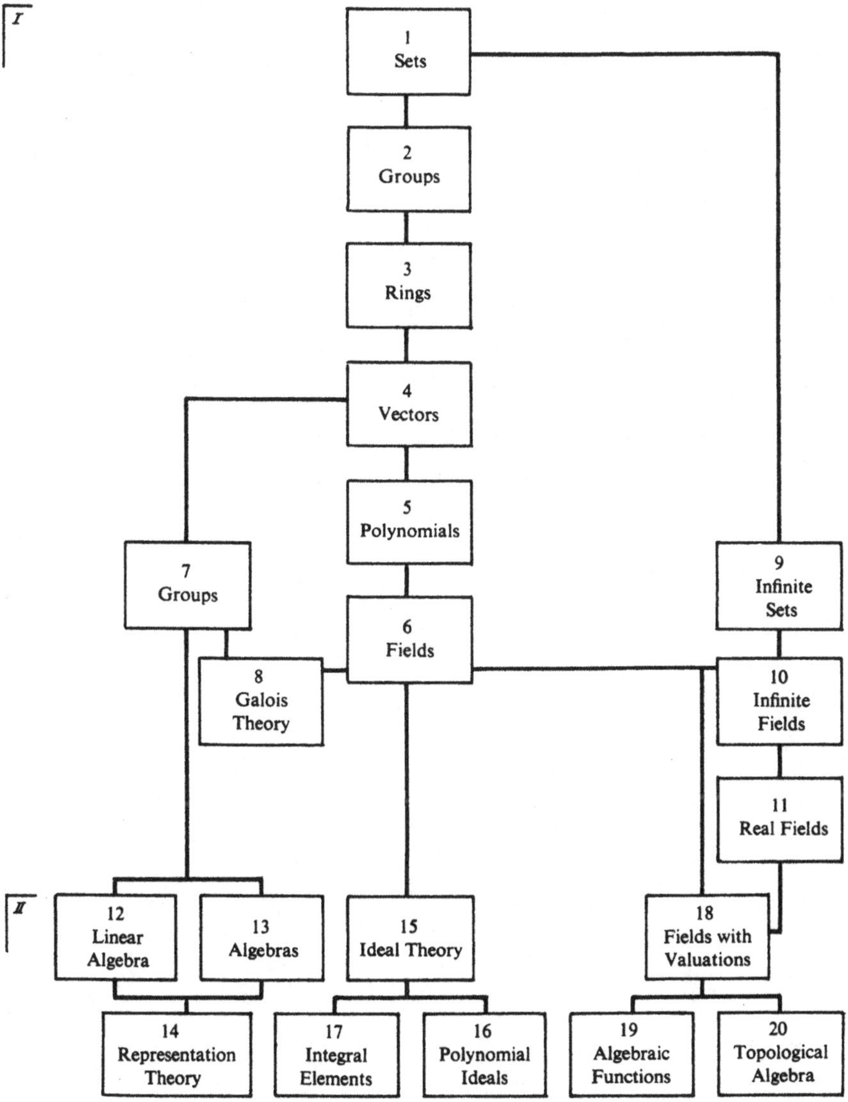

CONTENTS

Chapter 17

INTEGRAL ALGEBRAIC ELEMENTS 168

Chapter 18

FIELDS WITH VALUATIONS 191

Chapter 19

ALGEBRAIC FUNCTIONS OF ONE VARIABLE 223

Chapter 20

TOPOLOGICAL ALGEBRA 252

Chapter 12

LINEAR ALGEBRA

Linear algebra deals with modules and their homomorphisms and, in particular, with vector spaces and the linear transformations of vector spaces. In Section 12.3 the Fundamental Theorem of Abelian Groups is proved as an application of module theory. Section 12.7 deals with quadratic forms, and Section 12.8 with antisymmetric bilinear forms.

Chapter 12 is based entirely on the theory of groups with operators (Chapter 7).

12.1 MODULES OVER A RING

Let $\mathfrak{R}$ be a ring with identity element ε, and let $\mathfrak{M}$ be a right $\mathfrak{R}$-module, that is, an additive group with $\mathfrak{R}$ as operator domain. The elements of $\mathfrak{M}$ will be denoted by Latin letters and those of $\mathfrak{R}$ by Greek letters. The composition rules are those of an additive group and the following:

$$(a+b)\lambda = a\lambda+b\lambda$$

$$a(\lambda+\mu) = a\lambda+a\mu$$

$$a\cdot\lambda\mu = a\lambda\cdot\mu.$$

The distributive laws imply, as usual, the same laws for subtraction, the multiplicative properties of the minus sign, and the fact that a product is zero if a factor is zero (whether it is the zero element of $\mathfrak{R}$ or the zero element of $\mathfrak{M}$).

The fact that the multipliers are written on the right is entirely arbitrary. All theorems to be proved also hold with the multipliers written on the left.

The identity element of $\mathfrak{R}$ need not be the identity operator; $a\varepsilon$ may be different from a for certain a. (For example, all the composition rules are met if we put $a\lambda = 0$ for all a and all λ.) However, it is always the case that

$$a = (a-a\varepsilon)+a\varepsilon. \tag{12.1}$$

The first term $a-a\varepsilon$ is annihilated by the right factor ε; the second term is reproduced on multiplication by ε. The first terms form a submodule M_0 of $\mathfrak{M}$ which is annihilated by ε and therefore also by every element $\varepsilon\lambda$ of $\mathfrak{R}$; the second factors form a submodule $\mathfrak{M}_1$ for which ε is the identity operator. These two submodules have only the zero element in common, since for any other element annihilation and reproduction are mutually exclusive. The representation (12.1) shows, moreover, that $\mathfrak{M}$ is the direct sum $\mathfrak{M}_0 + \mathfrak{M}_1$. After the uninteresting part $\mathfrak{M}_0$ of $\mathfrak{M}$ is split off, we obtain a module for which ε is the identity operator. *We shall therefore assume in the following that the identity element of $\mathfrak{R}$ is also the identity operator for $\mathfrak{M}$.*

If, in particular, $\mathfrak{R}$ is a skew field, then $\mathfrak{M}$ is a vector space over $\mathfrak{R}$ in the sense of Section 4.1, Volume I.

The module $\mathfrak{M}$ is said to be *finite over* $\mathfrak{R}$ if its elements can be expressed linearly in terms of finitely many basis elements $u_1, \ldots, u_n$:

$$u_1\lambda_1 + \cdots + u_n\lambda_n. \tag{12.2}$$

In this case $\mathfrak{M}$ is the sum of the submodules $u_1\mathfrak{R}, \ldots, u_n\mathfrak{R}$:

$$\mathfrak{M} = (u_1\mathfrak{R}, \ldots, u_n\mathfrak{R}). \tag{12.3}$$

Instead of (12.3) we sometimes write for brevity:

$$\mathfrak{M} = (u_1, \ldots, u_n).$$

If in the representation (12.2) the coefficients $\lambda_1, \ldots, \lambda_n$ are uniquely determined by u, then $\mathfrak{M}$ is called a *module of linear forms* over $\mathfrak{R}$. In this case the sum (12.3) is direct:

$$\mathfrak{M} = u_1\mathfrak{R} + \cdots + u_n\mathfrak{R}.$$

Every finite-dimensional vector space is a module of linear forms, since by Section 4.1 we can always choose a linearly independent basis $(u_1, \ldots, u_n)$. By Section 4.2 the dimension n is independent of the choice of basis.

An operator homomorphism which maps a module of linear forms $\mathfrak{M} = (u_1, \ldots, u_m)$ into a module of linear forms $\mathfrak{N} = (v_1, \ldots, v_n)$ is called a *linear transformation of* $\mathfrak{M}$ *into* $\mathfrak{N}$. For such a transformation A, therefore, we have, as in Section 4.5,

$$A(x+y) = Ax + Ay$$

$$A(x\lambda) = (Ax)\lambda.$$

The transformation A is completely determined if the image of each basis element u_k,

$$Au_k = \sum u_i\alpha_{ik},$$

is given. The coefficients α_{ik} form a *matrix* of the transformation A.

If A is a one-to-one mapping of $\mathfrak{M}$ onto $\mathfrak{N}$, then there exists an inverse mapping A^{-1}. We then have

$$A^{-1}A = 1 \quad \text{and} \quad AA^{-1} = 1,$$

where 1 denotes the identity. In this case the mapping A and its matrix (α_{ik}) are called *invertible*.

In the following we shall often denote the linear transformation A and its matrix (α_{ik}) by the same letter A. This is not altogether logical, but it is practical.

12.2 MODULES OVER EUCLIDEAN RINGS. ELEMENTARY DIVISORS

We now require that the ring $\mathfrak{R}$ be commutative and Euclidean in the sense of Section 3.7. This means that to every ring element $a \neq 0$ there corresponds an "absolute value" $g(a)$ such that $g(ab) \geqq g(a)$ and also that a process of division is possible. According to Section 3.7, every ideal in $\mathfrak{R}$ is then a principal ideal.

Theorem: *Let $\mathfrak{M}$ be a module of linear forms over $\mathfrak{R}$ with basis $(u_1, \ldots, u_n)$. Then every submodule $\mathfrak{N}$ of $\mathfrak{M}$ is again a module of linear forms with at most n basis elements.*

Proof: For the null module $\mathfrak{M} = (0)$ the theorem is trivial. Suppose then that it is true for modules $\mathfrak{M}$ with $n-1$ basis elements.

If $\mathfrak{N}$ consists of linear forms in $u_1, \ldots, u_{n-1}$ only, then the theorem is true by the induction hypothesis. If $\mathfrak{N}$ contains a linear form $u_1\lambda_1 + \cdots + u_n\lambda_n$ with $\lambda_n \neq 0$, then the set of such λ_n forms a right ideal in $\mathfrak{R}$ which is thus a principal ideal (μ_n) with $\mu_n \neq 0$. Therefore $\mathfrak{N}$ contains a form $l = u_1\mu_1 + \cdots u_n\mu_n$; by subtracting an appropriate multiple $l\alpha$ of l from any other form $u_1\lambda_1 + \cdots + u_n\lambda_n$, the last coefficient λ_n can be eliminated. The linear forms of $\mathfrak{N}$ in $u_1, \ldots, u_{n-1}$ which then remain form a submodule, and by the induction hypothesis this submodule has a linearly independent basis $(l_1, \ldots, l_{m-1})$ with $m-1 \leqq n-1$. Clearly $\mathfrak{N}$ is generated by $l_1, \ldots, l_{m-1}, l$.

Now $l_1, \ldots, l_{m-1}$ are already linearly independent. If there were a linear dependence

$$l_1\beta_1 + \cdots + l_{m-1}\beta_{m-1} + l\beta = 0$$

with $\beta \neq 0$, then on equating coefficients of u_n it would follow that $\mu_n\beta = 0$, which is impossible.

Exercises

12.1 If $\mathfrak{M}$ is a module of integral linear forms and if the submodule $\mathfrak{N}$ is generated by finitely many linear forms $v_k = \sum u_i\alpha_{ik}$, then a basis $(l_1, \ldots, l_m)$ with the properties above can be constructed in a finite number of steps.

12.2 Using the basis $(l_1, \ldots, l_m)$ constructed in Exercise 12.1, give a method of determining whether a particular linear form $u_1\beta_1 + \cdots + u_n\beta_n$ is contained in the module $\mathfrak{N}$, that is, whether the linear system of diophantine equations

$$\sum \alpha_{ik}\xi_k = \beta_i$$

is solvable in terms of integers ξ_k.

Theorem on Elementary Divisors: *If $\mathfrak{N}$ is a submodule of the module of linear forms $\mathfrak{M}$, then there exists a basis $(u_1, \ldots, u_n)$ of $\mathfrak{M}$ and a basis $(v_1, \ldots, v_m)$ of $\mathfrak{N}$ such that*

$$v_i = u_i\varepsilon_i$$

$$\varepsilon_{i+1} \equiv 0(\varepsilon_i). \tag{12.4}$$

Proof: We start with arbitrary bases $(u_1, \ldots, u_n)$ of $\mathfrak{M}$ and $(v_1, \ldots, v_m)$ of $\mathfrak{N}$. Let

$$v_k = \sum u_i\alpha_{ik}. \tag{12.5}$$

In matrix notation (12.5) reads

$$(v_1 \ldots v_m) = (u_1 \ldots u_n)\cdot A. \tag{12.6}$$

By stepwise change of basis we shall now bring the matrix A to the desired diagonal form

$$\begin{pmatrix} \varepsilon_1 & 0 & \ldots & 0 \\ 0 & \varepsilon_2 & & \cdot \\ \cdot & & \cdot & \cdot \\ \cdot & & \cdot & \cdot \\ \cdot & & \cdot & \cdot \\ \cdot & & & \varepsilon_m \\ 0 & \cdot & \ldots & 0 \end{pmatrix}. \tag{12.7}$$

The changes permitted are as follows:

1. Interchange of two u or v; this effects an interchange of two rows or columns of A.
2. Replacement of u_i by $u_i + u_j\lambda$ $(j \neq i)$, whereby the ith row of A multiplied on the left by λ is subtracted from the jth row:

$$v_k = \sum u_i\alpha_{ik} = \cdots + (u_i + u_j\lambda)\alpha_{ik} + \cdots + u_j(\alpha_{jk} - \lambda\alpha_{ik}) + \cdots.$$

3. Replacement of v_k by $v_k - v_j\lambda$ $(j \neq k)$, whereby the jth column of A multiplied on the right by λ is subtracted from the kth column:

$$v_k - v_j\lambda = \sum u_i(\alpha_{ik} - \alpha_{ij}\lambda).$$

We transform the matrix A using (1), (2), and (3) until *the nonzero element of A with smallest absolute value has the least possible absolute value.* We bring this least element in the matrix to the site α_{11} by means of operation 1. If the other elements of the first column are now made as small as possible by subtracting appropriate multiples of the first row according to (2), then they become less than $|\alpha_{11}|$ in absolute value and are therefore zero. Similarly, using (3), the elements of the first row are made to vanish without altering the first column. After these operations all the elements in the entire matrix must be divisible by α_{11}, for if some α_{ik} were not divisibly by α_{11}, then by the division algorithm

$$\alpha_{ik} = \alpha_{11}\beta + \gamma, \qquad \gamma \neq 0, \qquad g(\gamma) < g(\alpha_{11}).$$

If now the first row is added to the ith row using (2) and the first column multiplied by β is subtracted from the kth column by means of (3), then the element γ with $g(\gamma) < g(\alpha_{11})$ appears at the place (ik); this contradicts the minimality of α_{11}.

Our matrix now has the appearance

$$\begin{pmatrix} \alpha_{11} & 0 \ldots 0 \\ 0 & \\ \cdot & \\ \cdot & A' \\ \cdot & \\ 0 & \end{pmatrix},$$

where all elements in A' are multiples of α_{11}. Using operations which leave the first row and column unaltered, we now proceed with A' as previously with A. The divisibility of all elements by α_{11} is hereby not destroyed. Finally A' acquires the form

$$\begin{pmatrix} \alpha_{22} & 0 \ldots 0 \\ 0 & \\ \cdot & \\ \cdot & A'' \\ \cdot & \\ 0 & \end{pmatrix},$$

where all elements of A'' are divisible by α_{22}. Continuing in this manner, we obtain after m steps the desired normal form (12.7). It is not possible that one of the matrices $A, A', A'', \ldots$ should consist solely of zeros before the form (12.7) is obtained, since this would imply that certain of the v_k were equal to zero; on the contrary, at each stage of the process the v form a linearly independent basis of $\mathfrak{N}$. This completes the proof of the theorem.

Remark 1: Operations 1 through 3 amount to multiplying the matrix A on the right or left by an invertible matrix with elements in $\mathfrak{R}$. Indeed, if $(u_1', \ldots, u_n') = (u_1, \ldots, u_n) \cdot B$ and $(v_1', \ldots, v_m') = (v_1, \ldots, v_m) \cdot C$ are new bases, then

$$(v_1' \ldots v_m') = (v_1 \ldots v_m)C = (u_1 \ldots u_n)AC = (u_1' \ldots u_n')B^{-1}AC.$$

The theorem on elementary divisors is therefore equivalent to the existence of two invertible matrices B and C such that $B^{-1}AC$ is a matrix having the form (12.7).

Remark 2: The reduction of the matrix A proceeds in precisely the same manner even if the v do not form a linearly independent system. In this case, however, one of the matrices A, A', A'' may become the zero matrix, and we obtain instead of the normal form (12.7) the more general form

$$B^{-1}AC = \begin{pmatrix} \varepsilon_1 & & & & 0 \\ & \cdot & & & \\ & & \cdot & & \\ & & & \cdot & \\ & & & \varepsilon_r & \\ 0 & & & & 0 \end{pmatrix}, \tag{12.8}$$

where r is the rank of A. The divisibility relations of the ε_i remain the same.
Remark 3: The k-rowed subdeterminants of the transformed matrix $D = B^{-1}AC$ are linear functions of the subdeterminants of A; similarly, those of $A = BDC^{-1}$ are linear functions of the subdeterminants of D. Hence, up to units the greatest common divisor δ_k of the k-rowed subdeterminants of A is the same as for D. We easily obtain for D the value

$$\delta_k = \varepsilon_1 \varepsilon_2 \ldots \varepsilon_k \qquad (k \leqq r).$$

Therefore

$$\delta_k = \delta_{k-1}\varepsilon_k \qquad (1 < k \leqq r). \tag{12.9}$$

The δ_k are called the *determinant divisors* of the matrix A. The ε_k are called the *elementary divisors* of A.[1] From (12.9) it now follows that *the elementary divisors are the quotients of two successive determinant divisors.*
Remark 4: In the next section we shall find in another way that the elementary divisors are uniquely determined up to units by the matrix A. It will be shown that the elementary divisors (insofar as they are not units) depend only on the factor module $\mathfrak{M}/\mathfrak{N}$, which is in turn determined by A.

Exercise

12.3 Any linear system of diophantine equations

$$\sum_1^n \alpha_{ik}\xi_k = \beta_i \qquad (i = 1, \ldots, m) \tag{12.10}$$

with integers α_{ik} and β_i can be brought to the form

$$\varepsilon_i \eta_i = \gamma_i \qquad (i = 1, \ldots, r;\ \varepsilon_i \neq 0)$$
$$0 = \delta_j \qquad (j = r+1, \ldots, m)$$

by unimodular transformation of the unknowns and equations. The conditions for the solvability of the system in terms of integers are

$$\gamma_i \equiv 0(\varepsilon_i); \quad \delta_j = 0.$$

The η_i with $i \leqq r$ can be determined; the other η_j are arbitrary. The ξ_k are linear integral functions of the arbitrary η_j.

12.3 THE FUNDAMENTAL THEOREM OF ABELIAN GROUPS

Let $\mathfrak{G}$ be an Abelian group with finitely many generators; $\mathfrak{G}$ is a module, and composition will be written additively. If $\mathfrak{G}$ has an operator domain $\mathfrak{R}$, then we

[1] They are often called invariant factors in the English literature (*Trans.*).

assume that $\mathfrak{R}$ has an identity element which is also the identity operator; if no operator domain is given, we take the ring of integers as operator domain, which always satisfies the condition mentioned. We shall now write the operators on the left of the module elements.

First let $\mathfrak{G}$ be cyclic: $\mathfrak{G} = (g)$. The set of μ in $\mathfrak{R}$ which annihilate g is a left ideal $\mathfrak{a}$ in $\mathfrak{R}$: $\mu_1 g = 0$ and $\mu_2 g = 0$ imply $(\mu_1 - \mu_2)g = 0$, and $\mu g = 0$ implies $\kappa\mu g = 0$ for every κ in $\mathfrak{R}$. To each λ of $\mathfrak{R}$ there corresponds a λg in $\mathfrak{G}$; since

$$(\lambda+\mu)g = \lambda g + \mu g$$

$$\lambda\mu \cdot g = \lambda \cdot \mu g,$$

this correspondence is an operator homomorphism with respect to $\mathfrak{R}$. It follows therefore by the homomorphism theorem that

$$\mathfrak{G} \cong \mathfrak{R}/\mathfrak{a},$$

or in words: *a cyclic $\mathfrak{R}$-module $\mathfrak{G}$ is isomorphic to the factor module of $\mathfrak{R}$ by the annihilating left ideal of $\mathfrak{G}$.*

In the case of an ordinary cyclic group $\mathfrak{G}$ we thus again obtain the result that $\mathfrak{G}$ is isomorphic to the additive group of integers or to the group of residue classes modulo an integer. If $n > 0$ is the basis element of the ideal $\mathfrak{a}$, then n is the order of the cyclic group (g) or the order of the group element g.

The theorem just proved holds without special conditions on $\mathfrak{R}$. However, if $\mathfrak{R}$ is commutative and Euclidean, as we shall assume in the following, then we can say still more. The ideal $\mathfrak{a}$ is then a principal ideal: $\mathfrak{a} = (\alpha)$. We assume that $\alpha \neq 0$ and, if possible, split α into two relatively prime factors:

$$\alpha = \rho\sigma$$

$$1 = \lambda\rho + \mu\sigma.$$

If we form the cyclic groups $\mathfrak{G}_1 = (\rho g)$ and $\mathfrak{G}_2 = (\sigma g)$, then $\mathfrak{G}_1$ is annihilated by σ and $\mathfrak{G}_2$ by ρ. Since

$$g = \lambda\rho g + \mu\sigma g,$$

$\mathfrak{G}$ is the sum of $\mathfrak{G}_1$ and $\mathfrak{G}_2$. The intersection $\mathfrak{G}_1 \cap \mathfrak{G}_2$ is annihilated by both ρ and σ and therefore by $\lambda\rho + \mu\sigma = 1$; hence $\mathfrak{G}_1 \cap \mathfrak{G}_2 = (0)$, and the sum is direct:

$$\mathfrak{G} = \mathfrak{G}_1 + \mathfrak{G}_2.$$

If σ or ρ can be split into further relatively prime factors, then $\mathfrak{G}_1$ or $\mathfrak{G}_2$ can be further decomposed. *The cyclic group $\mathfrak{G}$ finally becomes a direct sum of cyclic groups which are annihilated by powers of prime numbers.*[2] *The product of these prime powers is α.* Groups of this kind will be called *prime-power groups.*

[2]"Prime number" is short for "prime element of the ring $\mathfrak{R}$." In the case of ordinary Abelian groups this is an ordinary prime number.

We now proceed to the general case in which $\mathfrak{G}$ is an $\mathfrak{R}$-module with finitely many generators $g_1, \ldots, g_n$; the elements of $\mathfrak{G}$ then have the form

$$\lambda_1 g_1 + \cdots + \lambda_n g_n.$$

If we form the module of linear forms

$$\mathfrak{M} = (u_1, \ldots, u_n)$$

with indeterminates $u_1, \ldots, u_n$, then to each element $\sum \lambda_i g_i$ of $\mathfrak{G}$ there corresponds a linear form $\sum \lambda_i u_i$ of $\mathfrak{M}$; the correspondence is again a module homomorphism, and it follows from the homomorphism theorem that

$$\mathfrak{G} \cong \mathfrak{M}/\mathfrak{N},$$

where $\mathfrak{N}$ is the submodule of those linear forms $\sum \lambda_i u_i$ for which $\sum \lambda_i g_i = 0$.

We again assume that $\mathfrak{R}$ is Euclidean. By section 12.2 we can choose new bases $(v_1, \ldots, v_m)$ and $(u_1', \ldots, u_n')$ $(n \geqq m)$ for $\mathfrak{N}$ and $\mathfrak{M}$ such that

$$v_i = \varepsilon_i u_i' \qquad \text{for} \quad i = 1, \ldots, m$$

$$\varepsilon_{i+1} \equiv 0(\varepsilon_i).$$

The u' again correspond (under the homomorphism above) to elements $h_1, \ldots, h_n$ of $\mathfrak{G}$. All elements of $\mathfrak{G}$ have the form $\mu_1 h_1 + \cdots + \mu_n h_n$, and such an element is zero if and only if

$$\mu_1 \mu_1' + \cdots + \mu_n u_n' \equiv 0(v_1, \ldots, v_m),$$

that is,

$$\begin{aligned} \mu_1 &\equiv 0(\varepsilon_1), \qquad & \mu_{m+1} &= 0 \\ &\cdots & &\cdots \\ \mu_m &\equiv 0(\varepsilon_m), \qquad & \mu_n &= 0. \end{aligned}$$

A sum $\mu_1 h_1 + \cdots + \mu_n h_n$ is thus zero only if its individual terms are zero, and these are zero only if their coefficients μ_i are divisible by ε_i for $i = 1, \ldots, m$ and are zero for $i = m+1, \ldots, n$.

This may be expressed as the following theorem.

Theorem: *The group $\mathfrak{G}$ is the direct sum of cyclic groups $(h_1) + \cdots + (h_n)$, and the annihilating ideal of (h_i) is*

$$(\varepsilon_i) \qquad \text{for} \quad i = 1, \ldots, m$$

$$(0) \qquad \text{for} \quad i = m+1, \ldots, n.$$

This is the *Fundamental Theorem of Abelian Groups with Finitely Many Generators.*

In the case of ordinary Abelian groups the $|\varepsilon_i|$ are the orders of the cyclic groups $(h_1), \ldots, (h_m)$, and the other groups $(h_{m+1}), \ldots, (h_n)$ have infinite order.

Three supplements to the theorem are still required:

1. Elimination of units among the ε_i.
2. Further decomposition of the cyclic groups into prime-power groups.
3. Uniqueness.

1. Suppose that ε_1, for instance, is a unit; (ε_1) is then the unit ideal $\mathfrak{R}$ and hence $\mathfrak{R}h_1 = (0)$. The cyclic group $\mathfrak{R}h_1$ may therefore be omitted from the direct-sum decomposition $\mathfrak{R}h_1 + \cdots + \mathfrak{R}h_n$.

Let the annihilating ideals (ε_i), (0) which remain after elimination of units be written *in reverse order* as $\mathfrak{a}_1, \ldots, \mathfrak{a}_q$; then

$$\mathfrak{a}_i \equiv 0(\mathfrak{a}_{i+1}).$$

2. Those groups (h_i) whose annihilator is (0) are isomorphic to $\mathfrak{R}$. Those groups whose annihilator $(\varepsilon_i) \neq (0)$ can be further spilt into prime-power groups, as was demonstrated above. The annihilating prime powers are found by factoring the ε_i. The sum of all the groups in the decomposition of $\mathfrak{G}$ belonging to a prime number p form a group $\mathfrak{B}_p$ consisting of those elements of $\mathfrak{G}$ which are annihilated by a sufficiently high power p^{ϱ}. Hence *the groups $\mathfrak{B}_p$ are uniquely determined.* If $\mathfrak{U}$ denotes the sum of the groups with $\mathfrak{a} = (0)$, then we have

$$\mathfrak{G} = \sum_p \mathfrak{B}_p + \mathfrak{U}.$$

By further decomposition of the $\mathfrak{B}_p$ the prime-power groups are again obtained; these are determined uniquely up to isomorphism, as we shall soon see. In each $\mathfrak{B}_p$ there is a uniquely determined sequence of subgroups $\mathfrak{B}_{p,\varrho}, \mathfrak{B}_{p,\varrho-1}, \ldots, \mathfrak{B}_{p,0}$, where $\mathfrak{B}_{p,\nu}$ consists of those elements of $\mathfrak{B}_p$ which are annihilated by p^{ν}. The first group of this sequence is $\mathfrak{B}_p$; the last is the zero group.

The group $\mathfrak{U}$ is uniquely determined up to isomorphism, since

$$\mathfrak{U} \cong \mathfrak{G}/\sum_p \mathfrak{B}_p.$$

3. **Uniqueness Theorem:** *The annihilating ideals $\mathfrak{a}_1, \ldots, \mathfrak{a}_q$ with $\mathfrak{a}_i \equiv 0(\mathfrak{a}_{i+1})$ for a direct sum decomposition $\mathfrak{G} = \mathfrak{C}_1 + \cdots + \mathfrak{C}_q$ are uniquely determined by the module $\mathfrak{G}$ alone.* (Or equivalently: the groups $\mathfrak{C}_i$ are uniquely determined up to isomorphism.)

Proof: The asserted uniqueness is established once we have shown that it can be uniquely determined how many of the ideals $\mathfrak{a}_i$ are divisible by each prime power p^{σ} of $\mathfrak{R}$. Indeed, if p^{σ} divides precisely k of these ideals, then these must be the first k ideals $\mathfrak{a}_1, \ldots, \mathfrak{a}_k$ because of the divisibility property of these ideals. For every prime power p^{σ} we thus know not only how many but also which ideals are divisible by p^{σ}; this means that for each $\mathfrak{a}_i$ we know which prime powers divide it. Those ideals which are divisible by arbitrarily high powers are zero; the others are uniquely determined by their prime factorization.

If p^{σ} divides the annihilator of the cyclic group $\mathfrak{C}_i$, then

$$p^{\sigma-1}\mathfrak{C}_i/p^{\sigma}\mathfrak{C}_i$$

is a cyclic group with annihilator (p); it is therefore a simple group. On the other hand, if p^{σ} does not divide the annihilating ideal of $\mathfrak{C}_i$, then $p^{\sigma}\mathfrak{C}_i = p^{\sigma-1}\mathfrak{C}_i$ and therefore $p^{\sigma-1}\mathfrak{C}_i/p^{\sigma}\mathfrak{C}_i = (0)$. It follows that $p^{\sigma-1}\mathfrak{G}/p^{\sigma}\mathfrak{G}$ is the direct sum of the same number of simple groups as the number k of the $\mathfrak{a}_i$ divisible by p^σ. Thus, k is equal to the length of the composition series for $p^{\sigma-1}\mathfrak{G}/p^\sigma\mathfrak{G}$ and so is uniquely determined.

Exercises

12.4. Complete the last part of the above proof.

12.5. The group $\mathfrak{U}$ constructed in (2) above is a module of linear forms over the ring $\mathbb{Z}$ of integers, and the number of its cyclic summands is at the same time the rank of $\mathfrak{G}$ (rank = maximum number of linearly independent elements over $\mathfrak{R}$).

12.6. Give a second uniqueness proof using the length of the composition series of the uniquely determined groups constructed in (2) and their factor groups. The rank of the module $\mathfrak{U}$ (Exercise 12.5) may also be used.

12.4 REPRESENTATIONS AND REPRESENTATION MODULES

Let K be a skew field.

A *representation of a ring* $\mathfrak{o}$ *by linear transformations or by matrices in* K is a homomorphism

$$\mathfrak{o} \sim \mathfrak{O},$$

where $\mathfrak{O}$ is a ring of square matrices of degree r in K. If the homomorphism is an isomorphism, then the representation is said to be *faithful.*

A *representation module* of $\mathfrak{o}$ with respect to K is a *double module* $\mathfrak{M}$ having $\mathfrak{o}$ as left and K as right operator domain with the following properties.

1. $\mathfrak{M}$ is a module of linear forms over K.

$$\mathfrak{M} = u_1\mathsf{K} + \cdots + u_n\mathsf{K}.$$

2. For $a \in \mathfrak{o}$, $u \in \mathfrak{M}$, and $\lambda \in \mathsf{K}$,

$$a\cdot u\lambda = au\cdot\lambda. \tag{12.11}$$

The last condition states that multiplication with a is an operator homomorphism of the K-module $\mathfrak{M}$, that is, a linear transformation. The linear transformation is given by a square matrix $A = (\alpha_{ik})$:

$$a\cdot u_k = \sum u_j\alpha_{jk}$$

$$a\cdot\sum u_k\lambda_k = \sum\sum u_j\alpha_{jk}\lambda_k. \tag{12.12}$$

Thus, to each a of $\mathfrak{o}$ there corresponds a matrix A of K. As a consequence of the module postulates, to the sum and product of two elements a, b of $\mathfrak{o}$ there correspond the sum and product of the associated linear transformations and hence also of their matrices. *The correspondence $a \to A$ is therefore a representation of the ring* $\mathfrak{o}$.

Conversely, if a representation of a ring $\mathfrak{o}$ by linear transformations of a module $\mathfrak{M}$ of linear forms over K is given, then $\mathfrak{M}$ can be made a double module on defining the products $a \cdot u (a \in \mathfrak{o},\ u \in \mathfrak{M})$ by (12.12). We then find that all the properties of double modules and property (12.11) are satisfied, so that $\mathfrak{M}$ is a *representation module*.

Every representation module thus affords a representation of $\mathfrak{o}$ *by linear transformations or, after choosing a* K*-basis* $(u_1, \ldots, u_n)$, *by matrices in* K; *conversely, every representation defines a representation module.*

If we go from the basis $(u_1, \ldots, u_n)$ to another basis $(u_1', \ldots, u_n')$ by

$$(u_1' \ldots u_n') = (u_1 \ldots u_n)\, P,$$

then the same linear transformation is represented by the matrix

$$A' = P^{-1}AP.$$

To the ring elements a there now correspond new matrices A'; this is called an *equivalent representation.* Since the transition to an equivalent representation amounts to nothing more than the choice of another basis for the same representation module (or one operator isomorphic to it), we conclude: *to isomorphic representation modules there correspond equivalent representations, and conversely.*

A system of linear transformations of a module of linear forms $\mathfrak{M}$, in particular a representation of a ring, is said to be *reducible* if all the transformations of the system take a fixed linear subspace $\mathfrak{N} \neq (0),\ \neq \mathfrak{M}$ into itself. Then $\mathfrak{N}$ is called an *invariant subspace.* When it is a question of a representation of a ring $\mathfrak{o}$, if we interpret $\mathfrak{M}$ as a double module with respect to $\mathfrak{o}$ and K, then the invariant subspace $\mathfrak{N}$ admits all the elements of $\mathfrak{o}$ as left operators. From this it follows that *a representation of a ring is reducible if and only if the associated representation module contains a (double) submodule* $\mathfrak{N}$.

In order to investigate the form of the matrices of a reducible representation, we start with a K-basis for $\mathfrak{N}$ and enlarge it to a K-basis for $\mathfrak{M}$. Let then

$$\mathfrak{N} = v_1\mathsf{K} + \cdots + v_r\mathsf{K}$$

$$\mathfrak{M} = v_1\mathsf{K} + \cdots + v_r\mathsf{K} + w_1\mathsf{K} + \cdots + w_t\mathsf{K}.$$

The fact that a linear transformation takes the module $\mathfrak{N}$ into itself means that the transforms of the v are expressed in terms of the v alone:

$$v_j' = \sum v_i\rho_{ij}$$

$$w_j' = \sum v_i\sigma_{ij} + \sum w_i z_{ij}. \tag{12.13}$$

If we put $R = (\rho_{ij})$, $S = (\sigma_{ij})$, and $T = (\tau_{ij})$, then the transformation is represented by the matrix

$$A = \begin{pmatrix} R & S \\ 0 & T \end{pmatrix}. \tag{12.14}$$

It follows that *a system of matrices is reducible if and only if all the matrices of the system can simultaneously be brought to the form* (12.14) *by a transformation* $A' = P^{-1}AP$ *(choice of a new basis).*

Equations (12.13) imply

$$(v_1' \dots v_r') = (v_1 \dots v_r) \cdot R$$

$$(w_1' \dots w_t') \equiv (w_1 \dots w_t) \cdot T \pmod{\mathfrak{N}}. \tag{12.15}$$

This leads to the following observation.

If in the case of a reducible representation of a ring $\mathfrak{o}$ *we interpret the invariant submodule* $\mathfrak{N}$ *and the factor module* $\mathfrak{M}/\mathfrak{N}$ *themselves as representation modules, then the representations thus obtained are given by the components R and T of* (12.14).

If we take for $\mathfrak{N}$ a maximal invariant submodule $\mathfrak{M}_{l-1}$, in this again a maximal invariant submodule $\mathfrak{M}_{l-2}$, and so on until a composition series

$$\mathfrak{M} = \mathfrak{M}_l, \mathfrak{M}_{l-1}, \dots, \mathfrak{M}_0 = (0)$$

is obtained, then with an appropriate choice of basis the matrices of the representation have the following appearance:

$$\begin{pmatrix} R_{11} & \cdots & \cdots & R_{1l} \\ 0 & R_{22} & & \vdots \\ \vdots & & \ddots & \vdots \\ 0 & \cdots & 0 & R_{ll} \end{pmatrix}. \tag{12.16}$$

The diagonal blocks R_{ii} provide representations belonging to the composition factors $\mathfrak{M}_i/\mathfrak{M}_{i-1}$; since these composition factors are simple double modules (that is, they have no invariant submodules), the associated representations are *irreducible.* The process leading to (12.16) is called *reduction* of a representation. By the Jordan-Hölder theorem (Section 7.4), the composition factors are uniquely determined up to order and operator isomorphism. Hence *the irreducible components* R_{ii} *of the reduced representation* (12.16) *are uniquely determined up to order and equivalent representations.*

If the σ_{ij} are absent in (12.13), this means that $(w_1, \dots, w_t)$ as well as $(v_1, \dots, v_r)$ is an invariant submodule and hence that $\mathfrak{M}$ is a *direct sum of two invariant submodules* $\mathfrak{N}$ *and* $\mathfrak{Q}$. The matrix (12.14) then has the form

$$A = \begin{pmatrix} R & 0 \\ 0 & T \end{pmatrix},$$

where R belongs to the representation provided by $\mathfrak{N}$ and T to that provided by $\mathfrak{Q}$. The representation $a \to A$ is then said to decompose into the representations $a \to R$ and $a \to T$.

If the double module $\mathfrak{M}$ is completely reducible in the sense of Section 7.6, that is, a direct sum of simple double modules, then the representation provided by $\mathfrak{M}$ is given by matrices of the form

$$\begin{pmatrix} R_{11} & & & & 0 \\ & R_{22} & & & \\ & & \cdot & & \\ & & & \cdot & \\ & & & & \cdot \\ 0 & & & & R_{ll} \end{pmatrix}, \tag{12.17}$$

where the individual blocks give irreducible representations which need not all be distinct. Such a representation is said to be *completely reducible.*

The theory of a single matrix in the next section provides examples of the ideas developed above.

Exercises

12.7. If $\mathfrak{o}$ is a ring with identity and if in a representation of $\mathfrak{o}$ the identity of $\mathfrak{o}$ corresponds to the identity matrix, then for the representation module this means that the identity element is the identity operator. Show with the help of a theorem of Section 12.1 that every representation of $\mathfrak{o}$ decomposes into one in which the identity element corresponds to the identity matrix and one in which every element of $\mathfrak{o}$ is mapped onto the zero matrix:

$$A = \begin{pmatrix} S & 0 \\ 0 & 0 \end{pmatrix}.$$

12.8. A representation is completely reducible if and only if for every invariant subspace $\mathfrak{N}$ another such subspace $\mathfrak{Q}$ can be found which together with $\mathfrak{N}$ spans the space $\mathfrak{M}$:

$$\mathfrak{M} = \mathfrak{N} + \mathfrak{Q}.$$

12.9. If $(u_1', \ldots, u_n') = (u_1, \ldots, u_n)P$ is a homomorphism of the representation module into itself, then the matrix P commutes with all the matrices of the representation

$$AP = PA,$$

and conversely.

12.5 NORMAL FORMS OF A MATRIX IN A COMMUTATIVE FIELD

Let $\mathfrak{M} = (u_1, \ldots, u_n)$ be a module of linear forms over a commutative field K, and let

$$u_k \to v_k = \sum u_i \alpha_{ik}$$

be a linear transformation of $\mathfrak{M}$ into itself. By introducing a new basis

$$(u_1' \dots u_n') = (u_1 \dots u_n)\,P$$

(where P is thus an invertible matrix in K), we wish to bring the matrix $A = (\alpha_{ik})$ to a simplest possible normal form

$$A' = P^{-1}AP.$$

We interpret the powers of the matrix A as a representation of the powers of an indeterminate x and extend this representation to a representation of the polynomial ring $\mathsf{K}[x]$ by sending the polynomial

$$f(x) = \sum \alpha_\nu x^\nu$$

into the matrix

$$f(A) = \sum \alpha_\nu A^\nu.$$

This correspondence is a homomorphism, since the powers of A commute with one another and with the coefficients α_ν.

To this representation there belongs a representation module $\mathfrak{M}$ in which the product of a polynomial of $\mathsf{K}[x]$ with a u of $\mathfrak{M}$ is defined by

$$(\sum \alpha_\nu x^\nu)u = \sum \alpha_\nu A^\nu u.$$

The representation module $\mathfrak{M}$ is a double module with respect to $\mathsf{K}[x]$ and K; since the elements of K commute with one another and with all other elements, they may also be written on the left of the elements of $\mathfrak{M}$:

$$u\lambda = \lambda u.$$

Therefore $\mathfrak{M}$ may be interpreted simply as a $\mathsf{K}[x]$-module.

Since the polynomial ring $\mathsf{K}[x]$ is Euclidean, the fundamental theorem of Section 12.3 is applicable: the module $\mathfrak{M}$ is the direct sum of cyclic $\mathsf{K}[x]$-modules $(w_1), \dots, (w_r)$ whose annihilating ideals are either zero or are each generated by a polynomial of $\mathsf{K}[x]$. However, the case of a null ideal is excluded, since for each $w = w_\nu$ at most n of the quantities $w, xw, x^2w, \dots$ can be linearly independent; therefore there exists a polynomial $\sum \alpha_\nu x^\nu \neq 0$ with the property that

$$\sum \alpha_\nu x^\nu w = 0.$$

Each $w = w_\nu$ thus has an annihilating polynomial of least degree

$$f_\nu(x) = f(x) = x^k + \alpha_{k-1}x^{k-1} + \cdots + \alpha_0,$$

and

$$f_{\nu+1} \equiv 0(f_\nu).$$

The quantities $w, xw, \dots, x^{k-1}w$ are linearly independent over K and may therefore be used as a K-basis for the cyclic $\mathsf{K}[x]$-module $(w) = (w, xw, x^2w, \dots)$.

Thus

$$
\begin{aligned}
Aw &= xw \\
Axw &= x^2 w \\
Ax^{k-1}w &= x^k w = -\alpha_0 \cdot w - \alpha_1 \cdot xw - \cdots - \alpha_{k-1} \cdot x^{k-1} w.
\end{aligned}
$$

The transformation A of the module $(w, xw, \ldots)$ into itself is represented in the new basis by the matrix

$$
A_\nu = \begin{pmatrix} 0 & . & \ldots & 0 & -\alpha_0 \\ 1 & 0 & \ldots & . & -\alpha_1 \\ 0 & 1 & & . & . \\ . & & . & . & . \\ . & & . & . & . \\ . & & . & . & . \\ 0 & . & \ldots & 1 & -\alpha_{k-1} \end{pmatrix}. \tag{12.18}
$$

These matrices are called *companion matrices*; to each w_ν there corresponds a companion matrix A_ν of this type. Since $\mathfrak{M}$ is the direct sum of the (w_ν), we obtain the *first normal form* for the matrix A:

$$
A = \begin{pmatrix} A_1 & & & & \\ & A_2 & & & \\ & & . & & \\ & & & . & \\ & & & & A_r \end{pmatrix}, \tag{12.19}
$$

the blocks A_ν are here companion matrices of the type (12.18).

It follows from the uniqueness theorem of Section 12.3 that the polynomials $f_\nu(x)$, and hence also the companion matrices A_ν, are *uniquely determined* by the module $\mathfrak{M}$.

The blocks A_ν can be still further "reduced" by representing the modules (w_ν) as direct sums of cyclic modules whose annihilators are powers of prime polynomials. The form (12.19) is retained, but the companion matrices (12.18) now belong to powers of prime polynomials $(p(x))^\varrho$ (*second normal form*). The companion matrices are still uniquely determined up to their order in (12.19). The polynomials $(p(x))^\varrho$ are sometimes called *elementary divisors* of the matrix A. The word thus has a different meaning here from its meaning in Section 12.2. The relation between the two concepts will emerge in Section 12.6.

By means of the composition series for the cyclic modules (w_ν) we can still further reduce the normal form just obtained. We shall carry this out here for the case in which the prime polynomials $p(x)$ are linear; this is always the case if the field K is algebraically closed. Let then

$$
\begin{aligned}
p(x) &= x - \lambda \\
f(x) &= (x-\lambda)^\varrho.
\end{aligned}
$$

As basis elements we use

$$v_1 = (x-\lambda)^{\varrho-1}w$$
$$v_2 = (x-\lambda)^{\varrho-2}w$$
$$\dots$$
$$v_\varrho = w;$$

thus,

$$(x-\lambda)v_1 = 0$$
$$(x-\lambda)v_\mu = v_{\mu-1} \qquad (1<\mu\leqq\sigma)$$

or

$$Av_1 = xv_1 = \lambda v_1$$
$$Av_\mu = xv_\mu = \lambda v_\mu + v_{\mu-1}. \tag{12.20}$$

The "block" A_1 thus acquires the "reduced" form

$$A_1 = \begin{pmatrix} \lambda & 1 & 0 & \cdot & \cdot & \cdot & 0 \\ 0 & \lambda & 1 & & & & \cdot \\ \cdot & & \cdot & \cdot & \cdot & & \cdot \\ \cdot & & & \cdot & \cdot & \cdot & \cdot \\ \cdot & & & & \cdot & \cdot & \cdot \\ \cdot & & & & & \cdot & 1 \\ 0 & \cdot & \cdot & \cdot & \cdot & 0 & \lambda \end{pmatrix};$$

similarly, since to each w_ν there belongs a λ_ν,

$$A_\nu = \begin{pmatrix} \lambda_\nu & 1 & \cdot & \cdot & \cdot & 0 \\ \cdot & \cdot & \cdot & & & \cdot \\ \cdot & & \cdot & \cdot & & \cdot \\ \cdot & & & \cdot & \cdot & \cdot \\ \cdot & & & & \cdot & 1 \\ 0 & \cdot & \cdot & \cdot & \cdot & \lambda_\nu \end{pmatrix}.$$

If these blocks are again inserted in (12.19) the *third normal form* is obtained. The "characteristic roots" λ_ν and the degrees ρ_ν of the blocks are again *uniquely determined.*

All vectors v_μ belonging to the same root λ generate a module $\mathfrak{B}_\nu$ which is annihilated by a power of $x-\lambda$ (Section 12.3); this module is called (in the language of vectors) the *subspace belonging to the root* λ. The entire module $\mathfrak{M}$ is the direct sum of such subspaces. These subspaces further contain the sequence of subspaces mentioned in Section 12.3 which are annihilated by $(x-\lambda)^\varrho$, $(x-\lambda)^{\varrho-1}, \dots, 1$. The vectors w annihilated by $x-\lambda$, that is, those vectors for which

$$Aw = \lambda w,$$

are also called *eigenvectors* of the matrix A for the *eigenvalue* λ.[3]

[3]In place of these highly favored German-English hybrids, the terms "characteristic value (vector)", "proper value (vector)", and so on are sometimes used (*Trans.*).

The *completely reducible* case (cf. Section 12.4), in which the normal form (12.19) has the diagonal form

$$\begin{pmatrix} \lambda_1 & & & & 0 \\ & \lambda_2 & & & \\ & & \cdot & & \\ & & & \cdot & \\ & & & & \cdot \\ 0 & & & & \lambda_n \end{pmatrix}, \tag{12.21}$$

occurs if all the ϱ are equal to 1, that is, if all the polynomials $f_\nu(x)$ from which the $(p(x))^\varrho$ are obtained by prime factorization are free of multiple factors. Because of

$$f_{\nu+1} \equiv 0(f_\nu),$$

it is sufficient for this that the highest elementary divisor $f_r(x)$ have no multiple factors.

Methods for actually determining the characteristic roots and writing down the normal forms are given in the next section.

Exercises

12.10. The highest elementary divisor $f_r(x)$ may be characterized as the polynomial $f(x)$ of least degree with the property

$$f(x)\mathfrak{M} = 0 \qquad \text{or} \quad f(A) = 0.$$

12.11. For an arbitrary matrix A in the second or third normal form, determine the set of all matrices which commute with A (cf. Exercise 12.9).

12.6 ELEMENTARY DIVISORS AND CHARACTERISTIC FUNCTIONS

Under the transformation

$$A' = P^{-1}AP$$

the matrix $xE-A$ goes into

$$\begin{aligned} P^{-1}(xE-A)P &= xP^{-1}EP-P^{-1}AP \\ &= xE-A'. \end{aligned}$$

We wish to determine the elementary divisors of the matrix $xE-A$ in $\mathsf{K}[x]$ in the sense of Section 12.2. Since the elementary divisors are invariant under pre- and postmultiplication with arbitrary invertible matrices, they may be found from $xE-A'$ where A' has the first normal form of Section 12.5. By (12.18) and

(12.19), $xE-A'$ consists of blocks of the form

$$xE_1-A_1=\begin{pmatrix} x & 0 & . & . & . & 0 & \beta_0 \\ -1 & x & . & & & . & . \\ 0 & . & . & . & & . & . \\ & . & . & . & & . & . \\ & . & & . & . & . & . \\ . & & & . & . & x & \beta_{h-2} \\ 0 & . & . & . & 0 & -1 & x+\beta_{h-1} \end{pmatrix}.$$

To determine the elementary divisors, we must bring this matrix to diagonal form. If we multiply the second through hth rows by $x, x^2, \ldots, x^{h-1}$, respectively, and add these to the first row, we obtain

$$\begin{pmatrix} 0 & 0 & . & . & . & 0 & f(x) \\ -1 & x & . & & & . & \beta_1 \\ 0 & \cdot & . & . & & . & . \\ . & . & . & . & . & . & . \\ . & & . & . & . & . & . \\ . & & & . & \cdot & x & \beta_{h-2} \\ 0 & . & . & . & 0 & -1 & x+\beta_{h-1} \end{pmatrix}.$$

If the first row is brought to the bottom by interchanging rows, then only zeros remain under the main diagonal. By adding multiples of preceding columns to subsequent ones, it is very easy to replace everything above the main diagonal by zeros. As a result there remains

$$\begin{pmatrix} -1 & & & & & 0 \\ & -1 & & & & \\ & & . & & & \\ & & & . & & \\ & & & & . & \\ & & & & -1 & \\ 0 & & & & & f(x) \end{pmatrix}.$$

If all these blocks are now juxtaposed and rows and columns are interchanged until all the -1 occur first on the main diagonal, we obtain the desired diagonal form

$$\begin{pmatrix} -1 & & & & & & & & 0 \\ & -1 & & & & & & & \\ & & . & & & & & & \\ & & & . & & & & & \\ & & & & . & & & & \\ & & & & & -1 & & & \\ & & & & & & f_1(x) & & \\ & & & & & & & . & \\ & & & & & & & & . \\ 0 & & & & & & & & f_r(x) \end{pmatrix}.$$

The polynomials $f_\nu(x)$ together with a certain number of ones are thus the elementary divisors of $xE-A$.

The *characteristic polynomial* (*characteristic function*) *of A*

$$\chi(x) = \prod_1^r f_\nu(x)$$

annihilates the module $\mathfrak{M}$, since the factor $f_r(x)$ already does; thus

$$\chi(A) = 0. \qquad (12.22)$$

The characteristic polynomial is the greatest determinant divisor of $xE-A$ and is therefore equal to the determinant $|xE-A|$ up to a constant. It is immediate that the constant is equal to one, and hence

$$\chi(x) = |xE-A|. \qquad (12.23)$$

The characteristic equation (12.22) for the matrix A can also be derived directly from (12.23). Indeed,

$$xu_k = \sum u_i \alpha_{ik},$$

and elimination of all the u from this system of equations gives (since x and its powers commute with the α_{ik})

$$\begin{vmatrix} x-\alpha_{11} & -\alpha_{12}\cdots & -\alpha_{1n} \\ \cdot & \cdot & \cdot \\ \cdot & \cdot & \cdot \\ \cdot & \cdot & \cdot \\ -\alpha_{n1} & -\alpha_{n2}\cdots & x-\alpha_{nn} \end{vmatrix} \cdot u_j = 0$$

or

$$|xE-A|\, u_j = 0;$$

that is, $\chi(x) = |xE-A|$ annihilates all the u_j and hence the entire module $\mathfrak{M}$, Q.E.D.

According to the preceding remarks, the coefficients of the characteristic function $\chi(x)$ of A are invariant under the transformation

$$A \to P^{-1}AP.$$

The most important of these coefficients are the first and last:

1. The *trace* of A, which is the coefficient of $-x^{n-1}$.

$$S(A) = \sum \alpha_{ii};$$

2. The *norm* of A, which is the coefficient of $(-1)^n x^0$:

$$N(A) = |A|.$$

The roots of the characteristic function are the *characteristic roots* λ_ν; these were introduced in the preceding section as roots of the polynomial $f_\nu(x)$. This provides a practical method of determining the λ_ν and finding the normal forms of the preceding section: we first find the λ_ν as roots of

$$\chi(x) = |xE - A|,$$

and then determine the v_1 from the linear equations [cf. (12.20)]

$$Av_1 = \lambda_\nu v_1.$$

In the case of multiple roots ($\varrho > 1$) $v_2, \ldots, v_\varrho$ are easily found from (12.20); it may be necessary to replace the v_1 belonging to the same λ_ν by suitable linear combinations.

The equation $\chi(\lambda) = 0$ having the λ_ν as roots comes up repeatedly in applications; it is sometimes called the *secular equation* because it occurs in the theory of secular perturbations.

12.7 QUADRATIC AND HERMITIAN FORMS

Let K be a commutative field, and let Q be a quadratic form

$$Q(x_1, \ldots, x_n) = \sum_i q_i x_i^2 + \sum_{i<k} q_{ik} x_i x_k \tag{12.24}$$

with coefficients in K. If we put $q_i = q_{ii}$ and write $Q(x)$ instead of $Q(x_1, \ldots, x_n)$, then (12.24) can be more briefly written as

$$Q(x) = \sum_{i \leq k} q_{ik} x_i x_k.$$

We now form $Q(x+y)$, where y stands for a new sequence of indeterminates $y_1, \ldots, y_n$. Computation gives

$$Q(x+y) = Q(x) + Q(y) + B(x, y), \tag{12.25}$$

where $B(x, y)$ is a symmetric bilinear form

$$B(x, y) = \sum b_{ik} x_i y_k \tag{12.26}$$

with coefficients

$$b_{ii} = 2q_i$$

$$b_{ik} = b_{ki} = q_{ik} \qquad (i<k).$$

We call $B(x, y)$ the *polar form* of $Q(x)$.

If x is linearly transformed,

$$x_i = \sum \pi_{ij} x_j' \qquad (\pi_{ij} \in \mathsf{K}), \tag{12.27}$$

then $Q(x)$ is transformed into a new form $Q'(x)$:

$$Q(x) = Q'(x').$$

The matrix $P = (\pi_{ij})$ is here assumed to be nonsingular. The forms Q and Q' are called *rationally equivalent in the field* K. If the matrix P and its inverse P^{-1} belong to a ring $\mathfrak{R} \subseteq$ K, then the forms are called *equivalent in the ring* $\mathfrak{R}$ (for example, integrally equivalent if $\mathfrak{R} = \mathbb{Z}$ is the ring of integers).

If y is transformed just as x with the same coefficients π_{ij},

$$y_i = \sum \pi_{ij} y_j', \tag{12.28}$$

then $B(x, y)$ is transformed into a bilinear form $B'(x', y')$:

$$B(x, y) = B'(x', y').$$

From (12.25) now follows

$$Q'(x'+y') = Q'(x')+Q'(y')+B'(x', y'). \tag{12.29}$$

Thus, if B is the polar form of Q, then B' is the polar form of Q'. The construction of the polar form is thus invariant with respect to linear transformation of the variables.

Putting $y = x$ in (12.25), we obtain

$$4Q(x) = 2Q(x)+B(x, x)$$

or

$$2Q(x) = B(x, x). \tag{12.30}$$

If the characteristic of the field is not equal to 2, then the form $Q(x)$ can be regained from $B(x, x)$:

$$Q(x) = \tfrac{1}{2}B(x, x) = \tfrac{1}{2}\sum b_{ik}x_i x_k.$$

Putting $\tfrac{1}{2}b_{ik} = a_{ik}$, we can write the quadratic form as

$$Q(x) = \sum a_{ik}x_i x_k \qquad (a_{ik} = a_{ki}). \tag{12.31}$$

From the coefficients b_{ik} of the bilinear form $B(x, y)$ we may form the determinant

$$D = \begin{vmatrix} b_{11} & b_{12} & \cdots & b_{1n} \\ \cdot & \cdot & & \cdot \\ \cdot & \cdot & & \cdot \\ \cdot & \cdot & & \cdot \\ \cdot & \cdot & & \cdot \\ b_{n1} & b_{n2} & \cdots & b_{nn} \end{vmatrix} = \begin{vmatrix} 2q_1 & q_{12} & \cdots & q_{1n} \\ q_{12} & 2q_2 & \cdots & q_{2n} \\ \cdot & \cdot & & \cdot \\ \cdot & \cdot & & \cdot \\ \cdot & \cdot & & \cdot \\ q_{1n} & q_{2n} & \cdots & 2q_n \end{vmatrix}. \tag{12.32}$$

We call D the *determinant of the form* Q. If the characteristic of the base field is not 2, we may form the determinant Δ of the halved coefficients a_{ik}. This determinant is called the *discriminant* of the form Q. Clearly

$$D = 2^n\Delta. \tag{12.33}$$

We now investigate the behavior of the determinant D under the linear transformation (12.27). If (12.27) and (12.28) are substituted into (12.26), we obtain

$$B'(x', y') = \sum b_{ik}\pi_{ij}\pi_{kl}x_j'x_l',$$

and hence

$$b_{jl}' = \sum b_{ik}\pi_{ij}\pi_{kl}, \tag{12.34}$$

where the summation extends over repeated indices. Equation (12.34) can be written as the matrix equation

$$B' = P^t BP, \tag{12.35}$$

where P^t denotes the transpose of the matrix $P = (\pi_{ij})$.

Forming the determinants of both sides of (12.35), we obtain

$$D' = \{\mathrm{Det}(P)\}^2 \cdot D, \tag{12.36}$$

or in words: *The determinant D is multiplied by the square of the determinant of the transformation.*

We shall henceforth assume that the characteristic of the base field is not equal to 2. We replace the variables x_i by the coordinates c_i of a vector u and the y_i by the coordinates d_i of a vector v, and write

$$f(u, v) = \sum a_{ik}c_i d_k = \tfrac{1}{2}B(c, d);$$

in particular

$$f(u, u) = \sum a_{ik}c_i c_k = Q(c).$$

We now wish to bring the quadratic form $f(u, u)$ to simplest possible form by means of a linear transformation. To this end we choose a vector v_1 such that $f(v_1, v_1) \neq 0$, which is always possible if f is not identically zero. The equation $f(v_1, u) = 0$ then defines a subspace R_{n-1} of the vector space R_n which does not contain v_1. In this subspace we now choose, if possible, a vector v_2 such that $f(v_2, v_2) \neq 0$; the equation $f(v_2, u) = 0$ together with the preceding one then defines a subspace R_{n-2} in R_{n-1} which does not contain v_2. We continue in this manner until a subspace R_{n-r} is reached such that $f(u, u) = 0$ for all u in R_{n-r} and hence also $f(u, v) = 0$ for all u and v in R_{n-r}. It may happen that $r = n$; in this case R_{n-r} is the null vector space. Otherwise we choose basis vectors $v_{r+1}, \ldots, v_n$ arbitrarily in R_{n-r}. We thus have

$$\begin{aligned} f(v_i, v_k) &= 0 && (i \neq k) \\ f(v_i, v_i) &= \gamma_i \neq 0 && (i = 1, \ldots, r) \\ f(v_i, v_i) &= 0 && (i = r+1, \ldots, n). \end{aligned}$$

If every vector v is referred to the new basis vectors $v_1, \ldots, v_n$,

$$v = \sum v_i d_i,$$

then

$$f(v, v) = \sum\sum f(v_i, v_k) d_i d_k = \sum_1^r \gamma_i d_i^2 \tag{12.37}$$

The form f has thus been transformed, as is said, *to a sum of squares*.

The vectors w or R_{n-r} have the property that

$$f(w, u) = 0 \qquad \text{for every } u$$

and are hereby characterized. The space R_{n-r} and its dimension $n-r$ are thus invariants of the form f. The number of negative γ_i in (12.37) is called the *index of inertia* of f. We shall show that this index of inertia is also an invariant of f (*Sylvester's law of inertia*).

Suppose that in relation to other basis vectors the form f has the representation

$$f = \sum_1^r \gamma_i' d_i'^2.$$

Let us suppose that, say, $\gamma_1, \ldots, \gamma_h$ are positive and $\gamma_{h+1}, \ldots, \gamma_r$ negative and similarly that $\gamma_1', \ldots, \gamma_k'$ are positive and $\gamma_{k+1}', \ldots, \gamma_r'$ negative. If now $k > h$, then the linear equations

$$d_1 = 0, \ldots, d_h = 0, \qquad d_{k+1}' = 0, \ldots, d_r' = 0$$

would define a space of dimension greater than $n-r$. For a vector u of this space we would have $f(u, u) = \sum_{h+1}^r \gamma_i d_i^2 \leqq 0$ and also $f(u, u) = \sum_1^k \gamma_i' d_i'^2 \geqq 0$. Thus, $f(u, u) = 0$ and all d_i and $d_i' = 0$ so that u would lie in R_{n-r}. But this would imply that a space of dimension greater than $n-r$ were contained in a space of dimension $n-r$, which is impossible.

If all the γ_i in (12.37) are positive, then the form f is said to be *positive definite* in the case $r = n$ and *semidefinite* in the case $r < n$. The positive definite forms are characterized by the fact that they have a positive value for any vector $u \neq 0$, and the semidefinite forms by the fact that their value is always nonnegative.

It follows immediately from (12.37) that a positive definite form can be transformed to the *identity form*

$$E(u, u) = \sum d_i^2$$

by adjoining the quantities $\sqrt{\gamma_i}$ to the field K.

Hermitian forms are analogous to quadratic forms. To consider such forms, we adjoin to the ordered field K a square root θ of a negative quantity α of K, for example, $\theta = \sqrt{-1}$. To distinguish the elements of K from those of K(θ), we shall sometimes refer to the former as "real" elements, since in applications K is usually the field of real numbers and $\theta = \sqrt{-1}$.

Every element $c = a + b\theta$ has a conjugate $\bar{c} = a - b\theta$. The product $\bar{c}c = a^2 - b^2\theta^2$ is always real and nonnegative; it is zero only if $c = 0$.

A *Hermitian form* is defined by

$$H(u, u) = \sum \sum h_{ik} \bar{c}_i c_k \qquad (h_{ik} = \bar{h}_{ki}).$$

The value of the form H for any vector u is always real.

If we now form

$$H(u+\lambda v, u+\lambda v) = \sum\sum h_{ik}\bar{c}_i c_k + \lambda \sum\sum h_{ik}\bar{c}_i d_k + \bar{\lambda}\sum\sum h_{ik} d_i c_k + \lambda\bar{\lambda}\sum\sum h_{ik} d_i d_k,$$

we find as coefficients of λ the bilinear form

$$H(u, v) = \sum\sum h_{ik} \bar{c}_i d_k$$

with the property

$$H(v, u) = \overline{H(u, v)}.$$

Under a linear transformation of the c_i, whereby the $\bar{c}_i$ are naturally transformed by the conjugate transformation with matrix $\bar{P} = (\pi_{ij})$, the matrix H of the Hermitian form is transformed as follows:

$$H' = P^{\dagger}AP,$$

where $P^{\dagger} = \bar{P}^t$ is the conjugate transposed matrix.

Our previous considerations regarding the representation of quadratic forms as sums of squares continue to hold unaltered for Hermitian forms. As a normal form we find

$$H(u, u) = \sum_1^r \gamma_i \bar{c}_i c_i \qquad (\gamma_i \text{ real}). \tag{12.38}$$

The form H is again called *positive definite* if its values $H(u, u)$ are always positive except for $u = 0$, or if $r = n$ and $\gamma_1, \ldots, \gamma_n$ are all positive. Any positive definite form can be transformed to the *identity form*

$$E(u, u) = \sum \bar{c}_i c_i$$

by adjoining the square roots of the γ_i.

The following discussion applies to both Hermitian and quadratic forms. We shall speak of Hermitian forms, but one need only choose all elements in K and omit all bars in order to obtain the corresponding statements for quadratic forms.

We choose a fixed, preferably positive definite, Hermitian form $G(u, u)$ of rank n as *fundamental form* and denote its coefficient matrix (g_{ik}) by G. In particular, if $G(u, u)$ is the identity form, then G is the identity matrix E. Two vectors u and v are said to be *perpendicular* if $G(u, v) = 0$; in this case $G(v, u) = 0$ also. The vectors v perpendicular to a vector $u \neq 0$ form a linear subspace: *the space perpendicular to u.* If G is positive definite, then $G(u, u) \neq 0$, and hence u itself does not belong to its perpendicular space R_{n-1}. A system of n mutually perpendicular basis vectors $v_1, \ldots, v_n$ [as was used, for example, in deriving the normal form (12.38) for $G(u, u)$] is called a *complete orthogonal system* of vectors. The orthogonal system is said to be *normalized* if

$$G(v_j, v_j) = 1.$$

Those linear transformations A which have the property

$$G(Au, v) = G(u, Av) \qquad (\text{for all } u \text{ and } v) \tag{12.39}$$

are called *Hermitian symmetric* or simply *symmetric*. Written out explicitly, this condition reads:

$$\sum\sum\sum g_{il}\bar{a}_{ij}\bar{c}_j c_l = \sum\sum\sum g_{jk}\bar{c}_j a_{kl} c_l$$

or

$$\sum_i g_{il}\bar{a}_{ij} = \sum_k g_{jk}a_{kl}$$

or

$$A^\dagger G = GA. \tag{12.40}$$

In particular, if G is the identity form, then the symmetry condition is simply

$$A^\dagger = A \quad \text{or} \quad \bar{a}_{ik} = a_{ki},$$

which explains the name "symmetric."

Those linear transformations U which leave the fundamental form $G(u, u)$ invariant,

$$G(Au, Au) = G(u, u) \quad \text{or} \quad A^\dagger G A = G, \tag{12.41}$$

are called *unitary*, or in the real case *orthogonal*. We clearly then also have $G(Au, Av) = G(u, v)$. In particular, if $G = E$ (as may always be assumed in the positive definite case), then the condition reads:

$$A^\dagger A = E \quad \text{or} \quad A^\dagger = A^{-1} \quad \text{or} \quad AA^\dagger = E.$$

Writing this out explicitly, we obtain the "orthogonality conditions"

$$\sum \bar{a}_{ik}a_{il} = \delta_{kl} = \begin{cases} 0 & \text{for } k \neq l \\ 1 & \text{for } k = l \end{cases}$$

or equivalently

$$\sum a_{ik}\bar{a}_{jk} = \delta_{ij}.$$

A real orthogonal transformation with determinant 1 is called a *rotation*.

If a symmetric or unitary transformation A takes a nonzero vector u into a multiple of itself,

$$Au = \lambda u, \tag{12.42}$$

that is, if A leaves the line generated by u invariant, then A also leaves the space R_{n-1} perpendicular to u invariant.

Proof: If v belongs to R_{n-1}, then $G(u, v) = 0$ and, for a symmetric transformation A,

$$G(u, Av) = G(Au, v) = G(\lambda u, v) = \lambda G(u, v) = 0,$$

whereas, for a unitary transformation A,

$$\begin{aligned} G(u, Av) &= G(AA^{-1}u, Av) = G(A^{-1}u, v) \\ &= G(\lambda^{-1}u, v) = \lambda^{-1}G(u, v) = 0. \end{aligned}$$

A vector $u \neq 0$ with property (12.42) is called an *eigenvector* of the transformation A; λ is called the associated *eigenvalue*.

As we have already seen in Section 12.6, the eigenvalues are found from the "secular equation"

$$\chi(\lambda) = \begin{vmatrix} \lambda-\alpha_{11} & -\alpha_{12} & \cdots \\ -\alpha_{21} & \lambda-\alpha_{22} & \cdots \\ \cdots & & \end{vmatrix} = 0; \tag{12.43}$$

the associated eigenvectors are found from the linear equations

$$\sum \alpha_{ik} c_k = \lambda c_i, \tag{12.44}$$

which are equivalent to (12.42).

If we now assume that K is a real closed field (say the field of real numbers), and hence that $K(\theta)$ is algebraically closed (cf. Section 11.5, Volume I), then the secular equation (12.43) always has a root λ_1 in $K(\theta)$ to which there corresponds an eigenvector e_1. The space R_{n-1} perpendicular to e_1 is transformed into itself by A, and A in R_{n-1} is again symmetric or unitary if it was symmetric or unitary in R_n. By the same argument there then exists an eigenvector e_2 in R_{n-1} whose perpendicular space R_{n-2} in R_{n-1} is again invariant, and so on. *Continuing in this manner, we finally find a complete system of n linearly independent, mutually perpendicular eigenvectors* $e_1, \ldots, e_n$:

$$Ae_\nu = \lambda_\nu e_\nu.$$

Referred to the new basis $(e_1, \ldots, e_n)$, the matrix A acquires the diagonal form

$$A_1 = P^{-1}AP = \begin{pmatrix} \lambda_1 & & & & 0 \\ & \lambda_2 & & & \\ & & \cdot & & \\ & & & \cdot & \\ & & & & \cdot \\ 0 & & & & \lambda_n \end{pmatrix}. \tag{12.45}$$

This normal form is valid for both symmetric and unitary transformations by the preceding discussion.

Let us now normalize the e_ν by the condition $G(e_\nu, e_\nu) = 1$, which is always possible in a real closed field K, since the square roots of the positive quantities $G(e_\nu, e_\nu)$ are always contained in K. Then G becomes equal to the identity form E when referred to the basis given by the e_ν. If now the matrix A is symmetric, then A_1 must also be symmetric and therefore identical with $A_1{}^\dagger$. From this it follows that

$$\lambda_\nu = \bar{\lambda}_\nu \quad \text{or} \quad \lambda_\nu \in K.$$

The characteristic polynomial of the matrix A or A_1 is

$$\chi(x) = \prod_1^n (x - \lambda_n), \tag{12.46}$$

and hence: *the secular equation* $\chi(\lambda) = 0$ *of a symmetric matrix A has only real roots.*

If, in addition, the matrices A and G are real, then the eigenvectors e_ν are also real, as the solutions of the real equations (12.44). *A real symmetric matrix A can therefore be brought to the diagonal form* (12.45) *by means of real matrices.*

A Hermitian form

$$H(u, u) = G(u, Au) = G(Au, u)$$

is coupled with the symmetric transformation A in an invariant manner; its matrix is clearly

$$H = GA_j,$$

conversely, the matrix A is determined by

$$A = G^{-1}H.$$

Then $H = GA$ is brought to diagonal form along with A and G; the transformed form reads:

$$H(u, u) = \sum \bar{c}_\nu c_\nu \lambda_\nu.$$

We have thus proved the following.

Any pair of Hermitian forms G, H of which one, say G, is positive definite can be brought simultaneously by a single transformation to the form

$$G(u, u) = \sum \bar{c}_\nu c_\nu$$
$$H(u, u) = \sum \bar{c}_\nu c_\nu \lambda_\nu.$$

The λ_ν are the characteristic roots of the matrix $A = G^{-1}H$, or equivalently the roots of the secular equation

$$|\lambda g_{jk} - h_{jk}| = 0.$$

In particular, *any pair of real quadratic forms of which one is positive definite can be brought simultaneously by a real transformation to sums of squares*[4]:

$$G(u, u) = \sum c_\nu^2$$
$$H(u, u) = \sum c_\nu^2 \lambda_\nu.$$

Exercises

12.12. If r vectors $v_1, \ldots, v_r$ generate a space R_r, then the vectors perpendicular to them generate a space R_{n-r}, and the whole space R_n is the direct sum $R_r + R_{n-r}$.

12.13. If a symmetric or unitary transformation A leaves the space R_r invariant, then it also leaves the space R_{n-r} perpendicular to it invariant.

12.14. Every system of symmetric or unitary transformations is completely reducible.

12.15. The determinant D of a unitary transformation has modulus 1, that is, $D\bar{D} = 1$. The determinant of a real orthogonal transformation is ± 1.

12.16. The unitary and likewise the real orthogonal transformations of a vector space into itself each form a group.

[4]For a general treatment of the classification of pairs of quadratic forms, see L. E. Dickson, *Modern Algebraic Theories*, Chicago, 1926 (also in German by E. Bodewig, Leipzig, 1929).

12.8 ANTISYMMETRIC BILINEAR FORMS

A bilinear form in $x_1, \ldots, x_n$ and $y_1, \ldots, y_n$ with coefficients in a field K,

$$f(x, y) = \sum_{i,k} a_{ik} x_i y_k, \tag{12.47}$$

is called *antisymmetric* if it has the following two properties:

$$f(x, y) = -f(y, x) \tag{12.48}$$

$$f(x, x) = 0. \tag{12.49}$$

In terms of the coefficients this means that

$$a_{ik} = -a_{ki} \tag{12.50}$$

$$a_{ii} = 0. \tag{12.51}$$

If new variables x_j' and y_l' are introduced in place of x_i and y_k by the same linear transformation

$$x_i = \sum p_{ij} x_j'$$

$$y_k = \sum p_{kl} y_l',$$

then the form $f(x, y)$ is transformed into a new bilinear form

$$\begin{aligned} f'(x', y') &= \sum a_{ik} (\sum p_{ij} x_j')(\sum p_{kl} y_l') \\ &= \sum a_{jl}' x_j' y_l' \end{aligned}$$

which is again antisymmetric with coefficients given by

$$a_{jl}' = \sum p_{ij} a_{ik} p_{kl}$$

or in matrix notation

$$A' = P^t A P. \tag{12.52}$$

For the determinant D of the a_{ik}, we obtain from (12.52) the transformation formula

$$D' = D\Delta^2, \tag{12.53}$$

where Δ is the determinant of the transformation matrix.

Exercise

12.17. Show that (12.48) follows from (12.49).

We now wish to bring the form f to simplest possible normal form by means of an appropriate transformation matrix P. The transformation is accomplished in several steps.

If f is identically zero, then without transformation f already has the normal form

$$f_0 = 0.$$

If a coefficient is nonzero, then we may assume that $a_{12} \neq 0$. We now collect all terms in (12.47) containing x_1:

$$x_1(a_{12}y_2 + \cdots + a_{1n}y_n).$$

The terms with y_1 are then

$$-(a_{12}x_2 + \cdots + a_{1n}x_n)y_1.$$

In place of x_2 and y_2 we introduce the new variables x_2' and y_2':

$$x_2' = a_{12}x_2 + \cdots + a_{1n}x_n$$

$$y_2' = a_{12}y_2 + \cdots + a_{1n}y_n;$$

f is then written as a form in $x_1, x_2', x_3, \ldots, x_n$ and $y_1, y_2', y_3, \ldots, y_n$. The terms with x_1 and y_1 are now simply

$$x_1y_2' - x_2'y_1.$$

Suppose that the terms containing y_2' are

$$(x_1 + b_3x_3 + \cdots + b_nx_n)y_2'.$$

We introduce in place of x_1 and y_1 the new variables

$$x_1' = x_1 + b_3x_3 + \cdots + b_nx_n$$

$$y_1' = y_1 + b_3y_3 + \cdots + b_ny_n$$

and write f as a form in $x_1', x_2', x_3, \ldots, x_n$ and $y_1', y_2', y_3, \ldots, y_n$. There are now only two terms which contain x_1', x_2', y_1', or y_2', namely

$$x_1'y_2' - x_2'y_1'.$$

All other terms contain only $x_3, \ldots, x_n, y_3, \ldots, y_n$. If they are all zero, we have the normal form

$$f_1 = x_1'y_2' - x_2'y_1'.$$

Otherwise the process is repeated, whereby in place of x_3, x_4, y_3, y_4 the new variables x_3', x_4', y_3', y_4' are introduced and a term

$$x_3'y_4' - x_4'y_3'$$

is split off. In this manner we finally obtain, dropping the prime, a normal form

$$f_k = (x_1y_2 - x_2y_1) + \cdots + (x_{2k-1}y_{2k} - x_{2k}y_{2k-1}) \tag{12.54}$$

with

$$0 \leqq 2k \leqq n.$$

In the n-dimensional vector space of vectors $(c_1, \ldots, c_n)$ there is a linear subspace $\mathfrak{N}$ defined by the equations

$$f(c, y) = 0 \qquad \text{identically in the } y_k$$

or

$$\sum a_{ik} c_i = 0.$$

The dimension of this subspace is $n - r$, where r is the rank of the matrix A. This dimension is clearly an invariant of the form f under invertible linear transformations of the x_i and y_k. The number r is therefore also an invariant.

Computing the rank r for the normal form f_k, we obtain

$$r = 2k. \tag{12.55}$$

Since r is an invariant, the rank r is also an even number for the original form f. Hence we have the following.

The rank of an antisymmetric matrix A is an even number $2k$. It is equal to the number of terms in the normal form (12.54).

If n is odd, the rank is necessarily less than n, and hence the determinant D is then equal to zero. If, however, $n = 2m$ is even, then there exist forms with determinant $D \neq 0$, for example, the normal form f_m. Hence the determinant of an antisymmetric matrix with an even number of rows is not identically zero.

A *general antisymmetric bilinear form* is obtained by taking the a_{ik} with $i < k$ to be independent indeterminates and defining the others by (12.50) and (12.51). If n is even ($n = 2m$), then the determinant of this general form is different from zero as has just been shown. If this general form is transformed to normal form, the normal form (12.54) with $k = m$ is obtained. The coefficients of the transformation matrix are rational functions of the indeterminates a_{ik}, and the determinant D' of the normal form is 1. Hence, (12.53) implies

$$D = \Delta^{-2}, \tag{12.56}$$

where Δ is a rational function of the a_{ik} which may be expressed as a quotient of polynomials:

$$\Delta = \frac{G.}{H} \tag{12.57}$$

From (12.56) and (12.57) it follows that

$$DG^2 = H^2. \tag{12.58}$$

Thus, H^2 is divisible by G^2, and hence H is divisible by G:

$$H = GQ.$$

If this is substituted in (12.57) and (12.58), we find

$$\Delta = Q^{-1} \tag{12.59}$$

and

$$D = Q^2. \tag{12.60}$$

Since D is a form of degree $n = 2m$ it follows that Q is a form of degree m in the a_{ik}. If we carry out the computation for $n = 2$ and $n = 4$, we find

$$n = 2: \quad Q = a_{12}$$

$$n = 4: \quad Q = a_{12}a_{34} - a_{13}a_{24} + a_{14}a_{23}.$$

The general formula for Q was found by Pfaff. A proof may be found in a very instructive "Letter from the Nether World" by R. Lipschitz, *Ann. of Math.*, **69**, 247 (1959).

The group of linear transformations of the x_i and y_k which in the case $n = 2m$ takes the normal form f_m into itself is called the *complex group* or the *symplectic group*. For the structure of this group as well as that of the orthogonal and general linear groups, see J. Dieudonné, *Sur les Groupes Classiques*, Hermann, Paris, 1948.

Chapter 13

ALGEBRAS

A ring $\mathfrak{A}$ which is also a finite-dimensional vector space over a field P and satisfies the condition

$$(\alpha u)v = u(\alpha v) = \alpha(uv) \qquad \text{for} \quad \alpha \in \mathrm{P}$$

is called an *associative algebra* or a *hypercomplex system* over P. If the requirement of associativity is dropped, we have the more general idea of a (linear) *algebra*. There are two types of nonassociative algebras which deserve special mention.

1. *Alternative rings* in which the following restricted associative law holds:

$$a(ab) = (aa)b$$

$$b(aa) = (ba)a.$$

The oldest example of a genuine alternative ring is the algebra of Cayley numbers; for an account of this, see M. Zorn, "Alternativkörper und Quadratische Systeme", *Abh. math. Sem. Univ. Hamburg*, **9**, 395 (1933). Alternative rings are important in the axiomatics of plane geometry.[1] For recent investigations, see R. D. Schafer, "Structure and Representation of Non-Associative Algebras", *Bull. Amer. Math. Soc.*, **61**, 469 (1955).

2. *Lie rings*, in which the following composition laws hold:

$$ab - ba = 0$$

$$a \cdot bc + b \cdot ca + c \cdot ab = 0.$$

The infinitesimal generators of Lie groups satisfy these rules. In the fundamental work of E. Cartan[2] and H. Weyl[3], Lie rings were studied in connection

[1]R. Moufang, "Alternativkörper und Satz vom Vollständigen Vierseit," *Abh. Math. Sem. Univ. Hamburg*, **9**, 207; see also *Math. Ann.*, **110**, 416. For further discussion, see H. Freudenthal, "Zur Ebenen Oktavengeometrie," *Proc. Akad. Amsterdam*, **A 56**, 195 (1953); also **A 57**, 218, 363, and **A 58**, 151.

[2]E. Cartan, *These*, 1894. Also, H. Freudenthal, *Proc. Akad. Amsterdam*, **A 56** (1953).

[3]H. Weyl, "Darstellung Halbeinfacher Gruppen durch Lineare Transformation I–III", *Math. Z.*, **23**, 271 (1925), and **24**, 328, 789 (1926). Also, B. L. van der Waerden, *Math. Z.*, **37**, 446.

with Lie groups. For recent investigations, see E. Witt. *J. Reine u. Angew. Math.*, **177**, 152 (1937), and *Abh. Math. Sem. Univ. Hamburg*, **14**, 289 (1941); H. Freudenthal, *Proc. Akad. Amsterdam*, **A 57**, 369, 487 (1954); **A 59**, 511 (1956); **A61**, 379 (1958).

In this book we shall restrict ourselves to associative algebras of finite dimension over P. The word "algebra" is henceforth always meant in this restricted sense.

13.1 DIRECT SUMS AND INTERSECTIONS

Emmy Noether repeatedly emphasized in her lectures the importance of the relationship between the direct sum and intersection representations of modules. This idea weaves through her work in a striking manner. We shall now explain this relationship. We begin with multiplicative groups and then go over to the additive notation.

Let a group $\mathfrak{G}$ be the direct product of subgroups $\mathfrak{A}_1, \ldots, \mathfrak{A}_n$. This means:

1. Each $\mathfrak{A}_1$ is a normal subgroup of $\mathfrak{G}$;
2. The product $\mathfrak{A}_1 \ldots \mathfrak{A}_n$ is $\mathfrak{G}$;
3. If $\mathfrak{B}_i$ is the product of all the $\mathfrak{A}_j$ except $\mathfrak{A}_i$, then

$$\mathfrak{A}_i \cap \mathfrak{B}_i = \mathfrak{E},$$

where $\mathfrak{E}$ consists of the identity element alone.

From properties 1, 2, and 3 it follows by Section 7.6 that each element g of $\mathfrak{G}$ can be uniquely represented as a product $a_1 \ldots a_n$ $(a_i \in \mathfrak{A}_i)$ and that for $i \neq j$ each element of $\mathfrak{A}_i$ commutes with every element of $\mathfrak{A}_j$. From property 2 it further follows that

$$\mathfrak{A}_i\mathfrak{B}_i = \mathfrak{G}.$$

The group $\mathfrak{B}_i$ consists of all products $a_1 \ldots a_n$ in which the factor a_i is equal to e. From this it follows that the intersection of all the $\mathfrak{B}_i$ is equal to $\mathfrak{E}$ and that the intersection of all the $\mathfrak{B}_j$ with $j \neq i$ is equal to $\mathfrak{A}_i$. The $\mathfrak{B}_i$ thus have the following three properties, which in a certain sense are dual to properties 1, 2, and 3:

1′. Each $\mathfrak{B}_i$ is a normal subgroup of $\mathfrak{G}$;
2′. The intersection $\mathfrak{B}_1 \cap \ldots \cap \mathfrak{B}_n$ is $\mathfrak{E}$;
3′. If $\mathfrak{A}_i$ is the intersection of all the $\mathfrak{B}_j$ except $\mathfrak{B}_i$, then

$$\mathfrak{A}_i\mathfrak{B}_i = \mathfrak{G}.$$

If properties 1′, 2′, and 3′ are satisfied, then the identity group $\mathfrak{E}$ is called the *direct intersection* of $\mathfrak{B}_1, \ldots, \mathfrak{B}_n$. If, in property 2′, $\mathfrak{E}$ is replaced by a group $\mathfrak{D}$ while (1′) and (3′) continue to hold, then $\mathfrak{D}$ is called the direct intersection of $\mathfrak{B}_1, \ldots, \mathfrak{B}_n$. By forming the factor groups $\mathfrak{G}/\mathfrak{D}$ and $\mathfrak{B}_i/\mathfrak{D}$, this more general case can be immediately reduced to the case $\mathfrak{D} = \mathfrak{E}$.

We shall now assume properties 1′, 2′, and 3′ and prove properties 1, 2, 3.

If $\mathfrak{A}_i$ is defined by (3'), then it follows from (2') that

$$\mathfrak{A}_i \cap \mathfrak{B}_i = \mathfrak{E}. \tag{13.2}$$

As intersections of normal subgroups, the $\mathfrak{A}_i$ are again normal subgroups of $\mathfrak{G}$. We shall now show that their product is $\mathfrak{G}$ and that the product of all $\mathfrak{A}_j$ excepting $\mathfrak{A}_i$ is $\mathfrak{B}_i$.

Let g be an element of $\mathfrak{G}$. By (13.1) and (13.2), $\mathfrak{G}$ is the direct product of $\mathfrak{A}_i$ and $\mathfrak{B}_i$, and g can therefore be uniquely represented as the product

$$g = a_i b_i \qquad (a_i \in \mathfrak{A}_i,\ b_i \in \mathfrak{B}_i).$$

Further, each element of $\mathfrak{A}_i$ commutes with every element of $\mathfrak{B}_i$ and so, in particular, with every element of $\mathfrak{A}_j$ $(j \neq i)$. If we form the product

$$g' = a_1 \cdots a_n,$$

then

$$g^{-1}g' = b_i^{-1}a_i^{-1}a_1 \cdots a_n.$$

Since the a_j commute, this may be written:

$$g^{-1}g' = b_i^{-1}a_1 \cdots a_{i-1}a_{i+1} \cdots a_n.$$

All the factors on the right-hand side are contained in $\mathfrak{B}_i$, and therefore $g^{-1}g'$ is in $\mathfrak{B}_i$ for every i. From property 2' it now follows that

$$g^{-1}g' = e,$$

and thus $g' = g$. Thus each element g of $\mathfrak{G}$ can be represented as a product $a_1 \ldots a_n$. If g is in $\mathfrak{B}_i$, then the factor $a_i = e$; therefore, each element of $\mathfrak{B}_i$ can be represented as a product,

$$a_1 \cdots a_{i-1}a_{i+1} \cdots a_n.$$

Hence the product of all the $\mathfrak{A}_j$ is $\mathfrak{G}$, and the product of all the $\mathfrak{A}_j$ except $\mathfrak{A}_i$ is $\mathfrak{B}_i$. The $\mathfrak{A}_i$ therefore have properties 1, 2, and 3.

From (13.1) and (13.2) it follows by the first isomorphism theorem that

$$\mathfrak{G}/\mathfrak{B}_i \cong \mathfrak{A}_i.$$

In additive notation, what has been proved can be formulated as follows.

If a module $\mathfrak{G}$ *is the direct sum of submodules* $\mathfrak{A}_1, \ldots, \mathfrak{A}_n$ *and if* $\mathfrak{B}_i$ *is the sum of all the* $\mathfrak{A}_j$ *with the exception of* $\mathfrak{A}_i$, *then* $\{0\}$ *is the direct intersection of* $\mathfrak{B}_1, \ldots \mathfrak{B}_n$ *and* $\mathfrak{A}_i$ *is the intersection of all the* $\mathfrak{B}_j$ *except* $\mathfrak{B}_i$ *and conversely. Further,* $\mathfrak{G}/\mathfrak{B}_i \cong \mathfrak{A}_i$.

All this is also true for groups with operators. In ring theory applications $\mathfrak{G}$ is a ring with $\mathfrak{G}$ itself as left or right operator domain. The modules $\mathfrak{A}_i$ and $\mathfrak{B}_i$ are then left or right ideals in $\mathfrak{G}$. We are then concerned with a representation of a ring $\mathfrak{G}$ as a direct sum of left or right ideals $\mathfrak{A}_i$ and with a corresponding representation of the null ideal as a direct intersection of left or right ideals $\mathfrak{B}_i$.

The group theory notation will be retained, since in this theory each ring is interpreted as an additive group with itself as operator domain.

If the $\mathfrak{A}_i$ (and hence also the $\mathfrak{B}_i$) are two-sided ideals, then $\mathfrak{A}_i\mathfrak{A}_j$ is contained in both $\mathfrak{A}_i$ and $\mathfrak{A}_j$. However, for $i \neq j$, $\mathfrak{A}_i \cap \mathfrak{A}_j$ is the null ideal, and hence $\mathfrak{A}_i\mathfrak{A}_j = \{0\}$. We thus have the following statement.

If a ring $\mathfrak{G}$ is the direct sum of two-sided ideals $\mathfrak{A}_i$,

$$\mathfrak{G} = \mathfrak{A}_1 + \cdots + \mathfrak{A}_n, \tag{13.3}$$

then the $\mathfrak{A}_i$ are rings which annihilate one another:

$$\mathfrak{A}_i\mathfrak{A}_j = \{0\} \qquad \text{for} \quad i \neq j. \tag{13.4}$$

Conversely, if $\mathfrak{G}$, considered as an additive group, is the direct sum of rings $\mathfrak{A}_i$ which annihilate one another, then these $\mathfrak{A}_i$ are two-sided ideals in $\mathfrak{G}$. The proof is obvious. In this case the ring $\mathfrak{G}$ (or the algebra $\mathfrak{G}$) is said to be the *direct sum of the rings* (or algebras) $\mathfrak{A}_i$.

If (13.3) and (13.4) are satisfied, then the structure of the ring $\mathfrak{G}$ is determined by the structure of the rings $\mathfrak{A}_i$ in a simple manner. Indeed, if g and h are ring elements and if they are represented in accordance with (13.3) by the sums

$$g = g_1 + \cdots + g_n$$
$$h = h_1 + \cdots + h_n,$$

then it follows that

$$g+h = (g_1+h_1) + \cdots + (g_n+h_n)$$
$$gh = g_1h_1 + g_2h_2 + \cdots + g_nh_n;$$

that is, two elements are added or multiplied by adding or multiplying their components.

Exercise

13.1. If a ring with identity is the direct sum of left ideals,

$$\mathfrak{G} = \mathfrak{l}_1 + \cdots + \mathfrak{l}_n \tag{13.5}$$

and if the decomposition of the identity is given by

$$e = e_1 + \cdots + e_n, \tag{13.6}$$

then $\mathfrak{l}_i = \mathfrak{G}e_i$ and

$$e_i^2 = e_i \tag{13.7}$$

$$e_ie_j = 0 \qquad (i \neq j). \tag{13.8}$$

Conversely, if (13.6), (13.7), and (13.8) are satisfied and if we define

$$\mathfrak{l}_i = \mathfrak{G}e_i, \tag{13.9}$$

then $\mathfrak{G}$ is the direct sum of the left ideals $\mathfrak{l}_i$.

Similarly, if we make the definition

$$\mathfrak{r}_i = e_i\mathfrak{G}, \tag{13.10}$$

then $\mathfrak{G}$ is the direct sum of the right ideals $\mathfrak{r}_i$.

13.2 EXAMPLES OF ALGEBRAS

Example 1: An important example of an algebra is the *complete matrix ring* P_n consisting of $n \times n$ matrices with elements in P. This algebra has rank n^2. As basis elements we may choose the matrices C_{ik} having a 1 in the ith row and kth column and zeros elsewhere. Each matrix A with elements α_{ik} can be represented as the sum

$$\sum C_{ik}\alpha_{ik},$$

where the summation extends over all i and k from 1 to n. The rules for the multiplication of the basis elements C_{ik} are

$$C_{hi}C_{jk} = 0 \qquad (i \neq j)$$

$$C_{hi}C_{ik} = C_{hk}.$$

Example 2: *The algebra of quaternions.* Let $\mathfrak{A}$ be a four-dimensional vector space with basis elements e, j, k, l. Here e is taken to be the identity element, so that $e^2 = e$, $ej = j$, and so on. Further, let

$$j^2 = -e\alpha, \qquad k^2 = -e\beta,$$

where α and β are arbitrary elements of P, and let

$$jk = -kj = l.$$

It then follows that

$$\begin{aligned} l^2 &= jkjk &&= -jjkk &&= -e\alpha\beta \\ jl &= jjk &&= -e\alpha k &&= -k\alpha \\ lj &= -kjj &&= +ke\alpha &&= k\alpha \\ kl &= -kkj &&= +e\beta j &&= j\beta \\ lk &= jkk &&= -je\beta &&= -j\beta. \end{aligned}$$

The algebra $\mathfrak{A}$ so defined is called a *generalized quaternion algebra.* Its elements are

$$x = ex_0 + jx_1 + kx_2 + lx_3 \qquad (x_0, x_1, x_2, x_3 \text{ in } \mathsf{P}).$$

Identifying the element ex_0 with x_0, it is seen that P is imbedded in $\mathfrak{A}$.

The *norm* of an element x is defined as

$$\begin{aligned} N(x) = x\bar{x} &= (ex_0 + jx_1 + kx_2 + lx_3)(ex_0 - jx_1 - kx_2 - lx_3) \\ &= x_0{}^2\alpha x_1{}^2 + \beta x_2{}^2 + \alpha\beta x_3{}^2. \end{aligned}$$

If this quadratic form is indefinite (that is, if it is zero for values of x_i not all zero), then $x\bar{x}$ can vanish for $x \neq 0$, and $\mathfrak{A}$ has zero divisors. If the form is not indefinite, then every $x \neq 0$ has an inverse

$$x^{-1} = \bar{x}(x_0{}^2+\alpha x_1{}^2+\beta x_2{}^2+\alpha\beta x_3{}^2)^{-1},$$

and $\mathfrak{A}$ is then a skew field.

A *matrix representation* of the generalized quaternion algebra $\mathfrak{A}$ can be obtained by considering $\mathfrak{A}$ to be a double module with $\mathfrak{A}$ as left and $\Sigma = \mathsf{P}(j)$ as right operator domain. If we assume that $-\alpha$ is not a square in P, then

$$\Sigma = \mathsf{P}(j) = \mathsf{P}(\sqrt{-\alpha})$$

is a field. Then $\mathfrak{A}$ itself is a two-dimensional vector space over this field; we may take, say, e and $-k$ as basis elements. The vectors x are then

$$x = e(x_0+jx_1)+(-k)(-x_2+jx_3). \tag{13.11}$$

If the vectors x are multiplied on the left by an arbitrary element y, we obtain a linear transformation Y of the vector space $\mathfrak{A}$ which can be represented by a matrix. We denote this matrix by Y also. The columns of the matrix Y are obtained by multiplying the basis elements e and $-k$ on the left by y and writing the results again in the form (13.11). If, in particular, we take y equal to j, k, or l, then we obtain the matrices

$$J = \begin{pmatrix} j & 0 \\ 0 & -j \end{pmatrix}, \quad K = \begin{pmatrix} 0 & \beta \\ -1 & 0 \end{pmatrix}, \quad L = \begin{pmatrix} 0 & j\beta \\ j & 0 \end{pmatrix}. \tag{13.12}$$

Choosing $\alpha = \beta = 1$, we obtain Hamilton's quaternions

$$x = ex_0+jx_1+kx_2+lx_3$$

with the composition rules

$$\begin{aligned} j^2 &= k^2 = l^2 = -1 \\ jk &= l, \qquad kj = -l \\ kl &= j, \qquad lk = -j \\ lj &= k, \qquad jl = -k. \end{aligned}$$

If P is a real number field, then in the matrix representation j may be replaced by the imaginary unit i. We then obtain

$$J = \begin{pmatrix} i & 0 \\ 0 & -i \end{pmatrix}, \quad K = \begin{pmatrix} 0 & 1 \\ -1 & 0 \end{pmatrix}, \quad L = \begin{pmatrix} 0 & i \\ i & 0 \end{pmatrix}.$$

Example 3: *The group ring* of a finite group is obtained by taking the group elements $u_1, \ldots, u_n$ as the basis elements of an algebra. The associative law is necessarily satisfied.

Example 4: *The Grassmann exterior product.* We start with a vector space

$$\mathfrak{M} = u_1\mathsf{P}+\cdots+u_n\mathsf{P}$$

and consider the problem of defining an associative multiplication of vectors such that

$$uu = 0 \qquad \text{and} \qquad uv+vu = 0. \tag{13.13}$$

To this end, we first form purely formal products of basis vectors u_i in the natural order:

$$u_{ijk\ldots} = u_i u_j u_k \ldots \qquad (i<j<k\ldots);$$

the empty product is denoted by e. These 2^n products we take as basis elements of a vector space $\mathfrak{A}$. The elements of $\mathfrak{A}$ are thus all sums of the form

$$e\alpha + \sum_i u_i\alpha_i + \sum_{i<j} u_{ij}\alpha_{ij} + \cdots + u_{12\ldots n}\alpha_{12\ldots n}. \tag{13.14}$$

Arbitrary products

$$u_{abc\ldots} = u_a u_b u_c \ldots \tag{13.15}$$

are now defined as follows. If in (13.15) two indices are equal, then we set $u_{abc\ldots} = 0$. If all indices are distinct, they are brought into natural order $ijk\ldots$ by a permutation, and we put

$$u_{abc\ldots} = \varepsilon u_{ijk\ldots},$$

where $\varepsilon = +1$ if the permutation is even and -1 if the permutation is odd.

Finally, the product of two basis elements is defined by

$$u_{ijk\ldots}u_{pqr\ldots} = u_{ijk\ldots pqr\ldots}. \tag{13.16}$$

Two sums of the type (13.14) are multiplied by multiplying the individual terms according to (13.16) and adding the results. According to this definition, the product $u_a u_b \ldots$ is in fact equal to $u_{ab\ldots}$ as required by (13.15). The composition rules (13.13) are clearly satisfied, and the associative law for multiplication can be easily demonstrated.

With this definition of multiplication, the sums (13.14) form the *Grassmann algebra* $\mathfrak{A}$ going with the vector space $\mathfrak{M}$; the multiplication itself is called the *exterior product.* The vector space $\mathfrak{M}$ is imbedded in $\mathfrak{A}$. The notation $a \wedge b$ is frequently used for the exterior product of a and b.

An equivalent definition is obtained by first forming from $\mathfrak{M}$ the *tensor ring* which consists of all finite sums

$$e\beta + \sum u_i\beta_i + \sum u_{ij}\beta_{ij} + \sum u_{ijk}\beta_{ijk} + \cdots \tag{13.17}$$

in which the indices $i, j, \ldots$ are unrestricted. Two such sums are called equal if all their coefficients are equal. It is clear how such sums are added. Multiplication is defined by (13.16).

It is easily seen that addition and multiplication in the tensor ring are independent of the choice of basis.

In the tensor ring $\mathfrak{T}$ we now form the two-sided ideal $\mathfrak{J}$ generated by products uu where u runs through all vectors of $\mathfrak{M}$. The ideal also contains

$$(u+v)(u+v) - uu - vv = uv + vu.$$

A homomorphism of $\mathfrak{T}$ onto $\mathfrak{A}$ is obtained by sending each sum (13.17) into the same sum in $\mathfrak{A}$. Under this mapping the elements of $\mathfrak{J}$ are sent to the zero element. Conversely, if a sum (13.17) is sent by this mapping into zero, then it belongs to the ideal $\mathfrak{J}$. Indeed, the sum (13.17) may first be written as

$$e\beta + \sum u_i\beta_i + \sum u_iu_j\beta_{ij} + \sum u_iu_ju_k\beta_{ijk} + \cdots, \tag{13.18}$$

and by addition of elements of $\mathfrak{J}$ it can then be brought to the normal form

$$e\alpha + \sum_i u_i\alpha_i + \sum_{i<j} u_iu_j\alpha_{ij} + \sum_{i<j<k} u_iu_ju_k\alpha_{ijk} + \cdots,$$

which corresponds to an element (13.14) of $\mathfrak{A}$ with the same coefficients α, α_i, α_{ij}, ... If this element is zero, all these coefficients are zero, and the sum (13.18) therefore lies in $\mathfrak{J}$. Thus $\mathfrak{J}$ is precisely the kernel of the ring homomorphism $\mathsf{T} \to \mathfrak{A}$, and we have

$$\mathfrak{A} \cong \mathfrak{T}/\mathfrak{J}. \tag{13.19}$$

On the right-hand side of (13.19). $\mathfrak{T}$ and $\mathfrak{J}$ are independent of the choice of basis $(u_1, \ldots, u_n)$. Up to isomorphism, $\mathfrak{A}$ is therefore also independent of the choice of basis. An invariant definition of the Grassmann algebra $\mathfrak{A}$ is obtained if we define $\mathfrak{A}$ directly as $\mathfrak{T}\mathfrak{J}$.

Example 5: *Clifford algebras.* These algebras can be defined much the same as the Grassmann algebra. Let $Q(x)$ be a quadratic form in $x_1, \ldots, x_n$ with coefficients in P:

$$Q(x_1, \ldots, x_n) = \sum_i q_ix_i^2 + \sum_{i<j} q_{ij}x_jx_i.$$

For every vector $u = \sum u_i\gamma_i$ of $\mathfrak{M}$, the value of the form

$$Q(u) = Q(\gamma_1, \ldots, \gamma_n)$$

is then defined. Moreover, for any two vectors u and v the symmetric bilinear form

$$B(u, v) = Q(u+v) - Q(u) - Q(v)$$

is also defined. In particular

$$Q(u_i) = q_i$$

$$B(u_i, u_j) = B(u_j, u_i) = q_{ij} \qquad (i<j).$$

We now wish to define a multiplication of vectors such that

$$uu = Q(u) \tag{13.20}$$

$$uv + vu = B(u, v). \tag{13.21}$$

Here (13.21) is a consequence of (13.20):

$$\begin{aligned} uv + vu &= (u+v)(u+v) - uu - vv \\ &= Q(u+v) - Q(u) - Q(v) = B(u, v). \end{aligned}$$

In particular, we must have

$$u_i u_i = q_i \tag{13.22}$$

$$u_i u_j + u_j u_i = q_{ij} \qquad (i<j). \tag{13.23}$$

We again form a 2^n-dimensional vector space consisting of the sums

$$e a + \sum_i u_i \alpha_i + \sum_{i<j} u_{ij} \alpha_{ij} + \cdots + u_{12\ldots n} \alpha_{12\ldots n}. \tag{13.24}$$

Arbitrary products

$$u_a u_b u_c \ldots = u_{abc\ldots} \tag{13.25}$$

will now be defined. If the indices $a, b, c \ldots$ are distinct and in natural order, then $u_{abc\ldots}$ is again one of our basis vectors. In all other cases the product $u_a u_b u_c \ldots$ is transformed by means of relations (13.22) and (13.23). If, for example, bc is the first pair of adjacent indices for which $b<c$ does not hold, then we write the product (13.25) as

$$u_a(u_b u_c) \ldots$$

and reduce $u_b u_c$ by (13.22) or (13.23):

$$u_b u_b = q_b \qquad (c = b)$$

$$u_b u_c = -u_c u_b + q_{cb} \qquad (c < b).$$

The factors q_b and q_{cb} are written in front of the product. We thus have:

$$u_a(u_b u_b) \ldots = q_b u_a \ldots$$

$$u_a(u_b u_c) \ldots = -u_a u_c u_b \ldots + q_{cb} u_a \ldots.$$

After transformation the product contains either two fewer factors or one less inversion. Continuing in this manner, an expression of the form (13.24) is finally obtained.

Once this meaning of $u_{abc\ldots}$ has been settled, the product of two basis elements can again be defined by (13.16) and the associative law of multiplication demonstrated. This completes the definition of the Clifford algebra $\mathfrak{C}$ belonging to the form $Q(x)$. If Q is the zero form, then the Clifford algebra becomes the Grassmann algebra.

If, in (13.24), we consider only terms $u_{ij\ldots}$ which have an even number of indices,

$$e\alpha + \sum_{i<j} u_{ij}\alpha_{ij} + \sum_{i<j<k<l} u_{ijkl}\alpha_{ijkl} + \cdots,$$

we obtain a subalgebra called the *second Clifford algebra* $\mathfrak{C}_+$.

An invariant definition of the algebra $\mathfrak{C}$ is obtained if in the tensor ring $\mathfrak{T}$ we take the ideal $\mathfrak{J}$ generated by all expressions of the type

$$uu - Q(u)$$

and form the residue class ring $\mathfrak{T}/\mathfrak{J}$. Starting with this definition, Chevalley[4]

[4] C. Chevalley, *The Algebraic Theory of Spinors*, Columbia University Press, 1954.

developed the theory of Clifford algebras for arbitrary base fields. A simple proof that the invariant definition coincides with the definition given above can be found in the paper by B. L. van der Waerden, *Proc. Kon. Ned. Akad. Amsterdam*, **69**, 78.

The second Clifford algebra was used by Brauer and Weyl (*Amer. J. Math.*, **57**, 245) to represent orthogonal transformations (that is, linear transformations T with determinant 1 which leave the quadratic form Q invariant) in the form

$$Tu = sus^{-1}.$$

Here u runs through the vector space $\mathfrak{M}$, and the s are elements of $\mathfrak{C}_+$ which transform the space $\mathfrak{M}$ into itself:

$$s\mathfrak{M}s^{-1} = \mathfrak{M}.$$

We must hereby require that the characteristic of P be different from 2 and that the form Q be nonsingular. In the case of characteristic 2, the situation is somewhat more complicated (see Chevalley, Theorem II. 3.3).

Exercises

13.2. The second Clifford algebra of a binary quadratic form

$$Q(x_1, x_2,) = q_1x_1^2 + q_{12}x_1x_2 + q_2x_2^2,$$

which does not split into linear factors in the base field P, is the quadratic extension field in which the form splits.

13.3. The second Clifford algebra of the ternary quadratic form

$$Q(x_1, x_2, x_3) = q_1x_1^2 + q_2x_2^2 + q_3x_3^2$$

is an algebra of generalized quaternions.

13.3 PRODUCTS AND CROSSED PRODUCTS

PRODUCTS OF VECTOR SPACES

Let $\mathfrak{A}$ and $\mathfrak{B}$ be finite-dimensional vector spaces over a field P:

$$\mathfrak{A} = u_1\mathrm{P} + \cdots + u_m\mathrm{P}$$

$$\mathfrak{B} = v_1\mathrm{P} + \cdots + v_n\mathrm{P}.$$

We wish to define the product $\mathfrak{A} \times \mathfrak{B}$. To this end, from mn basis vectors w_{ik}, where i runs from 1 to m and k from 1 to n, we form a *product space*

$$\mathfrak{C} = \sum_{i,k} w_{ik}\mathrm{P}$$

and for each u of $\mathfrak{A}$ and each v of $\mathfrak{B}$ define a product

$$uv = (\sum u_i\alpha_i)(\sum v_k\beta_k) = \sum_{i,k} w_{ik}\alpha_i\beta_k.$$

In particular,

$$u_iv_k = w_{ik}.$$

All elements of $\mathfrak{C}$ are therefore expressions of the following type:

$$w = \sum_{i,k} u_iv_k\gamma_{ik} = \sum_{i,k} w_{ik}\gamma_{ik}. \tag{13.26}$$

These expressions are called *tensors of rank* 2, and the product space $\mathfrak{C}$ is called a *tensor space*.

Instead of (13.26) we may also write

$$w = \sum_i u_ib_i, \tag{13.27}$$

where the b_i are elements of $\mathfrak{B}$. The space $\mathfrak{C}$ is therefore the direct sum of the subspaces $u_i\mathfrak{B}$:

$$\mathfrak{C} = u_1\mathfrak{B} + \cdots + u_m\mathfrak{B}. \tag{13.28}$$

Formula (13.28) shows that the module $\mathfrak{C}$ is independent of the choice of basis in $\mathfrak{B}$. The elements w of $\mathfrak{C}$ may be written directly in the form (13.27), and their addition and multiplication by elements of P defined without introducing a basis in $\mathfrak{B}$.

We may similarly write, instead of (13.26),

$$w = \sum a_kv_k. \tag{13.29}$$

Therefore,

$$\mathfrak{C} = \mathfrak{A}v_1 + \cdots + \mathfrak{A}v_n, \tag{13.30}$$

and it is evident that the product space $\mathfrak{C}$ is also independent of the choice of basis in $\mathfrak{A}$. According to (13.28), the module $\mathfrak{C} = \mathfrak{A}\times\mathfrak{B}$ can also be formed if $\mathfrak{A}$ is a finite-dimensional vector space and $\mathfrak{B}$ is any P-module. Similarly, according to (13.30), $\mathfrak{C}$ can be formed if $\mathfrak{A}$ is an arbitrary P-module and $\mathfrak{B}$ is any finite dimensional vector space.

The module product $\mathfrak{A}\times\mathfrak{B}$ can also be defined without using bases. This invariant definition is even meaningful if P is a commutative ring with identity and $\mathfrak{A}$ and $\mathfrak{B}$ are arbitrary P-modules, as long as the identity element of P is the identity operator for $\mathfrak{A}$ and $\mathfrak{B}$. Since we shall here need only the case in which P is a field and $\mathfrak{A}$ or $\mathfrak{B}$ has finite dimension, we restrict ourselves to the definition initially given and for the general case refer the reader to N. Bourbaki's *Algèbre multilinéaire* (*Eléments de Mathématique*, Vol. II, Chap. 3; Actualités scient., 1044).

In exactly the same way the product space of three or more vector spaces

$$\mathfrak{A}\times\mathfrak{B}\times\mathfrak{C} = (\mathfrak{A}\times\mathfrak{B})\times\mathfrak{C} = \mathfrak{A}\times(\mathfrak{B}\times\mathfrak{C}) \tag{13.31}$$

can be formed.

PRODUCTS OF ALGEBRAS

If $\mathfrak{A}$ and $\mathfrak{B}$ are algebras over P, then the module $\mathfrak{C} = \mathfrak{A} \times \mathfrak{B}$ can be made an algebra by defining the products of basis elements w_{ik} as follows:

$$w_{ik} w_{jl} = (u_i v_k)(u_j v_l) = (u_i u_j)(v_k v_l). \tag{13.32}$$

This can be done independently of the basis in $\mathfrak{B}$ by writing the elements of $\mathfrak{C}$ in the form (13.27) and defining the product by

$$(\sum u_i b_i)(\sum u_j b_j') = \sum_{i,j} u_i u_j b_i b_j'.$$

This may also be expressed as follows. *The products $u_i u_j$ of basis elements of $\mathfrak{A}$ are formed exactly as they are defined in $\mathfrak{A}$, but $\mathfrak{B}$ instead of* P *is taken as the ring of coefficients.* The algebra so obtained is denoted by $\mathfrak{A}_{\mathfrak{B}}$. This same notation is used if $\mathfrak{B}$ is an arbitrary ring which contains P in its center. Therefore $\mathfrak{A}_{\mathfrak{B}}$ is quite generally an algebra with the same basis elements as $\mathfrak{A}$ but with $\mathfrak{B}$ as the ring of coefficients.

Clearly,

$$\mathfrak{A} \times \mathfrak{B} \cong \mathfrak{B} \times \mathfrak{A}.$$

An isomorphism between $\mathfrak{A} \times \mathfrak{B}$ and $\mathfrak{B} \times \mathfrak{A}$ is indeed obtained by mapping $u_i v_k$ onto $v_k u_i$ and extending this mapping linearly to the sums (13.26).

The product relations for complete matrix rings deserve special mention. Here $\mathfrak{A}_r$ shall always denote the ring of $r \times r$ matrices with coefficients in an arbitrary ring $\mathfrak{A}$. We have:

$$\mathfrak{A} \times \mathsf{P}_r \cong \mathfrak{A}_r \tag{13.33}$$

$$\mathsf{P}_r \times \mathsf{P}_s = \mathsf{P}_{rs}. \tag{13.34}$$

To prove (13.33) we need only note that the matrices $\mathfrak{C}_{ik}$ defined in Section 13.2 form a basis for P_r. To form $\mathfrak{A} \times \mathsf{P}_r$ we must take these same basis elements with now $\mathfrak{A}$ as the ring of coefficients, but this gives precisely $\mathfrak{A}_r$.

We obtain (13.34) as follows. If P_r is generated by the r^2 basis elements C'_{ik} and P_s by the s^2 basis elements C''_{jl}, then $\mathsf{P}_r \times \mathsf{P}_s$ is generated by the $r^2 s^2$ products

$$C_{ij,kl} = C'_{ik} C''_{jl},$$

which satisfy the composition rules

$$C_{ij,kl} \cdot C_{mn,pq} = \begin{cases} 0 & \text{if } k \neq m \quad \text{or} \quad l \neq n \\ C_{ij,pq} & \text{if } k = m \quad \text{and} \quad l = n. \end{cases}$$

If we denote the rs pairs of indices ij by an index J which runs from 1 to rs, then

$$C_{JK} C_{LM} = \begin{cases} 0 & \text{for } K \neq L \\ C_{JM} & \text{for } K = L, \end{cases}$$

and this gives the isomorphism to P_{rs}.

CROSSED PRODUCTS

Let Σ be a separable, finite, normal extension field of P. The Galois group $\mathfrak{G}$ of Σ (Section 8.1) consists of all the automorphisms S_i of Σ which leave the elements of P fixed. We shall assume that the Galois theory as well as the contents of Section 8.1 are known and, in particular, the fact that the order of $\mathfrak{G}$ is equal to the degree of the extension $n = (\Sigma : \mathsf{P})$.

We denote the image of an element β of Σ under the automorphism S by β^S. The product of S and T (first S and then T) will now be denoted by ST and thus

$$\beta^{ST} = (\beta^S)^T.$$

The *crossed product of a field* Σ *with its Galois field group* $\mathfrak{G}$, which was introduced by E. Noether, is defined as follows. A vector space

$$\mathfrak{A} = u_1\Sigma + \cdots + u_n\Sigma$$

is formed by assigning to each group element S_i a basis vector u_i. If the index is dropped and S_i is denoted simply by S, then the corresponding u_i is denoted by u_S. The vector space $\mathfrak{A}$ therefore consists of all sums

$$\sum_i u_i\beta_i = \sum_S u_S\beta_S. \tag{13.35}$$

The products βu_S are now defined by the formula

$$\beta u_S = u_S\beta^S \tag{13.36}$$

and the products $u_S u_T$ by

$$u_S u_T = u_{ST}\delta_{S,\,T}, \tag{13.37}$$

where the factors $\delta_{S,\,T}$ are for the time being arbitrary nonzero elements of Σ.

Using (13.36) and (13.37), two arbitrary sums (13.35) can be multiplied by multiplying the individual terms,

$$u_S\beta\cdot u_T\gamma = u_S u_T\beta^T\gamma = u_{ST}\delta_{S,\,T}\beta^T\gamma,$$

and adding the products so obtained.

In order that the multiplication defined in terms of the system of factors $\delta_{S,\,T}$, be associative, it is necessary that the $\delta_{S,\,T}$ satisfy the following *associativity condition*:

$$\delta_{S,\,TR}\delta_{T,\,R} = \delta_{ST,\,R}(\delta_{S,\,T})^R. \tag{13.38}$$

There is arbitrariness in the basis elements u_S and therefore also in the factors $\delta_{S,\,T}$, for the u_S may be replaced by

$$v_S = u_S\gamma_S \qquad (\gamma_S \neq 0 \text{ in } \Sigma). \tag{13.39}$$

The system of factors going with this new basis is then

$$\varepsilon_{S,\,T} = \frac{\gamma_S^{\,T}\gamma_T}{\gamma_{ST}}\delta_{S,\,T}. \tag{13.40}$$

Two systems of factors $\delta_{S,\,T}$ and $\varepsilon_{S,\,T}$ related by (13.40) are called *associated.* Associated systems of factors thus define the same algebra $\mathfrak{A}$.

If E is the identity element of $\mathfrak{G}$, then the arbitrary factor for u_E can always be so chosen that

$$u_E u_E = u_E,$$

and thus $\delta_{E,\,E} = 1$. From the associative laws

$$(u_E u_E)u_R = u_E(u_E u_R)$$

$$u_S(u_E u_E) = (u_S u_E)u_E$$

it then follows that u_E is the identity element of $\mathfrak{A}$. The products $u_E\beta$ may therefore be identified with the elements β of Σ.

Which elements $c = \sum u_S\gamma_S$ commute with all β of Σ? The condition $\beta c = c\beta$ gives

$$\sum u_S\beta^S\gamma_S = \sum u_S\gamma_S\beta,$$

and therefore, since the u_S are linearly independent,

$$(\beta^S - \beta)\gamma_S = 0.$$

For $S = E$ this condition is trivially satisfied. For $S \neq E$ there exists a β such that $\beta^S \neq \beta$ and therefore $\gamma_S = 0$. Hence

$$c = u_E\gamma_E = \gamma_E$$

is an element of Σ.

Thus Σ *is a maximal commutative subring of the algebra* $\mathfrak{A}$.

We now determine the *center* of the ring $\mathfrak{A}$, that is, the set of elements c of $\mathfrak{A}$ which commute with all elements of $\mathfrak{A}$:

$$ac = ca \qquad \text{for all } a.$$

If c is a center element, then c must first of all commute with all elements of Σ and it is therefore contained in Σ. We may therefore put $c = \gamma$. In order that γ commute with all the u_S, γ must, by (13.36), be left fixed by all automorphisms S. By the last theorem of Section 8.1, this is possible only if γ lies in the base field P. We therefore have the following statement.

The center of $\mathfrak{A}$ *is* P.

Algebras over P whose center is precisely P are called *central* over P. They were earlier called "normal," but this word already has too many meanings.

We next prove the following.

Theorem: *If in any ring containing* Σ *the relations* (13.36) *and* (13.37) *hold with* $\delta_{S,\,T} \neq 0$, *then either all the* u_S *are zero or they are linearly independent over* Σ.

Proof: If a u_S is linearly dependent on certain other linearly independent u_T, then (for this one S)

$$u_S = \sum_{T \neq S} u_T\gamma_T. \tag{13.41}$$

Multiplying (13.41) on the right by β^S gives

$$u_S\beta^S = \sum_T u_T\gamma_T\beta^S. \tag{13.42}$$

Multiplying (13.41) on the left by β, we obtain, by (13.36),

$$u_S\beta^S = \sum_T u_T\beta^T\gamma_T. \tag{13.43}$$

Since the u_T were assumed to be linearly independent, comparison of (13.42) and (13.43) gives

$$\beta^T\gamma_T = \gamma_T\beta^S$$

or

$$\gamma_T(\beta^T - \beta^S) = 0. \tag{13.44}$$

Since $T \neq S$, we can choose β so that $\beta^T \neq \beta^S$. Then, from (13.44), $\gamma_T = 0$. hTis holds for all T occurring in (13.41), and it therefore follows that $u_S = 0$. From (13.37) it now follows that $u_{ST} = 0$ for all T; that is, all u_S are zero, Q.E.D.

From the theorem just proved now follows: *$\mathfrak{A}$ is simple; that is, there are no two-sided ideals in $\mathfrak{A}$ except for $\{0\}$ and $\mathfrak{A}$ itself.*

Indeed, if $\mathfrak{m}$ is a two-sided ideal, then $\mathfrak{A}/\mathfrak{m}$ is a ring in which the residue classes $\bar{u}_S$ satisfy (13.36) and (13.37); all the $\bar{u}_S$ therefore either are zero or are linearly independent over Σ. In the first case $\mathfrak{m} = \mathfrak{A}$, and in the second case $\mathfrak{m} = \{0\}$.

Combining these results, we have: *The crossed product $\mathfrak{A}$ is a central simple algebra over* P.

CYCLIC ALGEBRAS

If the Galois group $\mathfrak{G}$ is cyclic, then the crossed product $\mathfrak{A}$ is called a *cyclic algebra*. All elements T of $\mathfrak{G}$ are in this case powers of a generator S,

$$T_k = S^k \qquad (k = 0, 1, \ldots, n-1),$$

and all the u_T can be chosen as powers of u_S,

$$u_T = (u_S)^k \qquad (k = 0, 1, \ldots, n-1). \tag{13.45}$$

This choice of the u_T is in agreement with the previous convention of choosing u_E to be the identity element of $\mathfrak{A}$:

$$u_E = (u_S)^0 = e.$$

The nth power of u_S is the product of the $(n-1)$th power and the first power. From this there follows by (13.27) that

$$(u_S)^n = e\delta, \tag{13.46}$$

where δ is an element of Σ. This single element determines the entire system of factors, since, for $i+k<n$,

$$(u_S)^i(u_S)^k = (u_S)^{i+k}$$

and, for $i+k \geqq n$,

$$(u_S)^i \cdot (u_S)^k = (u_S)^{i+k-n} \cdot (u_S)^n = (u_S)^{i+k-n} \cdot \delta.$$

The factors $\delta_{T,R}$ are therefore equal to 1 or δ depending on whether in $T = S^i$ and $R = S^k$ the sum of the exponents $i+k<n$ or $\geqq n$.

Multiplying (13.46) on the left or right by u_S gives

$$(u_S)^{n+1} = u_S\delta = \delta u_S.$$

From this it follows by (13.36) that

$$\delta = \delta^S.$$

Hence δ is invariant under the group $\mathfrak{G}$ and therefore lies in P. If this is the case, then all of the associativity conditions (13.41) are satisfied; δ therefore is subject to no conditions except $\delta \neq 0$ and $\delta \in$ P.

If u_S is replaced by $v_S = u_S\gamma$, then

$$(v_S)^n = (u_S\gamma)(u_S\gamma)\cdots(u_S\gamma)$$
$$= (u_S)^n . \gamma\gamma^S\gamma^{S^2} \ldots \gamma^{S^{n-1}}.$$

The product of all the conjugates of γ is the norm of γ over P. It follows that

$$(v_S)^n = e\varepsilon \qquad \text{with} \quad \varepsilon = \delta N(\gamma). \tag{13.47}$$

We thus have the following.

As the crossed product of a cyclic field Σ with its Galois group $\mathfrak{G}$, a cyclic algebra $\mathfrak{A}$ is completely determined by prescribing a single element $\delta \neq 0$ of the base field P. *This element can be multiplied by the norm of any arbitrary element $\gamma \neq 0$ of Σ without changing the algebra $\mathfrak{A}$.*

Following Hasse, the cyclic algebra $\mathfrak{A}$ is denoted by (δ, Σ, S).

If we take for P a field of characteristic $\neq 2$ and for Σ a quadratic extension field $\mathrm{P}(\sqrt{-\alpha})$ and if we put $\delta = -\beta$, then we obtain as the cyclic algebra precisely the generalized quaternion algebra of Example 2.

The cyclic algebras, in spite of their simple structure, are very general. Indeed, Brauer, Hasse, and Noether have proved a "Hauptsatz" (*J. Reine u. Angew. Math.*, **167**, 399) which states that every central division algebra over an algebraic number field of finite degree is a cyclic algebra.

Exercises

13.4. The center of a ring is again a ring.

13.5. The complete matrix ring P_n is a central simple algebra over P.

13.6. If all the factors $\delta_{S,T} = 1$, then the crossed product of Σ with $\mathfrak{G}$ is the product of Σ with the group ring of $\mathfrak{G}$.

13.4 ALGEBRAS AS GROUPS WITH OPERATORS. MODULES AND REPRESENTATIONS

An algebra $\mathfrak{A}$, considered as an Abelian group with respect to addition, admits two operator domains.

1. The field P. With this operator domain the admissible subgroups are *linear subspaces*.
2. The algebra $\mathfrak{A}$ itself, the elements of which can be considered left or right operators. The admissible subgroups in this case are the *right ideals*, the *left ideals*, and the *two-sided ideals*.

We shall agree once and for all, when considering (left, right, or two-sided) ideals in an algebra, always to include the field P as an operator domain. This means that the admissible *left ideals* are only those subgroups which with a also contain not only every ra (r in $\mathfrak{A}$) but also every $a\beta$ (β in P); a corresponding statement holds for right ideals. The admissible ideals are therefore always linear subspaces as well. Thus, two left ideals are *operator isomorphic* if there exists an isomorphism which, if it takes a into $\bar{a}$, also takes ra into $r\bar{a}$ and $a\beta$ into $\bar{a}\beta$. A left ideal is called *simple* or *minimal* if it contains no admissible left ideals except itself and the null ideal.

With this restriction of the ideal concept, the *maximal* and *minimal* conditions are satisfied for the ideals of an algebra.

Every nonempty set of (right, left, or two-sided) ideals contains a maximal ideal, that is, an ideal which is contained in no other ideal of the set, and a minimal ideal which contains no other ideal of the set.

For by the above agreement every ideal is also a linear subspace, and in every nonempty set of linear spaces of rank $\leqq n$ there exists a space of greatest and a space of least rank.

In order to develop the principal theorems of algebra theory under quite general conditions, we shall henceforth consider in this chapter not only algebras but also arbitrary rings $\mathfrak{o}$, for which we shall assume the maximal or minimal condition for left or right ideals as needed. A ring $\mathfrak{o}$ may be given an operator domain Ω (which assumes the role of P discussed earlier), the operators $\beta, \gamma, \ldots$ of which must satisfy the conditions:

$$(a+b)\beta = a\beta + b\beta \tag{13.48}$$

$$(ab)\beta = (a\beta)b = a(b\beta). \tag{13.49}$$

If such an operator domain is given, then the ideal concept will be restricted as above by the requirement that each ideal with a should also contain $a\beta$ (β in Ω). When we wish to emphasize this explicitly, we shall speak of *admissible* right or left ideals. Only for such ideals will the maximal or minimal condition be imposed.

We must investigate which of the ideal theory concepts, such as sums, products, and so on, remain meaningful for noncommutative rings with or without operator domains. First of all, it is clear that the *intersection* $\mathfrak{a} \cap \mathfrak{b}$ and the sum $(\mathfrak{a}, \mathfrak{b})$ of two admissible right or left ideals are again admissible right or left ideals. A product $\mathfrak{a} \cdot \mathfrak{b}$ (the set of all sums $\sum ab, a \in \mathfrak{a}, b \in \mathfrak{b}$) is, as is easily seen, an admissible right ideal if the second factor is an admissible right ideal, and an admissible left ideal if the first factor is an admissible left ideal. The other factor may be an entirely arbitrary set or even a single element of $\mathfrak{o}$; for example, $p\mathfrak{b}$, the set of all products pa $(a \in \mathfrak{b})$ is a right ideal if $\mathfrak{b}$ is a right ideal.

If $\mathfrak{a}$ is a left ideal and $\mathfrak{c}$ is an arbitrary set in $\mathfrak{o}$, then we can define the left quotients $\mathfrak{a}:\mathfrak{c}$ as the set of x in $\mathfrak{o}$ such that

$$x\mathfrak{c} \subseteqq \mathfrak{a}.$$

The left quotient is again a left ideal, since $x\mathfrak{c} \subseteqq \mathfrak{a}$ and $y\mathfrak{c} \subseteqq \mathfrak{a}$ imply $(x-y)\mathfrak{c} \subseteqq \mathfrak{a}$ and $x\mathfrak{c} \subseteqq \mathfrak{a}$ implies $rx\mathfrak{c} \subseteqq r\mathfrak{a} \subseteqq \mathfrak{a}$ for every r of $\mathfrak{o}$. If $\mathfrak{a}$ and $\mathfrak{c}$ are both left ideals, then $\mathfrak{a}:\mathfrak{c}$ is even a two-sided ideal, since $x\mathfrak{c} \subseteqq \mathfrak{a}$ implies $xr\mathfrak{c} \subseteqq \mathfrak{c} \subseteqq \mathfrak{a}$. The right quotient of two right ideals can be defined similarly, but we shall have no need for it.

In order to indicate how incisive the minimal condition is, we prove the following theorems.

Theorem: *If $\mathfrak{o}$ is a ring with the minimal condition for left ideals and a is an element of $\mathfrak{o}$ which is not a right zero divisor in $\mathfrak{o}$, then the equation $xa = b$ is solvable in $\mathfrak{o}$ for every b.*

Proof: In the set of left ideals $\mathfrak{o}a^n$ $(n = 1, 2, \ldots)$ there must be a minimal one, say $\mathfrak{o}a^m$. Since $\mathfrak{o}a^{m+1} \subseteqq \mathfrak{o}a^m$, the case $\mathfrak{o}a^{m+1} \subset \mathfrak{o}a^m$ must be excluded and therefore $\mathfrak{o}a^{m+1} = \mathfrak{o}a^m$. Thus, each product ba^m may also be written in the form ca^{m+1}:

$$ba^m = ca^{m+1}.$$

Since m factors a may be canceled on the right- and left-hand sides, it follows that

$$b = ca;$$

the equation $xa = b$ is therefore solvable.

Similarly we have the following.

Theorem: *If $\mathfrak{o}$ is a ring with the minimal condition for right ideals and a is not a left zero divisor, then $ax = b$ is solvable.*

There now follows from both theorems: *If $\mathfrak{o}$ is a ring without zero divisors with the minimal condition for both right and left ideals, then $\mathfrak{o}$ is a skew field.*

In particular, any algebra without zero divisors is a skew field. Such algebras are called *division algebras.*

Exercises

13.7. For a ring with identity the above restriction of the ideal concept to include P or Ω as operator domains is not essential: Every ideal admits multiplication with P or Ω.

13.8. The existence of a composition series for the left ideals of a ring $\mathfrak{o}$ is a necessary and sufficient condition for the validity of the maximal *and* minimal condition for these ideals.

In addition to ideals of $\mathfrak{o}$, we consider also $\mathfrak{o}$-modules, and for such modules $\mathfrak{M}$ we shall write the multipliers of $\mathfrak{o}$ on the left. We call these modules left $\mathfrak{o}$-modules. If $a, b, \ldots$ are elements of $\mathfrak{o}$ and $u, v, \ldots$ elements of $\mathfrak{M}$, then it is required that

$$a(u+v) = au+av \tag{13.50}$$

$$(a+b)u = au+bu \tag{13.51}$$

$$(ab)u = a(bu). \tag{13.52}$$

If the ring $\mathfrak{o}$ has an operator domain Ω, then it is required that $\mathfrak{M}$ also admit the operators of Ω (which we shall write on the right). It is hereby required that

$$(u+v)\beta = u\beta+v\beta \tag{13.53}$$

$$(au)\beta = a(u\beta) = (a\beta)u. \tag{13.54}$$

These modules are therefore *double modules* (left $\mathfrak{o}$- and right Ω-modules).

By *submodules* of a module $\mathfrak{M}$ we shall always mean admissible submodules, that is, those which admit the operators of $\mathfrak{o}$ and Ω. A module $\mathfrak{M}$ having no submodules except $\mathfrak{M}$ and $\{0\}$ is called *simple* or *minimal*. The ring $\mathfrak{o}$ is called *simple* if it is simple as a double module with $\mathfrak{o}$ as right and left operator domain (and possibly Ω as an additional operator domain), that is, if it has no (admissible) two-sided ideals except for $\mathfrak{o}$ and $\{0\}$.

Multiplication of the elements of $\mathfrak{M}$ by an element a of $\mathfrak{o}$ generates an endomorphism $\mathfrak{A}$ of the Ω-module $\mathfrak{M}$:

$$au = Au. \tag{13.55}$$

To each element a of $\mathfrak{o}$ there thus corresponds an endomorphism A. To the product ab there corresponds the product AB, and to the sum $a+b$ there corresponds the sum $A+B$, which is defined by

$$(A+B)u = Au+Bu. \tag{13.56}$$

The correspondence $a \rightarrow A$ is therefore a ring homomorphism. If for $\beta \in \Omega$ we define the endomorphism $A\beta$ by

$$(A\beta)u = (Au)\beta, \tag{13.57}$$

then to the product $a\beta$ there corresponds the product $A\beta$. The ring homomorphism $a \rightarrow A$ is therefore also an operator homomorphism with respect to Ω. A ring homomorphism with this property is called a *representation of* $\mathfrak{o}$ (by endomorphisms of the Ω-module $\mathfrak{M}$).

We have seen that every double module $\mathfrak{M}$ (with $\mathfrak{o}$ as left and Ω as right operator domain) provides a representation of $\mathfrak{o}$. Conversely, if a representation

$a \to A$ of $\mathfrak{o}$ by endomorphisms of an Ω-module $\mathfrak{M}$ is given and if we define the product au by (13.55), then $\mathfrak{M}$ becomes a double module with $\mathfrak{o}$ as left and Ω as right operator domain.

If $\mathfrak{o}$ is an algebra over a base field Ω, and thus a vector space over Ω, then we usually consider only modules $\mathfrak{M}$ which are also vector spaces over Ω; that is, the identity of Ω is the identity operator. The endomorphisms are then linear transformations of the vector space $\mathfrak{M}$, and we are concerned with *representations of* $\mathfrak{o}$ *by linear transformations.*

The kernel of the homomorphism $a \to A$ consists of elements a such that $a\mathfrak{M} = \{0\}$; that is, the kernel is the two-sided ideal $\{0\}:\mathfrak{M}$. If the kernel is the null ideal, and the homomorphism is thus an isomorphism, the representation is called *faithful.*

The representation $a \to A$ is called (as in section 12.4) *reducible* if the representation module $\mathfrak{M}$ has a submodule $\mathfrak{N}$ distinct from $\{0\}$ and $\mathfrak{M}$. If no such submodule exists, then $\mathfrak{M}$ is simple and the representation $a \to A$ is called irreducible.

If the module $\mathfrak{M}$ is completely reducible in the sense of Section 7.6, and is thus the direct sum of simple modules, then the representation is called *completely reducible.* The appearance of the matrices of a reducible and that of a completely reducible representation were given by (12.14) and (12.17).

Two representations of $\mathfrak{o}$ are called *equivalent* if they are given by isomorphic double modules. In the case of a finite-dimensional vector space this means that with the appropriate choice of basis the matrices of both representations are the same.

Although these relations are very simple, they are of the greatest importance for the structure and representation theory of algebras. In Section 13.2, Example 2 we obtained a representation of the quaternions in terms of two-by-two matrices by considering the quaternion algebra $\mathfrak{A}$ itself as a double module (with $\mathfrak{A}$ as left and Σ as right operator domain).

13.5 THE LARGE AND SMALL RADICALS

A (left or right) ideal $\mathfrak{a}$ is called *nilpotent* if some power $\mathfrak{a}^m$ is the null ideal $\{0\}$.

Lemma 1: *The sum* $(\mathfrak{a}, \mathfrak{b})$ *of two nilpotent left ideals is nilpotent.*

Proof: Let $\mathfrak{a}^m = \mathfrak{b}^n = 0$. Writing out the ideal $(\mathfrak{a}, \mathfrak{b})^{m+n-1}$, we obtain a sum of products each containing $m+n-1$ factors $\mathfrak{a}$ or $\mathfrak{b}$. In such a product the factor $\mathfrak{a}$ occurs at least m times or the factor $\mathfrak{b}$ occurs at least n times. In, say, the first case the product has the form

$$\ldots\mathfrak{a}\cdots\mathfrak{a}\cdots\mathfrak{a}\ldots$$

with at least m factors $\mathfrak{a}$. Since now $\mathfrak{o}\mathfrak{a} \subseteq \mathfrak{o}$, it follows that

$$\ldots\mathfrak{a}\cdots\mathfrak{a}\cdots\mathfrak{a}\cdots \subseteq \mathfrak{a}^m\cdots = \{0\}.$$

All products are therefore $\{0\}$, and we obtain

$$(\mathfrak{a}, \mathfrak{b})^{m+n-1} = \{0\}.$$

Lemma 2: *Every nilpotent left (or right) ideal is contained in a nilpotent two-sided ideal.*

Proof: Let $\mathfrak{l}$ be a nilpotent left ideal: $\mathfrak{l}^n = \{0\}$. Then $\mathfrak{l}\mathfrak{o}$ is also nilpotent:

$$(\mathfrak{l}\mathfrak{o})^n = \mathfrak{l}(\mathfrak{o}\mathfrak{l})^{n-1}\mathfrak{o} \subseteqq \mathfrak{l}\mathfrak{l}^{n-1}\mathfrak{o} = \mathfrak{l}^n\mathfrak{o} = \{0\}.$$

The right ideal $(\mathfrak{l}, \mathfrak{l}\mathfrak{o})$ generated by $\mathfrak{l}$ is therefore the sum of two nilpotent left ideals. It is thus itself a nilpotent left ideal and therefore a nilpotent two-sided ideal.

We now define the *small radical* $\mathfrak{N}$ of $\mathfrak{o}$ to be the union of all nilpotent two-sided ideals. By Lemma 2, all nilpotent left and right ideals also lie in this union; $\mathfrak{N}$ may therefore also be defined as the union of all nilpotent left (or right) ideals. It may also be said that a lies in $\mathfrak{N}$ if a generates a nilpotent left (or right) ideal.

If the ring $\mathfrak{o}$ is an algebra or, more generally, if $\mathfrak{o}$ satisfies the minimal condition for left ideals, then the small radical $\mathfrak{N}$ coincides with the large radical $\mathfrak{R}$, which will be defined presently. In this case we may therefore omit the adjective "small" and call $\mathfrak{R} = \mathfrak{N}$ the *radical* of the algebra $\mathfrak{A}$.

An algebra without radical, that is, an algebra whose radical is the null ideal, is called *semisimple*. The structure of semisimple algebras has been clarified by J. H. MacLagan-Wedderburn. His principal theorems state as follows.

Theorem: *Every semisimple algebra is the direct sum of simple algebras with identity, and every such simple algebra is isomorphic to a complete matrix ring over a skew field.*

E. Artin (*Abh. Math. Sem. Univ. Hamburg*, **5**, 245) extended the theorems of Wedderburn to rings with the minimal condition for left ideals. Without such a condition it was not possible to obtain simple structure theorems. This is because, as was then already suspected, the radical $\mathfrak{N}$ is too small. Various authors, including R. Baer and J. Levitzki, have defined larger radicals, but N. Jacobson was the first to obtain structure theorems for rings without radical by an appropriate definition of the radical $\mathfrak{R}$. The reader is referred to Jacobson's book *Structure of Rings* (1956) for a comprehensive treatment of the entire theory. We shall here restrict ourselves to several main theorems.

In his book Jacobson defines the radical $\mathfrak{R}$ of a ring $\mathfrak{o}$ as the set of all elements a which are represented by zero in every irreducible representation. He then proves that the radical $\mathfrak{R}$ can also be obtained as the intersection of certain maximal right ideals which he calls "modular" right ideals. We can use left ideals in place of right ideals; this makes no difference. We shall here use the modular, maximal left ideals for the definition of $\mathfrak{R}$.

A left ideal is called *modular* if there exists an element c of $\mathfrak{o}$ with the property that

$$ac \equiv a(\mathfrak{L}) \qquad \text{for all } a \in \mathfrak{o}. \tag{13.58}$$

In a certain sense the element c plays the role of a right identity mod $\mathfrak{L}$. The word "modular" comes from "module," an old word for identity element.

We now define the *large radical*, or briefly the *radical*, $\mathfrak{R}$ of a ring $\mathfrak{o}$ as the intersection of all modular, maximal left ideals $\mathfrak{L}$. If there are no modular, maximal left ideals except $\mathfrak{o}$ itself, then the radical is the entire ring and this ring is called a *radical ring*.

Let $\mathfrak{L}$ be a modular, maximal left ideal. The factor module $\mathfrak{o}/\mathfrak{L}$ is therefore simple and provides an irreducible representation. The kernel of this representation is the two-sided ideal

$$\mathfrak{B} = \mathfrak{L} : \mathfrak{o} \tag{13.59}$$

or the set of all a with the property that

$$a\mathfrak{o} \subseteqq \mathfrak{L}. \tag{13.60}$$

Property (13.60) is equivalent to

$$ab \in \mathfrak{L} \qquad \text{for all} \quad b \in \mathfrak{o}. \tag{13.61}$$

In particular it follows from (13.60) that $ac \in \mathfrak{L}$, and hence from (13.58) $a \in \mathfrak{L}$. This holds for all a in $\mathfrak{P}$; therefore,

$$\mathfrak{P} \subseteqq \mathfrak{L}. \tag{13.62}$$

To each $\mathfrak{L}$ there belongs a $\mathfrak{P} = \mathfrak{L} : \mathfrak{o}$. The intersection of all $\mathfrak{P}$ is, by (13.62), contained in the intersection of all the $\mathfrak{L}$ and therefore in the radical. We shall now show, conversely, that $\mathfrak{R}$ is contained in all the ideals $\mathfrak{P}$ and therefore also in their intersection.

Let a be an element of $\mathfrak{R}$. We must show that (13.61) is satisfied for arbitrary b and $\mathfrak{L}$ and thus that a lies in every left ideal

$$\mathfrak{L}' = \mathfrak{L} : b.$$

Now a lies in $\mathfrak{o}$ and in all modular, maximal left ideals of $\mathfrak{o}$. It therefore suffices to show that $\mathfrak{L}'$ is either $\mathfrak{o}$ or is modular and maximal in $\mathfrak{o}$.

For fixed b and $\mathfrak{L}$, to each element x of $\mathfrak{o}$ there correspond the product xb and therefore also the coset $xb + \mathfrak{L}$ modulo $\mathfrak{L}$. The correspondence is a module homomorphism. The kernel of this homomorphism is precisely $\mathfrak{L}' = \mathfrak{L} : b$; therefore, $\mathfrak{o}/\mathfrak{L}'$ is mapped isomorphically into $\mathfrak{o}/\mathfrak{L}$. Now $\mathfrak{o}/\mathfrak{L}$ is minimal, and there are thus two possibilities. Either $\mathfrak{o}/\mathfrak{L}'$ is mapped isomorphically onto zero and is therefore itself zero, or $\mathfrak{o}/\mathfrak{L}'$ is mapped isomorphically *onto* $\mathfrak{o}/\mathfrak{L}$. In the first case $\mathfrak{L}' = \mathfrak{o}$; in the second case $\mathfrak{L}'$ is modular and maximal in $\mathfrak{o}$ just as $\mathfrak{L}$.

Summarizing, we have the following.

Theorem 1: *The radical $\mathfrak{R}$ is equal to the intersection of the two-sided ideals $\mathfrak{P} = \mathfrak{L} : \mathfrak{o}$ and is thus itself a two-sided ideal.*

We now form the factor ring $\bar{\mathfrak{o}} = \mathfrak{o}/\mathfrak{R}$. To each modular, maximal left ideal $\mathfrak{L}$ of $\mathfrak{o}$ there corresponds a modular, maximal ideal $\bar{\mathfrak{L}} = \mathfrak{L}/\mathfrak{R}$ of $\bar{\mathfrak{o}}$. Hence we have the following.

Theorem 2: *The factor ring $\mathfrak{o}/\mathfrak{R}$ is a ring without radical; that is, the radical of $\mathfrak{o}/\mathfrak{R}$ is the null ideal.*

Rings without radical are called *semisimple*. Theorem 2 may therefore also be formulated as follows. *The factor ring of $\mathfrak{o}$ by the radical $\mathfrak{R}$ is semisimple.*

Exercises

13.9. Each left ideal $\mathfrak{L}'$ such that $\mathfrak{o}/\mathfrak{L}'$ is operator isomorphic to $\mathfrak{o}/\mathfrak{L}$ gives the same quotient:

$$\mathfrak{L}' : \mathfrak{o} = \mathfrak{L} : \mathfrak{o} = \mathfrak{P}$$

13.10. Let $\mathfrak{L}$ be a modular left ideal. Then $\mathfrak{J} = \mathfrak{L} : \mathfrak{o}$ is a two-sided ideal contained in $\mathfrak{L}$ which contains all two-sided ideals contained in $\mathfrak{L}$.

We shall later need the following theorem.

Theorem 3: *Any modular left ideal $\mathfrak{l} \neq \mathfrak{o}$ can be extended to a maximal left ideal $\mathfrak{L} \neq \mathfrak{o}$ (which is, of course, again modular).*

Proof: Let c be an element of $\mathfrak{o}$ with the property that

$$ac \equiv a(\mathfrak{l}) \qquad \text{for all} \quad a \text{ in } \mathfrak{o}. \tag{13.63}$$

The left ideal $\mathfrak{l}$ does not contain c. We consider the set of all left ideals $\mathfrak{l}'$ which contain $\mathfrak{l}$ but not c. From these left ideals $\mathfrak{l}'$ we choose a maximal one $\mathfrak{L}$. Such an ideal exists by Zorn's lemma (Section 9.2). This $\mathfrak{L}$ is modular, since it contains $\mathfrak{l}$. It is also maximal and $\neq 0$. Indeed, if $\mathfrak{L}'$ is an ideal properly containing $\mathfrak{L}$, then $\mathfrak{L}'$ contains the element c and therefore every element of $\mathfrak{o}$ by (13.63).

To investigate the relationship between the large and small radicals, we need the expedient of a new product.

13.6 THE STAR PRODUCT

The *star product* $a{*}b$ of two elements a and b of a ring $\mathfrak{o}$ is defined by

$$a{*}b = a+b-ab.$$

Jacobson writes $a \circ b$ for this and calls it "circle composition."

The star product is associative, and zero is the identity element of star multiplication:

$$0{*}a = a = a{*}0.$$

If $\mathfrak{o}$ has an identity element 1, then $a{*}b = c$ may be defined by

$$(1-a)\,(1-b) = 1-c.$$

A *left star inverse* z' of a given z is defined by

$$z'{*}z = 0 \qquad \text{or} \quad z'+z-z'z = 0$$

or, if an identity element is present, by

$$(1-z')\,(1-z) = 1.$$

An element z which has a left star inverse z' is called *left star-regular* (or left quasi-regular). The concepts of a right star inverse and right star regularity are similarly defined by the condition $z*z' = 0$.

An element z is called *star-regular* if there exists a z' which is both a left and right star inverse of z:

$$z'*z = 0 = z*z'.$$

Theorem 4: *Any nilpotent element z is star-regular.*
Proof: If $z^m = 0$ and if we put

$$z' = -z-z^2-\cdots-z^{m-1},$$

then $z'*z = 0 = z*z'$. Hence z is star-regular.

If all elements of a left ideal $\mathfrak{l}$ are left star-regular, then they are star-regular. Indeed, let z be an element of $\mathfrak{l}$ and z' be its left star inverse, so that

$$z' = z'z-z.$$

Then z' lies in $\mathfrak{l}$ and therefore has a left star inverse z''. Hence

$$z = 0*z = z''*z'*z = z''*0 = z'',$$

and thus

$$z*z' = z''*z' = 0;$$

that is, z' is not only a left star inverse, but also a right star inverse of z.

A left or right ideal all of whose elements are star-regular is called star-regular. By what has just been proved, it suffices for a left ideal that all elements be left star-regular. The same is true for a right ideal if all the elements are right star-regular.

Theorem 5: *The radical $\mathfrak{R}$ is a star-regular left ideal which contains all star-regular left ideals.*
Proof: Let z be an element of $\mathfrak{R}$. We must show that z has a left star inverse. We form the set of all elements

$$xz-x,$$

where x runs through the ring $\mathfrak{o}$. This set is a modular left ideal, where z assumes the role formerly played by c. If this left ideal contains z, then this means that there exists an x with the property that

$$z = xz-x.$$

From this it follows that $x*z = 0$, and therefore x is a left star inverse of z. If the modular left ideal does not contain z, then it $\neq \mathfrak{o}$ and can therefore be extended to a modular, maximal left ideal $\mathfrak{L} \neq \mathfrak{o}$ by Theorem 3. Now z lies in $\mathfrak{R}$ and $\mathfrak{R}$ is the intersection of all modular, maximal left ideals; z is therefore contained in $\mathfrak{L}$. Then all elements

$$x = xz-(xz-x)$$

also lie in $\mathfrak{L}$; that is, $\mathfrak{L}$ is equal to $\mathfrak{o}$, although it was assumed that $\mathfrak{L} \neq \mathfrak{o}$.

Each element z of $\mathfrak{R}$ therefore has a left star inverse, and hence $\mathfrak{R}$ is a star-regular left ideal.

Now let $\mathfrak{l}$ be a star-regular left ideal. We wish to show that $\mathfrak{l}$ is contained in every modular, maximal left ideal $\mathfrak{L}$ and therefore also in $\mathfrak{R}$.

If $\mathfrak{l}$ were not contained in $\mathfrak{L}$, then $(\mathfrak{L}, \mathfrak{l})$ would be the entire ring $\mathfrak{o}$:

$$(\mathfrak{L}, \mathfrak{l}) = \mathfrak{o}. \tag{13.64}$$

Since $\mathfrak{L}$ is modular, there exists an element c with the property that

$$ac \equiv a(\mathfrak{L}) \qquad \text{for all} \quad a \in \mathfrak{o}. \tag{13.65}$$

This element c must, by (13.64), admit a representation of the form $y+z$, where y is in $\mathfrak{L}$ and z is in $\mathfrak{l}$. From this it follows that

$$c \equiv z(\mathfrak{L}). \tag{13.66}$$

Since z lies in $\mathfrak{l}$, z has a star inverse z':

$$z+z'-z'z = 0. \tag{13.67}$$

From (13.66) and (13.67) it follows that

$$c+z'-z'c \equiv 0(\mathfrak{L}),$$

and hence, by (13.65),

$$c \equiv 0(\mathfrak{L}),$$

which is impossible.

The equality of the left and right radicals follows easily from Theorem 5. Indeed, if the right radical $\mathfrak{R}'$ is defined as the intersection of all modular, maximal right ideals, then $\mathfrak{R}'$ is a star-regular two-sided ideal which is therefore contained in $\mathfrak{R}$ by Theorem 5. Similarly, $\mathfrak{R}$ is contained in $\mathfrak{R}'$ and therefore $\mathfrak{R} = \mathfrak{R}'$. The radical $\mathfrak{R}$ can therefore be defined as the intersection of all modular, maximal left or right ideals or as the union of all star-regular left or right ideals.

A left or right ideal is called a *nil ideal* if all its elements are nilpotent. It follows immediately from Theorem 4 that every nil ideal is star-regular. From this and Theorem 5 we obtain the following.

Theorem 6: *All nil ideals are contained in the radical* $\mathfrak{R}$.

In particular, all nilpotent ideals are contained in $\mathfrak{R}$. Their union is the small radical $\mathfrak{N}$. We therefore have the following.

Theorem 7: *The large radical* $\mathfrak{R}$ *contains the small radical* $\mathfrak{N}$.

Exercise

13.11. A left or right identity of a ring $\mathfrak{o}$ cannot be star-regular and is therefore never contained in the radical $\mathfrak{R}$.

13.7 RINGS WITH MINIMAL CONDITION

We henceforth assume that the minimal condition for left ideals holds in $\mathfrak{o}$. Under this assumption we first prove the following theorem.

Theorem 8: *The radical $\mathfrak{R}$ is nilpotent.*

Proof: In the sequence of powers $\mathfrak{R}^m$ there exists a minimal ideal $\mathfrak{R}^n$. Since $\mathfrak{R}^{2n}$ is contained in $\mathfrak{R}^n$, it follows that

$$\mathfrak{R}^{2n} = \mathfrak{R}^n$$

or, setting $\mathfrak{R}^n = \mathfrak{S}$, that $\mathfrak{S}^2 = \mathfrak{S}$. We wish to show that $\mathfrak{S} = \{0\}$.

If $\mathfrak{S} \neq \{0\}$, we consider the set of all left ideals $\mathfrak{I}$ such that

$$\mathfrak{I} \subseteq \mathfrak{S} \tag{13.68}$$

$$\mathfrak{S}\mathfrak{I} \neq \{0\}. \tag{13.69}$$

This set is nonempty, since $\mathfrak{S}$ is such a left ideal. There thus exists a minimal ideal $\mathfrak{I}_m$ with the properties (13.68) and (13.69). By (13.69), there exists a b in $\mathfrak{I}_m$ such that $\mathfrak{S}b \neq \{0\}$. The left ideal $\mathfrak{S}b$ is contained in $\mathfrak{I}_m$ and has properties (13.68) and (13.69); therefore $\mathfrak{S}b = \mathfrak{I}_m$. There thus exists an element z in $\mathfrak{S}$ such that $zb = z$. Since z lies in $\mathfrak{R}$, z has a left star inverse z' by Theorem 5:

$$z + z' - z'z = 0. \tag{13.70}$$

Multiplying this equation on the right by b, it follows that $b = 0$, which contradicts the assumption that $\mathfrak{S}b \neq \{0\}$. Hence $\mathfrak{S} = \{0\}$, and thus $\mathfrak{R}^n = \{0\}$.

The small radical $\mathfrak{N}$ contains all nilpotent two-sided ideals, and thus $\mathfrak{N} \supseteq \mathfrak{R}$. On the other hand, $\mathfrak{N} \subseteq \mathfrak{R}$ by Theorem 7. We thus have the following.

Theorem 9: *The small radical $\mathfrak{N}$ is equal to the radical $\mathfrak{R}$.*

Since $\mathfrak{R}$ contains all nil ideals by Theorem 6, there follows the next theorem.

Theorem 10: *All nil ideals are nilpotent.*

The factor ring $\mathfrak{o}/\mathfrak{R}$ is semisimple by Theorem 2. If the minimal condition for left ideals holds in $\mathfrak{o}$, then it naturally also holds in $\mathfrak{o}/\mathfrak{R}$. We now investigate quite generally the structure of semisimple rings with the minimal condition for left or right ideals.

Theorem 11: *Every semisimple ring $\mathfrak{o}$ with the minimal condition for left ideals is the direct sum of simple left ideals $\mathfrak{l}_i$.*

Proof: The radical of $\mathfrak{o}$, that is, the null ideal, is by definition the intersection of the modular, maximal left ideals $\mathfrak{L}$. We shall first show that the null ideal is the intersection of finitely many of these $\mathfrak{L}$.

We consider the set consisting of all intersections of finitely many modular, maximal left ideals $\mathfrak{L}$. In this set there exists a minimal ideal

$$\mathfrak{l} = \mathfrak{L}_1 \cap \cdots \cap \mathfrak{L}_m.$$

If $\mathfrak{l} \neq \{0\}$, there would exist an $\mathfrak{L}_{m+1}$ whose intersection with $\mathfrak{l}$ were a proper subset of $\mathfrak{l}$. This contradicts the minimal property of $\mathfrak{l}$. Hence $\mathfrak{l} = \{0\}$ and

$$\{0\} = \mathfrak{L}_1 \cap \cdots \cap \mathfrak{L}_m \tag{13.71}$$

If in this intersection representation an $\mathfrak{L}_i$ occurs which contains the intersection of the others, then this $\mathfrak{L}_i$ is redundant in (13.71). If all the redundant $\mathfrak{L}_i$

are deleted in (13.71), there remains finally and irreducible intersection representation

$$\{0\} = \mathfrak{L}_1 \cap \cdots \cap \mathfrak{L}_n, \tag{13.72}$$

in which no $\mathfrak{L}_i$ contains the intersection $\mathfrak{l}_i$ of the others. The sum $(\mathfrak{L}_i, \mathfrak{l}_i)$ is then an ideal properly containing $\mathfrak{L}_i$ and is therefore equal to $\mathfrak{o}$ since $\mathfrak{L}_i$ is maximal:

$$(\mathfrak{L}_i, \mathfrak{l}_i) = \mathfrak{o}. \tag{13.73}$$

Equations (13.72) and (13.73) state that $\{0\}$ is the direct intersection of the maximal left ideals $\mathfrak{L}_i$. From this it follows by Section 13.1 that $\mathfrak{o}$ is the direct sum of the left ideals $\mathfrak{l}_i$:

$$\mathfrak{o} = \mathfrak{l}_1 + \cdots + \mathfrak{l}_n. \tag{13.74}$$

From Section 13.1 we also have the operator isomorphisms

$$\mathfrak{l}_i \cong \mathfrak{o}/\mathfrak{L}_i, \tag{13.75}$$

so that the $\mathfrak{l}_i$ are simple, since the factor modules $\mathfrak{o}/\mathfrak{L}_i$ are simple. This completes the proof.

By (13.74) each element a of $\mathfrak{o}$ can be uniquely represented as a sum

$$a = a_1 + \cdots + a_n \qquad (a_i \in \mathfrak{l}_i). \tag{13.76}$$

In (13.76) we may select a particular term a_i and write, in place of (13.76),

$$a = a_i + b_i \qquad (a_i \in \mathfrak{l}_i, b_i \in \mathfrak{L}_i). \tag{13.77}$$

The term a_i is called the $\mathfrak{l}_i$-component of a. The correspondence $a \rightarrow a_i$ is an operator homomorphism with kernel $\mathfrak{L}_i$. Two elements a and a' are congruent (mod $\mathfrak{L}_i$) if and only if they have the same $\mathfrak{l}_i$-component a_i.

A ring element c with the property $c^2 = c$ is called *idempotent*. We now prove the following.

Theorem 12: *With the hypotheses and notation of Theorem* 11:

Each $\mathfrak{l}_i$ is generated by an idempotent element e_i:

$$\mathfrak{l}_i = \mathfrak{o}e_i, e_i^2 = e_i;$$

The elements e_i are mutually annihilating:

$$e_i e_k = 0 \qquad \text{for} \quad i \neq k; \tag{13.78}$$

The $\mathfrak{l}_i$-component a_i of any a is obtained by multiplication with e_i:

$$a_i = ae_i; \tag{13.79}$$

The sum

$$e = e_1 + \cdots + e_n \tag{13.80}$$

is the identity element of $\mathfrak{o}$.

Proof: Since $\mathfrak{L}_i$ is modular, there exists an element c_i of $\mathfrak{o}$ with the property

$$ac_i \equiv a(\mathfrak{L}_i) \qquad \text{for all } a. \tag{13.81}$$

We now decompose c_i as in (13.77):

$$c_i = e_i + f_i. \tag{13.82}$$

From this it follows that ac_i has the decomposition

$$ac_i = ae_i + af_i. \tag{13.83}$$

From congruence (13.81) it follows that ac_i and a have the same $\mathfrak{l}_i$-component. By (13.83), therefore,

$$a_i = ae_i. \tag{13.84}$$

This completes the proof of (13.79).

If a runs through the ring $\mathfrak{o}$, then a_i runs through the ideal $\mathfrak{l}_i$; hence

$$\mathfrak{l}_i = \mathfrak{o}e_i. \tag{13.85}$$

Putting $a = e_i$ in (13.84) gives

$$e_i = e_i^2. \tag{13.86}$$

Putting $a = e_k$ in (13.84) gives

$$0 = e_k e_i \qquad (k \neq i). \tag{13.87}$$

This completes the proof of assertions a, b, and c.

Putting

$$e = e_1 + \cdots + e_n, \tag{13.88}$$

it follows from (13.84) that

$$ae = ae_1 + \cdots + ae_n = a_1 + \cdots + a_n = a; \tag{13.89}$$

that is, e is a right identity of $\mathfrak{o}$. It remains only to show that e is also a left identity.

The elements $a - ea$ form a right ideal $\mathfrak{r}$. For any b we have $be = b$ and thus

$$b(a-ea) = ba - bea = ba - ba = 0.$$

In particular,

$$(a-ea)^2 = 0.$$

Thus $\mathfrak{r}$ is a null ideal and is therefore contained in the radical by Theorem 6; but then $\mathfrak{r}$ is the null ideal, and so for all a we have

$$a - ea = 0;$$

that is, e is a left identity.

A ring which as a left module is completely reducible—that is, the direct sum of simple left ideals—is called *completely left-reducible.* We may now combine Theorems 11 and 12 as follows.

Every semisimple ring with the minimal condition for left ideals is completely left-reducible and has an identity element.

The converse of this theorem is also true.

Theorem 13: *Every completely left-reducible ring with a right identity is semisimple and satisfies the minimal condition for left ideals.*

Proof: Let

$$\mathfrak{o} = \mathfrak{l}_1 + \cdots + \mathfrak{l}_n \tag{13.90}$$

be the decomposition of $\mathfrak{o}$ into simple left ideals, and let $\mathfrak{L}_i$ be the sum of all the $\mathfrak{l}_j$ except $\mathfrak{l}_i$. Then $\mathfrak{o}/\mathfrak{L}_i \cong \mathfrak{l}_i$, and $\mathfrak{L}_i$ is therefore maximal. If e is the identity element of $\mathfrak{o}$, then $ae = a$ for all a; each $\mathfrak{L}_i$ is therefore modular. The intersection of the $\mathfrak{L}_i$ is $\{0\}$ by Section 13.1, and hence $\mathfrak{o}$ is semisimple.

By section 7.6, $\mathfrak{o}$ has a composition series of length n. By Section 7.4, there is a composition series which includes any left ideal $\mathfrak{l}$. The section of this composition series from $\mathfrak{l}$ to $\{0\}$ has length $m \leqq n$; this number m is called the *length of* $\mathfrak{l}$. A proper subideal $\mathfrak{l}'$ of $\mathfrak{l}$ has a smaller length than $\mathfrak{l}$, since we can pass a composition series for $\mathfrak{l}$ through $\mathfrak{l}'$. In every nonempty set of left ideals there exists a left ideal $\mathfrak{l}''$ of smallest length. This $\mathfrak{l}''$ is a minimal ideal of the set, since any proper subideal $\mathfrak{l}'''$ would have smaller length. The minimal condition for left ideals therefore holds in $\mathfrak{o}$.

13.8 TWO-SIDED DECOMPOSITIONS AND CENTER DECOMPOSITION

In Section 13.7 we investigated the direct-sum decomposition of a ring $\mathfrak{o}$ in terms of left ideals under particular assumptions; we now wish to see what can be said about decomposition in terms of two-sided ideals.

Theorem 14: *If a ring $\mathfrak{o}$ with identity can be represented as the direct sum of directly irreducible two-sided ideals distinct from the null ideal:*

$$\mathfrak{o} = \mathfrak{a}_1 + \cdots + \mathfrak{a}_n, \tag{13.91}$$

then the ideals $\mathfrak{a}_1$ are uniquely determined.

Proof: If there is a second decomposition

$$\mathfrak{o} = \mathfrak{c}_1 + \cdots + \mathfrak{c}_m,$$

then

$$\mathfrak{c}_1 = \mathfrak{o}\mathfrak{c}_1 = (\mathfrak{a}_1\mathfrak{c}_1, \mathfrak{a}_2\mathfrak{c}_1, \ldots, \mathfrak{a}_n\mathfrak{c}_1).$$

The sum on the right-hand side is direct, since

$$\mathfrak{a}_1\mathfrak{c}_1 \subseteqq \mathfrak{a}_1, \ldots, \mathfrak{a}_n\mathfrak{c}_1 \subseteqq \mathfrak{a}_n.$$

Since $\mathfrak{c}_1$ is directly indecomposable, all the products $\mathfrak{a}_i\mathfrak{c}_1$ must be zero with the exception of a single one, say, $\mathfrak{a}_1\mathfrak{c}_1$. Thus,

$$\mathfrak{c}_1 = \mathfrak{a}_1\mathfrak{c}_1 \subseteqq \mathfrak{a}_1.$$

In the same manner we show that $\mathfrak{a}_1$ is contained in some $\mathfrak{c}_i$,and hence

$$\mathfrak{c}_1 \subseteqq \mathfrak{a}_1 \subseteqq \mathfrak{c}_i;$$

from this it follows that $i = 1$ and $\mathfrak{c}_1 = \mathfrak{a}_1$. Each $\mathfrak{c}_i$ is therefore equal to some $\mathfrak{a}_i$.

This uniqueness does not hold in the case of one-sided decompositions.

We now prove: *If* $\mathfrak{o}$ *is the direct sum of two-sided ideals* $\mathfrak{a}_i$, *then the center* $\mathfrak{Z}$ *of* $\mathfrak{o}$ *is the direct sum of the centers* $\mathfrak{Z}_i$ *of the rings* $\mathfrak{a}_i$.

$$\mathfrak{Z} = \mathfrak{Z}_1 + \cdots + \mathfrak{Z}_n.$$

Proof: Let $z = z_1 + \cdots + z_n$ be an element of the center, and let $x = x_1 + \cdots + x_n$ be an arbitrary element of $\mathfrak{o}$. Then $zx = xz$ and hence

$$z_1x_1 + \cdots + z_nx_n = x_1z_1 + \cdots + x_nz_n. \tag{13.92}$$

From this it follows that $z_ix_i = x_iz_i$ for all x_i in $\mathfrak{a}_i$ so that z_i lies in the center of $\mathfrak{a}_i$. Conversely, if each z_i lies in the center of $\mathfrak{a}_i$, then (13.92) holds for all x; hence $zx = xz$, and z lies in the center of $\mathfrak{o}$.

The preceding considerations hold for arbitrary rings $\mathfrak{o}$. We now assume that $\mathfrak{o}$ is semisimple and that the minimal condition for left ideals is satisfied. Then $\mathfrak{o}$ is completely left-reducible:

$$\mathfrak{o} = \mathfrak{l}_1 + \cdots + \mathfrak{l}_n \tag{13.93}$$

and has an identity element:

$$e = e_1 + \cdots + e_n \qquad (e_i \in \mathfrak{l}_i). \tag{13.94}$$

If now $\mathfrak{a}$ is a two-sided ideal, then each $\mathfrak{a}e_i$ is a left ideal contained in $\mathfrak{l}_i$ and is therefore equal to either $\mathfrak{l}_i$ or $\{0\}$. The $\mathfrak{l}_i$ can be ordered so that

$$\mathfrak{a}e_1 = \mathfrak{l}_1, \ldots, \mathfrak{a}e_m = \mathfrak{l}_m, \mathfrak{a}e_{m+1} = \{0\}, \ldots, \mathfrak{a}e_n = \{0\}.$$

Then $\mathfrak{l}_1, \ldots, \mathfrak{l}_m$ are contained in $\mathfrak{a}$, and therefore $\mathfrak{l}_1 + \cdots + \mathfrak{l}_m$ is also contained in $\mathfrak{a}$. Each element a of $\mathfrak{a}$ is equal to

$$a = ae = ae_1 + \cdots + ae_n.$$

In this sum $ae_{m+1}, \ldots, ae_n$ are all zero, and the sum thus reduces to

$$a = ae_1 + \cdots + ae_m.$$

Hence, $\mathfrak{a} \subseteq \mathfrak{l}_1 + \cdots + \mathfrak{l}_m$, and thus

$$\mathfrak{a} = \mathfrak{l}_1 + \cdots + \mathfrak{l}_m, \tag{13.95}$$

or in words: *Every two-sided ideal* $\mathfrak{a}$ *is the sum of a certain number of the* $\mathfrak{l}_i$.

For the $\mathfrak{l}_i$ occurring in (13.95) we have

$$\mathfrak{a}\mathfrak{l}_i = \mathfrak{a}\mathfrak{o}e_i = \mathfrak{a}e_i = \mathfrak{l}_i,$$

and for the $\mathfrak{l}_k$ not occurring in (13.95), on the other hand,

$$\mathfrak{a}\mathfrak{l}_k = \mathfrak{a}\mathfrak{o}e_k = \mathfrak{a}e_k = \{0\}.$$

The $\mathfrak{l}_i$ occurring in (13.95) can therefore also be characterized by the fact that they are not annihilated by $\mathfrak{a}$:

$$\mathfrak{a}\mathfrak{l}_i \neq \{0\}.$$

If an $\mathfrak{l}_i$ has this property, then all the $\mathfrak{l}_j$ which are operator isomorphic to $\mathfrak{l}_i$ likewise have the property $\mathfrak{a}\mathfrak{l}_j \neq \{0\}$. Thus, if an $\mathfrak{l}_i$ occurs in (13.95), then all the $\mathfrak{l}_j$ isomorphic to it also occur.

Suppose that $\mathfrak{l}_1, \ldots, \mathfrak{l}_g$ are isomorphic to $\mathfrak{l}_1$ and the others are not. We assert: $\mathfrak{a}_1 = \mathfrak{l}_1 + \cdots + \mathfrak{l}_g$ *is a two-sided ideal.*

Proof: For any $b \in \mathfrak{o}$.

$$\begin{aligned} \mathfrak{a}_1 b = \mathfrak{a}_1 be &= \mathfrak{a}_1(be_1 + \cdots + be_n) \\ &\subseteq (\mathfrak{a}_1 be_1, \ldots, \mathfrak{a}_1 be_g, \ldots, \mathfrak{a}_1 be_n) \\ &\subseteq (\mathfrak{l}_1, \ldots, \mathfrak{l}_g, 0, \ldots, 0) = \mathfrak{a}_1. \end{aligned}$$

Hence $\mathfrak{a}_1$ is a right ideal and therefore a two-sided ideal.

In this manner a two-sided ideal $\mathfrak{a}_i$ can be formed from any class of isomorphic $\mathfrak{l}_j$. Let the ideals so formed by $\mathfrak{a}_1, \mathfrak{a}_2, \ldots, \mathfrak{a}_r$.

Now every two-sided ideal $\mathfrak{a}$ is a sum such as (13.95), and if it contains $\mathfrak{l}_i$ it also contains all $\mathfrak{l}_j$ isomorphic to $\mathfrak{l}_i$. We thus have the following.

Every two-sided ideal is a sum of a certain number of the two-sided ideals $\mathfrak{a}_1, \ldots, \mathfrak{a}_r$. *These are minimal two-sided ideals. The ring* $\mathfrak{o}$ *is the direct sum of the* $\mathfrak{a}_k$:

$$\mathfrak{o} = \mathfrak{a}_1 + \cdots + \mathfrak{a}_r. \tag{13.96}$$

The last assertion follows directly from (13.93).

By Section 12.8, all the $\mathfrak{a}_i$ are rings which are mutually annihilating:

$$\mathfrak{a}_i\mathfrak{a}_k = \{0\} \quad \text{for} \quad i \neq k. \tag{13.97}$$

From (13.96) and (13.97) it follows that every left or right ideal of the ring $\mathfrak{a}_i$ is also a left or right ideal in $\mathfrak{o}$. For left ideals $\mathfrak{l}$ the proof goes as follows:

$$\begin{aligned} \mathfrak{o}\mathfrak{l} &= (\mathfrak{a}_1 + \cdots + \mathfrak{a}_r)\mathfrak{l} \\ &\subseteq (\mathfrak{a}_1\mathfrak{l}, \ldots, \mathfrak{a}_r\mathfrak{l}) \\ &\subseteq (0, \ldots, \mathfrak{a}_i\mathfrak{l}, 0, \ldots, 0) \subseteq \mathfrak{l}; \end{aligned}$$

for right ideals the proof is similar. Hence every two-sided ideal in $\mathfrak{a}_i$ is a two-sided ideal in $\mathfrak{o}$. But since $\mathfrak{a}_i$ is a minimal two-sided ideal, there exist no two-sided ideals in $\mathfrak{a}_i$ except $\mathfrak{a}_i$ and $\{0\}$. Thus each $\mathfrak{a}_i$ is a simple ring with identity element e_i. We thus obtain the next theorem.

Theorem 15: *Every semisimple ring* $\mathfrak{o}$ *with the minimal condition for left ideals is the direct sum of simple rings with identity.*

In the case of an algebra $\mathfrak{o}$, this is the first half of Wedderburn's theorem mentioned in Section 13.5.

We now investigate the structure of simple rings with identity.

13.9 SIMPLE AND PRIMITIVE RINGS

Let $\mathfrak{o}$ be a simple ring with right identity e:

$$ae = a \qquad \text{for all } a. \tag{13.98}$$

Equation (13.98) states that the null ideal is a modular left ideal. By Theorem 3, there thus exists a modular, maximal left ideal $\mathfrak{L} \neq \mathfrak{o}$. The factor module $\mathfrak{o}/\mathfrak{L}$ is simple and provides an irreducible representation. The kernel of this representation is a two-sided ideal $\mathfrak{P}$ which, by Equation (13.62), is contained in $\mathfrak{L}$ and therefore $\neq \mathfrak{o}$. Since $\mathfrak{o}$ is simple, $\mathfrak{P} = \{0\}$; this means that the representation provided by $\mathfrak{o}/\mathfrak{L}$ is faithful.

A ring which has a faithful irreducible representation is called *primitive*. We thus have the following.

Theorem 16: *A simple ring with identity is primitive.*

We now investigate whether the converse is also true.

Let $\mathfrak{o}$ be a primitive ring and let $\mathfrak{M}$ be a simple $\mathfrak{o}$-module providing a faithful representation of $\mathfrak{o}$. Let u be an element of $\mathfrak{M}$ which is not annihilated by $\mathfrak{o}$. Then $\mathfrak{o}u$ is a submodule of $\mathfrak{M}$ which is not null and therefore equal to $\mathfrak{M}$. The mapping $x \to xu$ defines a homomorphism of $\mathfrak{o}$ onto $\mathfrak{M}$ whose kernel is a left ideal $\mathfrak{L}$ of $\mathfrak{o}$. The factor module $\mathfrak{o}/\mathfrak{L}$ is isomorphic to $\mathfrak{M}$ and therefore simple; this means that $\mathfrak{L}$ is maximal. Since $\mathfrak{o}u = \mathfrak{M}$, u must have a representation of the form cu:

$$u = cu.$$

From this it follows that $au = acu$ for all a of $\mathfrak{o}$. The mapping $x \to xu$ therefore assigns the same element of $\mathfrak{M}$ to the elements a and ac. It follows that

$$a \equiv ac(\mathfrak{L});$$

that is, $\mathfrak{L}$ is modular.

Because of the isomorphism $\mathfrak{M} \cong \mathfrak{o}/\mathfrak{L}$, the representation provided by $\mathfrak{M}$ is equivalent to that provided by $\mathfrak{o}/\mathfrak{L}$. The kernel of the latter representation is the two-sided ideal

$$\mathfrak{P} = \mathfrak{L} : \mathfrak{o}.$$

Since the representation is faithful, it follows that $\mathfrak{P} = \{0\}$. The radical $\mathfrak{R}$ of $\mathfrak{o}$ is contained in $\mathfrak{P}$ by Theorem 1; therefore $\mathfrak{R} = \{0\}$ and $\mathfrak{o}$ is semisimple. We thus have the following.

Theorem 17: *A primitive ring* $\mathfrak{o}$ *is semisimple.*

In the first part of the proof the fact that the representation is faithful was not used. Only the facts that the module $\mathfrak{M}$ is simple and that all elements of $\mathfrak{M}$ are not annihilated by $\mathfrak{o}$ were used. We thus have the following theorem for arbitrary rings.

Theorem 18: *Every simple* $\mathfrak{o}$*-module* $\mathfrak{M}$ *which is not annihilated by* $\mathfrak{o}$ *is isomorphic to the factor module* $\mathfrak{o}/\mathfrak{L}$ *of* $\mathfrak{o}$ *by a modular, maximal left ideal* $\mathfrak{L}$. *If* $\mathfrak{P}$ *is* *the kernel of the representation provided by* $\mathfrak{M}$, *then the radival* $\mathfrak{R}$ *is contained in* $\mathfrak{P}$; *that is, all elements of* $\mathfrak{R}$ *are represented by the zero element.*

We now return to the study of primitive rings. Since $\mathfrak{o}$ is semisimple, it follows, assuming the minimal condition for the left ideals, that $\mathfrak{o}$ is a direct sum of minimal left ideals:

$$\mathfrak{o} = \mathfrak{l}_1 + \cdots + \mathfrak{l}_n.$$

At least one $\mathfrak{l}_i$ is not contained in $\mathfrak{L}$, since otherwise the sum $\mathfrak{o} = \mathfrak{l}_1 + \cdots + \mathfrak{l}_n$ would lie in $\mathfrak{L}$, which is not the case. The sum $(\mathfrak{l}_i, \mathfrak{L})$ is then equal to $\mathfrak{o}$, since $\mathfrak{L}$ is maximal; the intersection $\mathfrak{l}_i \cap \mathfrak{L}$ is null, since $\mathfrak{l}_i$ is minimal. We have therefore the isomorphism

$$\mathfrak{o}/\mathfrak{L} \cong \mathfrak{l}_i.$$

The module $\mathfrak{M}$ is thus isomorphic to a simple left ideal $\mathfrak{l}_i$, and the representation provided by $\mathfrak{M}$ is equivalent to that provided by $\mathfrak{l}_i$.

By section 13.8, $\mathfrak{o}$ is a direct sum of two-sided ideals $\mathfrak{a}_\nu$ which, except for a single one, are all represented by zero in the representation provided by $\mathfrak{l}_i$. If the representation is faithful, there can be only one $\mathfrak{a}_\nu$; this means that $\mathfrak{o}$ is itself a simple ring with identity. We thus have the following theorem.

Theorem 19: *Every primitive ring with the minimal condition for left ideals is simple and has an identity element.*

Combining Theorem 16 with Theorem 19, it has been shown that for rings with the minimal condition, and thus in particular for algebras, the properties "primitive" and "simple with identity" are equivalent.

Jacobson has clarified the structure of primitive rings in general (without the minimal condition). Every primitive ring $\mathfrak{o}$ can be imbedded in the ring $\mathfrak{D}$ of linear transformations of a vector space in such a manner that $\mathfrak{D}$ is the closure of $\mathfrak{o}$ in a particular topology in $\mathfrak{D}$.[5] Here the vector space will be constructed, and the imbedding of $\mathfrak{o}$ in $\mathfrak{D}$ will be given.

In this construction the endomorphism ring of an $\mathfrak{o}$-module plays an important role. The *endomorphisms L* of an $\mathfrak{o}$-module $\mathfrak{M}$ are defined as mappings of $\mathfrak{M}$ into itself with the properties

$$L(u+v) = Lu+Lv \tag{13.99}$$

$$L(au) = a(Lu). \tag{13.100}$$

Property (13.100) implies that the mapping L commutes with the transformation A of the representation $a \to A$ provided by $\mathfrak{M}$:

$$LA = AL \qquad \text{for all } A.$$

If $\mathfrak{M}$ also has a right operator domain, then the property

$$L(u\beta) = (Lu)\beta \tag{13.101}$$

for all β in Ω is required in addition to properties (13.99) and (13.100). If, for example, Ω is a field and $\mathfrak{M}$ is a vector space over this field, then properties (13.99),

[5] N. Jacobson, *Structure of Rings* (1956), Chapter II.

(13.100), and (13.101) imply that the endomorphisms L are linear transformations of the vector space $\mathfrak{M}$ which commute with all the linear transformations A of the representation $a \to A$.

If the sum and product of endomorphisms are defined according to Section 6.9 by

$$(L+M)u = Lu+Mu$$

$$(LM)u = L(Mu),$$

then the endomorphisms form a ring, the *left endormorphism ring* of the module $\mathfrak{M}$.

In the following it will often be expedient to write the endomorphisms as right operators λ, μ, . . . and define their product by

$$u(\lambda\mu) = (u\lambda)\mu.$$

In place of (13.99), (13.100), and (13.101) we then have

$$(u+v)\lambda = u\lambda+v\lambda \tag{13.102}$$

$$(au)\lambda = a(u\lambda) \tag{13.103}$$

$$(u\beta)\lambda = (u\lambda)\beta \quad \text{for} \quad \beta \in \Omega. \tag{13.104}$$

The right endomorphisms likewise form a ring, the *right endomorphism ring* of the $\mathfrak{o}$-module $\mathfrak{M}$. Whenever in this and the following section we speak simply of the *endomorphism ring* of a module, the right endomorphism ring is always meant. It is *anti-isomorphic* to the left endomorphism ring; that is, to each left endomorphism L there corresponds a right endomorphism λ in a one-to-one manner, to the sum $L+M$ there corresponds the sum $\lambda+\mu$, and to the product LM there corresponds the product $\mu\lambda$.

The endomorphism ring of a simple $\mathfrak{o}$*-module is a skew field* K.

The endomorphism ring naturally has an identity, the identity automorphism ι. We must therefore only show that each endomorphism $\lambda \neq 0$ has an inverse λ^{-1}. The endomorphism λ maps $\mathfrak{M}$ onto a submodule $\mathfrak{M}\lambda$. If $\lambda \neq 0$, this submodule is not the null module, and it must therefore be equal to $\mathfrak{M}$. The set of elements mapped by λ onto 0 is also a submodule of $\mathfrak{M}$. If $\lambda \neq 0$, this submodule is not $\mathfrak{M}$ and must therefore be the null module. The endomorphism λ therefore maps $\mathfrak{M}$ isomorphically *onto* itself. It then has an inverse automorphism λ^{-1}, which is what we wanted to show.

The skew field K is called the *endomorphism field* of the simple $\mathfrak{o}$-module $\mathfrak{M}$. Since the identity element ι of K is the identity operator, $\mathfrak{M}$ is a vector space over K. The relations

$$a(u+v) = au+av$$

$$a(u\lambda) = (au)\lambda$$

imply that the elements a of $\mathfrak{o}$ generate linear transformations A of the vector space $\mathfrak{M}$. The mapping $a \to A$ is a ring homomorphism. If the representation is faithful, then the correspondence $a \to A$ is an isomorphism and the ring $\mathfrak{o}$ is imbedded in the ring $\mathfrak{O}$ of linear transformations of the vector space $\mathfrak{M}$.

Exercises

13.12. In a completely reducible representation of a ring $\mathfrak{o}$ the radical $\mathfrak{R}$ is always represented by zero.

13.13. A simple ring which is not primitive is nothing other than a simple additive group; all the products ab are zero.

13.14. A simple algebra without identity is a one-dimensional vector space $a_1\mathrm{P}$ with $a_1{}^2 = 0$.

13.10 THE ENDOMORPHISM RING OF A DIRECT SUM

Let $\mathfrak{M} = \mathfrak{M}_1 + \cdots + \mathfrak{M}_n$ be the direct sum of n simple modules. We wish to investigate the endomorphism ring of $\mathfrak{M}$.

If an element u of $\mathfrak{M}$ is decomposed into its $\mathfrak{M}_i$-components:

$$u = u_1 + \cdots + u_n, \tag{13.105}$$

then each of the mappings $u \to u_i$ is an endomorphism κ_i. The sum of these endomorphisms is the identity ι:

$$\iota = \kappa_1 + \cdots + \kappa_n. \tag{13.106}$$

Every endomorphism can be decomposed as follows:

$$\mu = \iota\mu\iota = \left(\sum \kappa_h\right) \mu\left(\sum \kappa_i\right)$$
$$= \sum_{h,i} \kappa_h\mu\kappa_i.$$

If we put

$$\kappa_h\mu\kappa_i = \mu_{hi}, \tag{13.107}$$

then

$$\mu = \sum_{h,i} \mu_{hi}. \tag{13.108}$$

Each of the endomorphisms μ_{hi} maps $\mathfrak{M}_h$ into $\mathfrak{M}_i$ and all the other $\mathfrak{M}_k$ $(k \neq h)$ onto zero. We may also say that μ_{hi} is a *homomorphism of* $\mathfrak{M}_h$ *into* $\mathfrak{M}_i$. The n^2 homomorphisms μ_{hi} in (13.108) may be chosen arbitrarily; their sum is always an endomorphism μ, and every μ can be obtained in this manner. The decomposition of μ into homomorphisms μ_{hi} of $\mathfrak{M}_h$ into $\mathfrak{M}_i$ is unique, since (13.107) follows immediately from (13.108) on multiplying on the left with κ_h and on the right with κ_i.

If $\mu = \sum \mu_{hi}$ and $\nu = \sum \nu_{hi}$ are two endomorphisms, then their sum and product can easily be formed. It is to be noted that $\mu_{hi}\nu_{jk}$ is zero if $i \neq j$. We obtain

$$\mu+\nu = \sum_{h,i} (\mu_{hi}+\nu_{hi}) \tag{13.109}$$

$$\mu\nu = \sum_{h,k} \left(\sum_{i} \mu_{hi}\nu_{ik}\right). \tag{13.110}$$

The μ_{hi} can be arranged in the form of a matrix (μ_{hi}). To each endomorphism μ there thus corresponds a matrix of homomorphisms μ_{hi} which may be chosen arbitrarily; to the sum $\mu+\nu$ there corresponds the matrix sum, by (13.109), and to the product $\mu\nu$ there corresponds the matrix product, by (13.110).

In general, many homomorphisms μ_{hi} are equal to zero. Indeed, we have the following theorem.

If M_h is mapped homomorphically into $\mathfrak{M}_i$ and the mapping is not the null operator, then it is an isomorphism of M_h onto $\mathfrak{M}_i$.

Proof: The kernel of the homomorphism is a submodule of $\mathfrak{M}_h$ and hence is equal to $\{0\}$ if $\mathfrak{M}_h$ is not mapped onto zero. The image module is a submodule of $\mathfrak{M}_i$ and hence is equal to $\mathfrak{M}_i$ if it is not $\{0\}$.

From this theorem it follows that $\mu_{hi} = 0$ except when $\mathfrak{M}_h \simeq \mathfrak{M}_i$. If we partition the modules $\mathfrak{M}_i$ into isomorphism classes and enumerate them so that $\mathfrak{M}_1, \ldots, \mathfrak{M}_q$ are mutually isomorphic, likewise $\mathfrak{M}_{q+1}, \ldots, \mathfrak{M}_{q+r}$, and so on, then the matrices (μ_{hi}) evidently decompose into square blocks of q, r, ... rows and columns while the entries outside these blocks are all zero:

$$\left(\begin{array}{c|c|c}
\begin{array}{ccc} \mu_{11} & \cdots & \mu_{1q} \\ \vdots & & \vdots \\ \mu_{q1} & \cdots & \mu_{qq} \end{array} & & \\
\hline
 & \begin{array}{cc} \mu_{q+1,\, q+1} & \cdots \\ \vdots & \\ \mu_{q+r,\, q+1} & \cdots \end{array} & \\
\hline
 & & \ddots
\end{array}\right).$$

If we write arbitrary elements in the first block and zeros in all the others, then a matrix ring E_1 is obtained which is a subring of the original matrix ring E; similarly, writing zeros everywhere except in the second block gives a ring E_2, and so on. It is clear that each element of E can be uniquely represented as a sum of elements of $\mathsf{E}_1, \mathsf{E}_2, \ldots$ and that the elements of $\mathsf{E}_1, \mathsf{E}_2, \ldots$ annihilate one another. Hence *the ring E is the direct sum of mutually annihilating rings $\mathsf{E}_1, \mathsf{E}_2, \ldots$.*

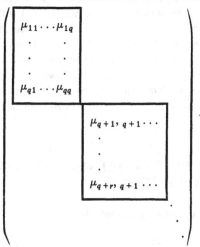

To determine the structure of E we have now only to study the structure of a single E_i, say E_1. To the elements of E_1 there correspond the $q \times q$ matrices

$$\begin{pmatrix} \mu_{11} \cdots \mu_{1q} \\ \cdot \quad \cdot \\ \cdot \quad \cdot \\ \cdot \quad \cdot \\ \mu_{q1} \cdots \mu_{qq} \end{pmatrix} \tag{13.111}$$

of the first block.

Here μ_{11} is an element of the endomorphism field K_1 of $\mathfrak{M}_1$. The other elements μ_{hi} do not belong to this skew field, but are rather homomorphisms of $\mathfrak{M}_h$ into $\mathfrak{M}_i$. However, we can map them in a well-defined way onto the elements of $\mathfrak{K}_1$ by choosing q fixed isomorphisms

$$\mu_1, \ldots, \mu_q$$

which map $\mathfrak{M}_1, \ldots, \mathfrak{M}_q$ onto $\mathfrak{M}_1$. For μ_1 we choose the identity automorphism. To each μ_{hi} we now assign the element

$$\lambda_{hi} = \mu_h^{-1}\mu_{hi}\mu_i \tag{13.112}$$

which belongs to $\mathfrak{K}_1$ (since μ_h^{-1} maps $\mathfrak{M}_1$ onto $\mathfrak{M}_h$, μ_{hi} maps $\mathfrak{M}_h$ into $\mathfrak{M}_i$, and μ_i maps $\mathfrak{M}_i$ again onto $\mathfrak{M}_1$); it is clear that to a sum $\mu_{hi}+\nu_{hi}$ corresponds again a sum and to the product $\mu_{hi}\nu_{ik}$ again a product as in (13.110). Thus, to each matrix (13.111) there corresponds a unique matrix (λ_{hi}) with elements in K_1, to the sum corresponds the sum, and to the product there corresponds the product. Hence, E_1 is isomorphic to the ring of all $q \times q$ matrices with elements in the skew field K_1, the automorphism skew field of the simple module $\mathfrak{M}_1$.

Combining these results, we obtain the following.

Structure Theorem for Endomorphism Rings: *The endomorphism ring of a completely reducible module $\mathfrak{M}$ is the direct sum of complete matrix rings E_i over skew fields K_i.*

13.11 STRUCTURE THEOREMS FOR SEMISIMPLE AND SIMPLE RINGS

We begin with a ring $\mathfrak{o}$ having a right identity e:

$$ae = a \qquad \text{for all} \quad a.$$

We consider $\mathfrak{o}$ as a module with $\mathfrak{o}$ itself as left operator domain and seek to determine the endomorphisms μ of this module. These are mappings of $\mathfrak{o}$ onto itself with the properties

$$(a+b)\mu = a\mu + b\mu$$

$$(ab)\mu = a(b\mu).$$

In the case $b = e$, the last property gives

$$a\mu = a(e\mu).$$

The endomorphism μ is therefore generated by right multiplication with an element $d = e\mu$. Conversely, every such right multiplication is an endomorphism:

$$(a+b)d = ad+bd$$
$$(ab)d = a(bd).$$

The endomorphisms μ thus correspond in one-to-one fashion to the elements d of $\mathfrak{o}$. Sums correspond to sums and products to products. We thus have the following.

If a ring $\mathfrak{o}$ *with right identity e is considered as a module with* $\mathfrak{o}$ *as left operator domain, then the right endomorphism ring of this module is isomorphic to the ring* $\mathfrak{o}$ *itself.*

As an application of this theorem, we determine the structure of semisimple rings with the minimal condition for left ideals. By Theorem 11, such a ring is the direct sum of simple left ideals:

$$\mathfrak{o} = \mathfrak{l}_1 + \cdots + \mathfrak{l}_n. \tag{13.113}$$

The endomorphism ring of such a direct sum is a direct sum of complete matrix rings over skew fields by Section 13.10. On the other hand, $\mathfrak{o}$ has an identity element by Section 13.7. The endomorphism ring is therefore isomorphic to the ring $\mathfrak{o}$ itself. We thus have the following.

Structure Theorem for Semisimple Rings: *Every semisimple ring* $\mathfrak{o}$ *with the minimal condition for left ideals is isomorphic to a direct sum of complete matrix rings over skew fields.*

If the ring $\mathfrak{o}$ is simple, then there can be only one matrix ring in the direct sum. We thus have the following.

Structure Theorem for Simple Rings: *A simple ring with identity which satisfies the minimal condition for left ideals is isomorphic to a complete matrix ring* $\mathfrak{K}_n$ *over a skew field* K.

The degree n of the matrices is hereby equal to the number of left ideals in the decomposition (13.113). Since $\mathfrak{o}$ is simple, all the $\mathfrak{l}_i$ are isomorphic. The skew field K is the endomorphism field of a left ideal $\mathfrak{l}_i$.

If, in particular, $\mathfrak{o}$ is a simple algebra over a field P, then the elements β of P generate endomorphisms $x \to x\beta$ of the left ideal $\mathfrak{l}_i$; P can therefore be imbedded in the endomorphism field K. Further, for each endomorphism λ of K,

$$(x\beta)\lambda = (x\lambda)\beta,$$

and thus β commutes with each λ of K; that is, P is contained in the center of K. Since the matrix ring K_n has only finite rank over P, K also has only finite rank over P; that is, K is a division algebra over P. We thus obtain the following.

Wedderburn Theorem: *Every simple algebra is isomorphic to a complete matrix ring over a division algebra.*

When in the future we speak of a *simple algebra*, we shall always mean a simple algebra with identity and thus a complete matrix ring K_n over a skew field K. The multiples $e\beta$ of the identity element will always be identified with the elements β of P.

Exercises

13.15. A direct sum of complete matrix rings over skew fields is semisimple.
13.16. A complete matrix ring over a skew field is primitive and simple.
13.17. A commutative semisimple ring with minimal condition is a direct sum of fields.

13.12 THE BEHAVIOR OF ALGEBRAS UNDER EXTENSION OF THE BASE FIELD

Let $\mathfrak{A}$ be a semisimple algebra over the base field P. We wish to examine the behavior of $\mathfrak{A}$ when the base field is extended to a field Λ, that is, to see which properties of $\mathfrak{A}$ are hereby preserved and which are lost. The investigation proceeds as follows: $\mathfrak{A}$ is assumed first to be a commutative field, then a skew field, then a simple algebra, and finally a semisimple algebra; the next more complicated case is always reduced back to the preceding case. All the rings considered are assumed to have identity elements.

1. *If $\mathfrak{A}$ is a finite separable extension field of* P, *then $\mathfrak{A}_\Lambda$ has no radical regardless of how the field Λ is chosen; if, on the other hand, $\mathfrak{A}$ is inseparable, then, for appropriate choice of Λ, $\mathfrak{A}_\Lambda$ has a radical.*

Proof: If $\mathfrak{A}$ is separable, θ is a primitive element of $\mathfrak{A}$ (Section 6.10), $\varphi(z)$ is the irreducible polynomial with root θ, and n is the degree of $\varphi(z)$, then, by Section 6.3,

$$\mathfrak{A} = \mathsf{P}(\vartheta) = \mathsf{P}+\vartheta\mathsf{P}+\cdots+\vartheta^{n-1}\mathsf{P} \cong \mathsf{P}[z]/\varphi(z)$$

and hence after extending the base field

$$\mathfrak{A}_\Lambda = \Lambda+\vartheta\Lambda+\cdots+\vartheta^{n-1}\Lambda \cong \Lambda[z]/\varphi(z).$$

Since $\varphi(z)$ has no multiple factors in $\Lambda[z]$, there does not exist a polynomial $f(z)$ such that some power of $f(z)$ is congruent to zero modulo $\varphi(z)$ unless $f(z)$ itself is congruent to zero modulo $\varphi(z)$; that is, there is no nilpotent element in $\Lambda[z]/\varphi(z)$ except zero. By Theorem 8, the radical of $\mathfrak{A}_\Lambda$ consists of nilpotent elements. Since there are no nilpotent elements except zero, the radical is the null ideal. This means then that $\mathfrak{A}_\Lambda$ is semisimple.

On the other hand, if $\mathfrak{A}$ is inseparable and ϑ is an inseparable element of $\mathfrak{A}$, then $\mathfrak{A}$ has the subfield $\mathsf{P}(\vartheta)$ and $\mathfrak{A}_\Lambda$ has the subring $\Lambda(\vartheta)$ which is isomorphic to $\Lambda[z]/\varphi(z)$ as above. If Λ is appropriately chosen, $\varphi(z)$ has multiple roots in Λ, and there exists in $\Lambda[z]$ a polynomial $f(z)$ which is itself not divisible by $\varphi(z)$

while some power of $f(z)$ is divisible by $\varphi(z)$. There thus exists a nonzero nilpotent element in $\Lambda[z]/\varphi(z)$; hence there is also such an element in $\Lambda(\vartheta)$, and this element generates a nil ideal in $\mathfrak{A}_\Lambda$, since in a commutative ring every nilpotent element generates a nil ideal. This completes the proof of the theorem.

Since the roles of $\mathfrak{A}$ and Λ are interchangeable, the first part of the theorem may be formulated as follows. *If at least one of the fields $\mathfrak{A}$ and Λ is finite and separable over* P, *then $\mathfrak{A} \times \Lambda$ is semisimple.* Since $\mathfrak{A} \times \Lambda$ is moreover commutative, we have: $\mathfrak{A} \times \Lambda$ *is the direct sum of fields* (cf. Exercise 13.17).

2. We now proceed to the case where $\mathfrak{A}$ is a skew field K. This case can be reduced to the preceding commutative case by means of the following *reduction theorem.*

Reduction Theorem: *If* K *is a skew field over* P *with center* $\mathsf{Z} \supseteqq \mathsf{P}$, *if further* Λ *is an algebra over* P, *and if we put* $\mathfrak{R} = \mathsf{K} \times \Lambda$ *and* $\mathfrak{Z} = \mathsf{Z} \times \Lambda$, *then every two-sided ideal* $\mathfrak{a}$ *in* $\mathfrak{R}$ *is generated by a two-sided ideal of* $\mathfrak{Z}$.

The reduction theorem is more easily understood if it is formulated as a *module theorem.*

Let K be a skew field which admits certain automorphisms σ. Let $\mathfrak{M}$ be a module over K of finite rank:

$$\mathfrak{M} = z_1\mathsf{K} + \cdots + z_q\mathsf{K}.$$

The automorphisms σ of K induce automorphisms of $\mathfrak{M}$ according to the following definition:

$$\sigma(z_1\kappa_1 + \cdots + z_q\kappa_q) = z_1(\sigma\kappa_1) + \cdots + z_q(\sigma\kappa_q).$$

We now assert: *every submodule* $\mathfrak{a}$ *of* $\mathfrak{M}$ *which admits the automorphisms* σ *has a* K*-basis, each of the elements of which is taken into itself by the automorphisms.*

Proof: If $(z_1, \ldots, z_r)$ is a K-basis for $\mathfrak{a}$, then by Section 4.2 it can be enlarged to a K-basis of $\mathfrak{M}$ by adjoining certain z_i, say $z_{r+1}, \ldots, z_q$. Each element of $\mathfrak{M}$ is then congruent modulo $\mathfrak{a}$ to a linear form in $z_{r+1}, \ldots, z_q$ with coefficients in K. In particular, for $i = 1, 2, \ldots, r$,

$$z_i \equiv \sum_{k=r+1}^{q} z_k\gamma_k i \quad (\text{mod } \mathfrak{a}).$$

If we put

$$l_i = z_i - \sum_{k=r+1}^{q} z_k\gamma_{ki},$$

then the l_i are linearly independent elements of $\mathfrak{a}$. Indeed, any linear relation between the l_i implies the same relation between $z_1, \ldots, z_r$ and these are linearly independent. Thus, $l_1, \ldots, l_r$ form a K-basis for $\mathfrak{a}$. If an automorphism σ is now applied to l_i, we obtain:

$$\sigma l_i = z_i - \sum_{r+1}^{q} z_k(\sigma\gamma_{ki}) \tag{13.114}$$

This σl_i must again belong to $\mathfrak{a}$, and it is therefore a linear combination of the original l_i:

$$\sigma l_i = \sum l_j \alpha_j = \sum_1^r z_j \alpha_j - \sum_{r+1}^q z_k \sum_j \gamma_{kj} \alpha_j. \tag{13.115}$$

On comparing (13.114) and (13.115), we find that $\alpha_j = 0$ with the exception of $\alpha_i = 1$. Hence, $\sigma l_j = l_i$ as asserted.

In order to obtain the reduction theorem from the module theorem, we take the inner automorphisms $\kappa \to \beta\kappa\beta^{-1}$ of K for the automorphisms in the module theorem. Transformation with β affects a sum $z_1\kappa_1 + \cdots + z_q\kappa_q$ in the required manner: it leaves the z_i unchanged and takes κ_i into $\beta\kappa_i\beta^{-1}$. A two sided ideal $\mathfrak{a}$ in $\mathsf{K} \times \Lambda$ is also a two-sided K-module and admits therefore the automorphisms $\mathfrak{a} \to \beta\mathfrak{a}\beta^{-1}$. Thus $\mathfrak{a}$ has a basis consisting of elements $\sum z_i\kappa_i$ which are taken individually into themselves by transformation with β; this means that the coefficients κ_i belong to the center Z of K. These basis elements therefore belong to $\mathfrak{Z} = \mathsf{Z} \times \Lambda$, and this completes the proof of the reduction theorem.

Remark: The reduction theorem remains true if an arbitrary skew field Ω is taken in place of Λ, provided that K has finite rank over P. Indeed, if $\mathfrak{a}$ is a two-sided ideal of $\mathfrak{K} = \mathsf{K} \times \Omega$, then $\mathfrak{a}$ as well as $\mathfrak{K}$ has finite rank over Ω and therefore has a finite Ω-basis $(a_1, \ldots, a_s)$. The basis elements, when expressed in the form $\sum \omega_i\kappa_i$, contain only finitely many ω_i which generate a finite submodule Λ of Ω. The module theorem may be applied to the product $\mathfrak{M} = \mathsf{K} \times \Lambda$ and the submodule $\mathfrak{a} \cap \mathfrak{M}$; we thus find a module basis for $\mathfrak{a} \cap \mathfrak{M}$, and hence an ideal basis for $\mathfrak{a}$, which is invariant under the inner automorphisms of K and therefore belongs to $\mathsf{Z} \times \Omega$.

Henceforth K and Λ shall be division algebras over P or, in particular cases, finite extension fields of P. From the reduction theorem we obtain the following.

If $\mathsf{Z} \times \Lambda$ *is simple, then* $\mathsf{K} \times \Lambda$ *is also simple. If* $\mathsf{Z} \times \Lambda$ *is semisimple, and thus the direct sum of simple algebras, then* $\mathsf{K} \times \Lambda$ *is the direct sum of the same number of simple algebras and is thus itself semisimple.*

Naturally, Λ may be replaced by its center in the same way that K is replaced by its center Z.

If the product of the centers of K *and* Λ *is simple or semisimple, then* $\mathsf{K} \times \Lambda$ *is likewise simple or semisimple. In particular,* $\mathsf{K} \times \Lambda$ *is semisimple if one of the centers is separable over* P.

If K is central over P, that is, if $\mathsf{Z} = \mathsf{P}$, then $\mathsf{Z} \times \Lambda = \Lambda$ is a division algebra and hence simple.

If one of the two division algebras K *or* Λ *is central over* P, *then* $\mathsf{K} \times \Lambda$ *is simple.*

3. The transition from division algebras to simple algebras, that is, to complete matrix rings $\mathfrak{A} = \mathsf{K}_r$, is easy. If Λ is an arbitrary skew field over P, then

$$\mathfrak{A} \times \Lambda = \mathsf{K}_r \times \Lambda = \mathsf{K} \times \mathsf{P}_r \times \Lambda \cong (\mathsf{K} \times \Lambda) \times \mathsf{P}_r.$$

If now $\mathsf{K} \times \Lambda$ is semisimple and is thus the direct sum of complete matrix rings, then to obtain $\mathfrak{A} \times \Lambda$ these matrix rings must be multiplied by P_r, that is, the degree of the matrices must be multiplied by r. Nothing is changed as far as the simplicity or semisimplicity of $\mathsf{K} \times \Lambda$ is concerned.

The center of $\mathfrak{A} = \mathsf{K}_r$ is equal to the center Z of K. We thus obtain the following.

If the center Z *of* $\mathfrak{A} = \mathsf{K}_r$ *is separable over* P, *then* $\mathfrak{A} \times \Lambda$ *is semisimple. If* $\mathfrak{A}$ *is central over* P, *that is,* $\mathsf{Z} = \mathsf{P}$, *then* $\mathfrak{A} \times \Lambda$ *is simple however the division algebra* Λ *is chosen.*

From the remark following the reduction theorem it follows that the last result is also true for infinite skew fields Λ.

4. A semisimple algebra $\mathfrak{A}$ is the sum of simple algebras $\mathfrak{A}', \mathfrak{A}'', \ldots$. The product $\mathfrak{A} \times \Lambda$ is obtained on multiplying the individual simple summands by Λ. If we take Λ to be a commutative field, we obtain the following.

A semisimple algebra remains semisimple under every separable extension of the base field. If the centers of the simple algebras $\mathfrak{A}', \mathfrak{A}'', \ldots$ *are all separable over* P, *then the semisimplicity is preserved under arbitrary extension of the base field.*

5. We have seen that the behavior of simple algebras under extension of the base field depends entirely on the behavior of the underlying division algebra. We shall now investigate the behavior of central division algebras somewhat further.

By what was proved in part 3 above, a central division algebra remains central and simple under every extension of the base field. It need not remain a skew field, but may go over into a matrix ring over a skew field. In this case we say that the extension of the base field effects a *decomposition* of the division algebra (a decomposition into simple left ideals).

If $\mathsf{K} \neq \mathsf{P}$ *is a central division algebra, then there always exists a field extension which effects a decomposition of* K.

Let β be an element of K which does not belong to P; β is a root of an irreducible polynomial $\varphi(x)$ of $\mathsf{P}[x]$. The polynomial $\varphi(x)$ splits in an appropriate field Λ; for example, we may take $\Lambda \cong \mathsf{P}(\beta)$ such that a linear factor is split off from $\varphi(x)$. It was previously shown that $\Lambda \times \mathsf{P}(\beta) \cong \Lambda[x]/\varphi(x)$; $\Lambda \times \mathsf{P}(\beta)$ therefore has zero divisors, and thus the larger ring $\Lambda \times \mathsf{K}$ also has zero divisors. This ring is therefore no longer a skew field; it can thus only be a matrix ring $\mathsf{K}_{r'}$ with $r' > 1$.

Let the symbol $(\mathsf{K} : \mathsf{P})$ denote the rank of K over P. Equating the rank over Λ of the left and right members of $\mathsf{K} \times \Lambda = \mathsf{K}'_{r'}$, we obtain

$$(\mathsf{K} : \mathsf{P}) = r'^2 \cdot (\mathsf{K}' : \Lambda).$$

The rank of K' over Λ is thus smaller than that of K over P. If $\mathsf{K}' \neq \Lambda$, then the skew field K' can also be decomposed by a further extension of the field Λ. Then $\mathsf{K}'_{r'}$ goes over into a matrix ring of degree $r'r''$. This process cannot be continued indefinitely, since the ranks of the skew fields always become smaller. A *complete decomposition* is finally obtained in which the division algebra K has become a matrix ring over Λ:

$$\mathsf{K} \times \Lambda \cong \Lambda_m.$$

A field Λ which accomplishes this is called a *splitting field* of the division algebra K. The above proof shows that there is always a splitting field of finite degree over P. The rank relation above now becomes

$$(\mathsf{K} : \mathsf{P}) = m^2.$$

The rank of a division algebra K over its center P is thus *always a square number* m^2. The number m—that is, the degree of the matrices after complete decomposition—is called the *index* of the division algebra K.

A splitting field of K is also a splitting field for K_r and conversely, since $\mathsf{K} \times \Lambda$ and $\mathsf{K}_r \times \Lambda$ are complete matrix rings over the same skew field.

Exercises

13.18. A product of two simple algebras over P, of which one is a central algebra is simple.

13.19. An algebraically closed extension field Ω of P is a splitting field for all central simple algebras over P.

Chapter 14

REPRESENTATION THEORY OF GROUPS AND ALGEBRAS

14.1 STATEMENT OF THE PROBLEM

Let $\mathfrak{G}$ be a group. A *representation of* $\mathfrak{G}$ *in a field* K is a group homomorphism which to each group element a assigns a linear transformation A of an n-dimensional vector space over K (or, what amounts to the same thing, an $n \times n$ matrix A). The dimension n is called the *degree* of the representation. The representation is said to be faithful if it is an isomorphism.

Similarly, a *representation of a ring* $\mathfrak{o}$ *in* K is a ring homomorphism $a \rightarrow A$, where the A are again linear transformations of an n-dimensional vector space. This definition agrees with that given in Section 12.4. It was shown there that to each representation of $\mathfrak{o}$ in K there corresponds a double module ($\mathfrak{o}$ on the left and K on the right), the *representation module*, and, conversely, every such representation module provides a representation. To isomorphic representation modules there correspond equivalent representations, and conversely. The representation is reducible if the representation module possesses a proper submodule distinct from $\{0\}$, and it is irreducible if the module is simple.

If $\mathfrak{o}$ is an algebra over P, then it is required of the representation that the base field P be contained in the field K and that $a \rightarrow A$ imply $a\beta \rightarrow A\beta$ for all β of P. For the representation module $\mathfrak{M}$ this means that

$$(a\beta)u = (au)\beta \qquad \text{for} \quad a \in \mathfrak{o}, \quad \beta \in \mathrm{P}, \quad u \in \mathfrak{M}.$$

The principal problem is to find all representations of a given group or algebra. The representation problem for finite groups can be reduced immediately to that for algebras by forming the group ring (Section 13.2)

$$\mathfrak{o} = a_1\mathrm{K} + \cdots + a_h\mathrm{K}$$

whose basis elements $a_1, \ldots, a_h$ are the elements of $\mathfrak{G}$. If $a_i \rightarrow A_i$ is the group representation, then

$$\sum a_i\beta_i \rightarrow \sum A_i\beta_i$$

is a representation of the group ring $\mathfrak{o}$, as is easily verified. Conversely, every representation of the group ring $\mathfrak{o}$ in K assigns, in particular, linear transformations to the basis elements $a_1, \ldots, a_h$. Thus: *Every representation of a finite group in a commutative field* K *can be obtained from a representation of the group ring.*

In the representation theory of algebras it is usually required that the representation field K be the same as the base field P. The general case can be reduced back to this special case by extending the algebra $\mathfrak{o}$ to $\mathfrak{o}_K$. If in the original representation the matrices $A_1, \ldots, A_h$ are assigned to the basis elements $a_1, \ldots, a_h$ of $\mathfrak{o}$, then the matrix $\sum A_i\beta_i$ may be assigned to an element $\sum a_i\beta_i$ $(\beta_i \in K)$ and the representation of $\mathfrak{o}$ is thereby extended to a representation of $\mathfrak{o}_K$. Thus *every representation of* $\mathfrak{o}$ *in a commutative field* K *can be obtained from a representation of* $\mathfrak{o}_K$.

A further restriction of the problem is obtained if the ring $\mathfrak{o}$ has an identity element. We may then always assume that this identity element 1 is also the identity operator for the representation module; that is, to it is assigned the identity matrix in the representation. Otherwise, by Section 12.1, the representation module is a direct sum $\mathfrak{M}_0+\mathfrak{M}_1$, where $\mathfrak{M}_0$ is annihilated by $\mathfrak{o}$ and 1 is the identity operator for $\mathfrak{M}_1$. The representation decomposes into two components of which the first consists only of null matrices, and is thus of no interest; the second provides a representation in which the identity element is the identity operator.

An especially important representation of an algebra is the *regular representation*, which is obtained by interpreting $\mathfrak{o}$ itself as a representation module ($\mathfrak{o}$ on the left and P on the right). The submodules are precisely the left ideals of $\mathfrak{o}$. The regular representation is completely reducible if the ring is completely left-reducible.

14.2 REPRESENTATION OF ALGEBRAS

In Section 13.9 (Theorem 18) we saw that the radical $\mathfrak{R}$ of an algebra $\mathfrak{o}$ is represented by zero in every irreducible representation. The same naturally holds also for every completely reducible representation, since any such representation is obtained from a sequence of irreducible representations. Any completely reducible representation of $\mathfrak{o}$ may therefore be interpreted as a representation of the semisimple algebra $\mathfrak{o}/\mathfrak{R}$.

The following theorem shows how all representations of a semisimple algebra—or, more generally, of a semisimple ring with the minimal condition for left ideals—are obtained. By Section 13.7 such a ring $\mathfrak{o}$ has an identity element and is completely left-reducible; that is, it is the direct sum of simple left ideals. Every representation of $\mathfrak{o}$ is provided by an $\mathfrak{o}$-module $\mathfrak{M}$.

Principal Theorem: *Let* $\mathfrak{o}$ *be a completely left-reducible ring with identity, and let* $\mathfrak{M}$ *be an* $\mathfrak{o}$*-module with finite basis. Let the identity of* $\mathfrak{o}$ *be the identity operator* *for $\mathfrak{M}$. Then $\mathfrak{M}$ is the direct sum of simple $\mathfrak{o}$-modules. Each of these is isomorphic to a simple left ideal of $\mathfrak{o}$.*

Proof: The ring $\mathfrak{o}$ is the direct sum of simple left ideals by hypothesis:

$$\mathfrak{o} = \mathfrak{l}_1 + \cdots + \mathfrak{l}_r. \tag{14.1}$$

The module $\mathfrak{M}$ is assumed to have a finite $\mathfrak{o}$-basis $(u_1, \ldots, u_s)$. This implies that

$$\mathfrak{M} = (\mathfrak{o}u_1, \ldots, \mathfrak{o}u_s). \tag{14.2}$$

Substituting (14.1) in (14.2) gives

$$\mathfrak{M} = (\ldots, \mathfrak{l}_i u_k, \ldots). \tag{14.3}$$

Those modules $\mathfrak{l}_i u_k$ which are null may be omitted from the sum on the right-hand side of (14.3). If $\mathfrak{l}_i u_k \neq 0$, then $x \to x u_k$ defines an operator isomorphism of $\mathfrak{l}_i$ onto $\mathfrak{l}_i u_k$. The modules $\mathfrak{l}_i u_k$ which are not null are thus isomorphic to $\mathfrak{l}_i$ and are therefore simple. If one of the $\mathfrak{l}_i u_k$ is contained in the sum of the other such modules, then this term may be omitted from the sum. This process is continued until each remaining term $\mathfrak{l}_i u_k$ has only the zero element in common with the sum of the other such terms. The sum is then direct.

The theorem also holds if a right multiplication domain Ω with the usual properties

$$(au)\beta = a(u\beta) = (a\beta)u \qquad (\beta \in \Omega)$$

is also given for $\mathfrak{o}$ and $\mathfrak{M}$. In application to the representation theory of algebras, Ω is the coefficient field P of the algebra $\mathfrak{o}$ and at the same time the representation field. If $\mathfrak{M}$ is a vector space of finite dimension over P, then $\mathfrak{M}$ automatically has a finite $\mathfrak{o}$-basis as required in the theorem.

When applied to semisimple algebras, the theorem states that every representation of such an algebra is completely reducible, and each of its irreducible components occurs as a component in the regular representation. The irreducible components of the regular representation are, according to (14.1), provided by the simple left ideals $\mathfrak{l}_i$.

By Section 13.8, a semisimple algebra $\mathfrak{o}$ is the direct sum of simple algebras $\mathfrak{a}_\nu$:

$$\mathfrak{o} = \mathfrak{a}_1 + \cdots + \mathfrak{a}_s. \tag{14.4}$$

The $\mathfrak{a}_\nu$ can be decomposed further into minimal left ideals $\mathfrak{l}_i$. All the $\mathfrak{l}_i$ occurring in a single $\mathfrak{a}_\nu$ are isomorphic and thus provide the same representation. The $\mathfrak{l}_i$ occurring in $\mathfrak{a}_\nu$ are annihilated by every $\mathfrak{a}_\mu$ with $\mu \neq \nu$:

$$\mathfrak{a}_\mu \mathfrak{l}_i \subseteq \mathfrak{a}_\mu \mathfrak{a}_\nu = \{0\}.$$

All such $\mathfrak{a}_\mu$ are therefore represented by zero in the representation provided by $\mathfrak{l}_i$. Only $\mathfrak{a}_\nu$ is faithfully represented. Indeed, the kernel of the representation of $\mathfrak{a}_\nu$ is a two-sided ideal in $\mathfrak{a}_\nu$; since $\mathfrak{a}_\nu$ is simple and is not represented by zero alone, the kernel must be the null ideal.

We now investigate the representation of a simple algebra which is provided by an arbitrary simple left ideal of this algebra.

By Section 13.11, a simple algebra $\mathfrak{o}$ with identity is isomorphic to a complete matrix ring over a skew field Δ. If c_{ik} denote the matrices which were introduced in Section 13.2 as C_{ik}, then

$$\mathfrak{o} = c_{11}\Delta + c_{12}\Delta + \cdots + c_{nn}\Delta.$$

A minimal left ideal $\mathfrak{l}$ is given by

$$\mathfrak{l} = c_{11}\Delta + c_{21}\Delta + \cdots + c_{n1}\Delta.$$

The base field P, which is also to be the representation field, is contained in the center of Δ, and Δ has finite rank over P.

We first consider the case $\Delta = \mathsf{P}$. The basis $(c_{11}, c_{21}, \ldots, c_{n1})$ of $\mathfrak{l}$ can serve to determine the matrices of the representation explicitly. If $a = \sum_{i,\,k=1}^{n} c_{ik}\alpha_{ik}$ is an element of $\mathfrak{o}$, then

$$ac_{k1} = \sum_{i=1}^{n} c_{ik}c_{k1}\alpha_{ik} = \sum_{i=1}^{n} c_{i1}\alpha_{ik};$$

thus, *in the representation provided by* $\mathfrak{l}$ *to the element* a *there corresponds the matrix* (α_{ik}). The isomorphism of $\mathfrak{o}$ to the complete matrix ring of matrices (α_{ik}) is thus precisely that irreducible representation which is provided by a minimal left ideal.

It is to be noted that in the case $\Delta = \mathsf{P}$ the matrices of the representation always from the *complete* matrix ring of degree n. This may also be expressed by saying that among the matrices of the representation n^2 of them are linearly independent.

If now Δ is a proper extension field of P:

$$\Delta = \lambda_1\mathsf{P} + \cdots + \lambda_r\mathsf{P},$$

then we first form the regular representation of Δ in P, whereby the matrix defined by

$$\beta\lambda_j = \sum \lambda_i\beta_{ij}, \qquad B = (\beta_{ij})$$

is assigned to each β of Δ. We then form

$$\begin{aligned} \mathfrak{l} &= c_{11}\Delta + \cdots + c_{n1}\Delta \\ &= (c_{11}\lambda_1\mathsf{P} + \cdots + c_{11}\lambda_r\mathsf{P}) + \cdots + (c_{n1}\lambda_1\mathsf{P} + \cdots + c_{n1}\lambda_r\mathsf{P}). \end{aligned}$$

If we represent an element $c_{ik}\cdot\beta$ of $\mathfrak{o}$ with respect to this basis, we obtain

$$c_{ik}\beta \rightarrow \begin{pmatrix} 0\ldots 0\ldots 0 \\ \cdot \quad \cdot \quad \cdot \\ \cdot \quad \cdot \quad \cdot \\ \cdot \quad \cdot \quad \cdot \\ 0\ldots B\ldots 0 \\ \cdot \quad \cdot \quad \cdot \\ \cdot \quad \cdot \quad \cdot \\ \cdot \quad \cdot \quad \cdot \\ 0\ldots 0\ldots 0 \end{pmatrix},$$

where the zeros stand for r-rowed matrices and B occupies the kth position in the ith row of matrices. On summing, it follows that

$$\sum_{i,\,k=1}^{n} c_{ik}\alpha_{ik} \rightarrow \begin{pmatrix} A_{11}\ldots A_{1n} \\ \cdot \qquad \cdot \\ \cdot \qquad \cdot \\ \cdot \qquad \cdot \\ A_{n1}\ldots A_{nn} \end{pmatrix}, \tag{14.5}$$

where the A_{ik} are again the matrices which correspond to α_{ik} in the regular representation of Δ.

From the form of the irreducible representation provided by $\mathfrak{l}$ it can also be found how this representation decomposes on extension of the base field P to a commutative extension field Ω. In this extension Δ goes over into a system $\Delta_\Omega = \Delta\times\Omega$ and the left ideal $\mathfrak{l} = c_{11}\Delta_\Omega + \cdots + c_{n1}\Delta_\Omega$ becomes

$$\mathfrak{l}_\Omega = c_{11}\Delta_\Omega + \cdots + c_{n1}\Delta_\Omega.$$

If now Δ_Ω is reducible and thus contains a proper left ideal $\mathfrak{l}'$, then $\mathfrak{l}_\Omega$ also contains a proper ideal

$$\mathfrak{L}' = c_{11}\mathfrak{l}' + \cdots + c_{n1}\mathfrak{l}'.$$

Thus, if Δ_Ω decomposes into left ideals $\mathfrak{l}'$, then $\mathfrak{l}_\Omega$ decomposes into the same number of left ideals $\mathfrak{L}'$. *The reducibility or decomposition of the irreducible representation of* $\mathfrak{o}$ *provided by* $\mathfrak{l}$ *on extension of* P *to* Ω *is thus completely determined by the reducibility or decomposition of the algebra* Δ_Ω *into left ideals.*

If $\Delta \neq \mathsf{P}$, then by Section 13.12 the field Ω can always be chosen so that Δ_Ω contains zero divisors; it is therefore no longer a skew field and thus contains at least one proper left ideal. The representation provided by $\mathfrak{l}$ which is irreducible in P then becomes reducible in Ω. In the case $\Delta = \mathsf{P}$, on the other hand, the representation provided by $\mathfrak{l}$ is *absolutely irreducible*; that is, it remains irreducible under any extension of the base field. *Thus* $\Delta = \mathsf{P}$ *is the necessary and sufficient condition that the representation irreducible in* P *be absolutely irreducible.*

If the algebra $\mathfrak{o}$ is not simple but only semisimple, and thus a direct sum of simple algebras $\mathfrak{a}_1 + \cdots + \mathfrak{a}_s$, and if $\mathfrak{l}$ is a left ideal of $\mathfrak{a}_\nu$, then to find the representation provided by $\mathfrak{l}$ of an element a of $\mathfrak{o}$ we must first write a as a sum $a_1 + \cdots + a_s$, pick out the component a_ν from this sum, and form the matrix corresponding to this a_ν according to (14.5). The other components $a_1, \ldots, a_{\nu-1}, a_{\nu+1}, \ldots, a_s$ annihilate the ideal $\mathfrak{l}$ and are thus represented by zero.

If $\mathfrak{a}_1, \ldots, \mathfrak{a}_s$ are, say, complete matrix rings of degrees $n_1, \ldots, n_s$ over the skew fields $\Delta_1, \ldots, \Delta_s$ and, further, if r_ν is the rank of Δ_ν and $\mathfrak{D}_\nu$ is the irreducible representation provided by a left ideal of $\mathfrak{a}_\nu$, then the rank h of $\mathfrak{o}$ is equal to the sum of the ranks of $\mathfrak{a}_1, \ldots, \mathfrak{a}_s$ and hence

$$h = \sum_1^s n_\nu^2 r_\nu. \tag{14.6}$$

Further, by (14.5) the degree of the representation $\mathfrak{D}_\nu$ is equal to

$$g_\nu = n_\nu r_\nu. \tag{14.7}$$

Finally, $\mathfrak{a}_\nu$ decomposes into n_ν equivalent left ideals $\mathfrak{l}$; the regular representation therefore contains the representation $\mathfrak{D}_\nu$ precisely n_ν times.

In particular, if all the $\mathfrak{D}_\nu$ are absolutely irreducible, then all $r_\nu = 1$; (14.6) and (14.7) then simplify to

$$h = \sum_1^s n_\nu^2; \qquad g_\nu = n_\nu. \tag{14.8}$$

14.3 REPRESENTATIONS OF THE CENTER

In an irreducible representation the center of an algebra $\mathfrak{o}$ must be represented by matrices which commute with all other matrices of the representation. If the base field is algebraically closed and the ring of matrices of the representation is thus a complete matrix ring, then its center consists only of multiples of the identity matrix E; the center of $\mathfrak{o}$ is thus represented by matrices of the form $E\alpha$. The same holds for absolutely irreducible representations, since for these representations the base field can be extended to an algebraically closed field without destroying the irreducibility. Hence, *in an absolutely irreducible representation of an algebra $\mathfrak{o}$ the elements of the center are represented by multiples of the identity matrix.*

If $\mathfrak{o}$ is itself commutative, and thus is its own center, then all matrices of an absolutely irreducible representation have the form $E_n\lambda$. It then follows from the irreducibility that the representations must be of first degree. Thus, *the absolutely irreducible representations of a commutative algebra are of first degree.*

A representation of $\mathfrak{o}$ of first degree is a homomorphism of $\mathfrak{o}$ into the representation field K. If K is commutative, then two equivalent representations are in fact equal; for if $A = (\alpha)$ is a matrix of the representation and λ is an element of K, then

$$\lambda^{-1}(\alpha)\lambda = (\lambda^{-1}\alpha\lambda) = (\alpha).$$

It thus follows that *the number of inequivalent representations of a commutative algebra* $\mathfrak{o}$ *of first degree in a commutative field* K *is equal to the number of distinct homomorphisms from* $\mathfrak{o}$ *into* K.

Let us now return to noncommutative algebras and suppose that $\mathfrak{o}$ is semisimple. Then $\mathfrak{o}$ is the direct sum of simple algebras:

$$\mathfrak{o} = \mathfrak{a}_1 + \cdots + \mathfrak{a}_s,$$

and the center $\mathfrak{Z}$ of $\mathfrak{o}$ is the sum of precisely the same number of fields:

$$\mathfrak{Z} = \mathfrak{Z}_1 + \cdots + \mathfrak{Z}_s \qquad (\mathfrak{Z}_\nu \text{ the center of } \mathfrak{a}_\nu).$$

The number of inequivalent, irreducible representations of $\mathfrak{o}$, and likewise of $\mathfrak{Z}$, is equal to the number s of the two-sided components of $\mathfrak{o}$ or $\mathfrak{Z}$; for every such representation $\mathfrak{D}_\nu$ of $\mathfrak{o}$ is provided by a left ideal of $\mathfrak{a}_\nu$, and every such representation $\mathfrak{D}_\nu'$ of $\mathfrak{Z}$ is provided by a $\mathfrak{Z}_\nu$. *There are thus the same number of inequivalent, irreducible representations of* $\mathfrak{o}$ *as of* $\mathfrak{Z}$, *and to each irreducible representation* $\mathfrak{D}_\nu$ *of* $\mathfrak{o}$ *in which all the* $\mathfrak{a}_1, \ldots, \mathfrak{a}_s$ *with the exception of* $\mathfrak{a}_\nu$ *are represented by zero there corresponds a representation* $\mathfrak{D}_\nu'$ *of* $\mathfrak{Z}$ *in which all the* $\mathfrak{Z}_1, \ldots, \mathfrak{Z}_s$ *with the exception of* $\mathfrak{Z}_\nu$ *are represented by zero.*

In particular, if $\mathfrak{o}$ is the sum of complete matrix rings over P, then the fields $\mathfrak{Z}_\nu$ are of rank 1 and are isomorphic to P; in this case the number s of irreducible representations of $\mathfrak{o}$ is thus equal to the rank of the center $\mathfrak{Z}$. The relation between the irreducible representations $\mathfrak{D}_\nu$ of $\mathfrak{o}$ and the irreducible representations (of first degree) of $\mathfrak{Z}$ is very simple in this case. Indeed, in the representation $\mathfrak{D}_\nu$ each center element z is represented by a matrix of the form $E\alpha$, where E denotes the identity matrix of degree n_ν. To each z there thus corresponds a particular α (for given ν), and we may write:

$$\alpha = \Theta_\nu(z).$$

The function Θ_ν affords a homomorphism of the center; that is,

$$\Theta_\nu(y+z) = \Theta_\nu(y) + \Theta_\nu(z)$$
$$\Theta_\nu(yz) = \Theta_\nu(y)\Theta_\nu(z)$$
$$\Theta_\nu(z\beta) = \Theta_\nu(z)\cdot\beta.$$

Under this homomorphism all the $\mathfrak{Z}_1, \ldots, \mathfrak{Z}_s$ with the exception of $\mathfrak{Z}_\nu$ are represented by zero; that is, the homomorphism Θ_ν is precisely the first-degree representation of the center previously denoted by $\mathfrak{D}_\nu'$.

The representation Θ_ν is known as soon as a P-basis for the module $\mathfrak{Z}_\nu$ is given; the identity element e_ν of the field $\mathfrak{Z}_\nu$ may be taken as such a basis. If each element z of $\mathfrak{Z}$ is written in the form

$$z = \sum_{\nu=1}^{s} e_\nu \beta_\nu, \tag{14.9}$$

then

$$z e_\nu = e_\nu^2 \beta_\nu = e_\nu \beta_\nu;$$

and $E\beta_\nu$ is thus the representing matrix, that is,

$$\Theta_\nu(z) = \beta_\nu.$$

For (14.9) we may now also write:

$$z = \sum_{\nu=1}^{s} e_\nu \Theta_\nu(z), \tag{14.10}$$

or in words: *the coefficients $\Theta_\nu(z)$ in the expansion of a center element z in terms of the idempotent elements e_ν of the center give at the same time the homomorphisms or representations of first degree of the center.*

Exercises

14.1. The number of representations of first degree of a commutative algebra $\mathfrak{o}$ in an algebraically closed extension field Ω of P is equal to the rank of $\mathfrak{o}_\Omega/\mathfrak{R}$ over P, where $\mathfrak{R}$ denotes the radical of $\mathfrak{o}_\Omega$.

14.2. If K is a commutative field over P, then the number of first-degree representations of K in $_\Omega$ is equal to the reduced field degree of K over P. Here $\mathfrak{R} = \{0\}$ if and only if K is separable over P.

14.4 TRACES AND CHARACTERS

The *trace of an element a in the representation $\mathfrak{D}$*, written

$$S_{\mathfrak{D}}(a) \qquad \text{or simply} \quad S(a),$$

is defined to be the trace $S(A)$ of the matrix A corresponding to a in the representation $\mathfrak{D}$. The trace $S_{\mathfrak{D}}$, considered as a function of the element a for fixed $\mathfrak{D}$, is called the *trace of the representation $\mathfrak{D}$*.

The relation

$$S(P^{-1}AP) = S(A)$$

implies that *equivalent representations have the same trace.*

The trace is a *linear function*; that is,

$$S(a+b) = S(a)+S(b)$$
$$S(a\beta) = S(a)\beta.$$

Traces of absolutely irreducible representation (or, what is the same thing, traces of irreducible representations in an algebraically closed field Ω) are called *characters.*[1] The character of an element a in the νth irreducible representation

[1] Many authors also use the word "character" for reducible representations and then speak of "compound characters." This designation is avoided here, since it does not coincide with the older meaning of the word "character" in the special case of Abelian groups and since, moreover, the word "trace" conveys the meaning just as clearly.

$\mathfrak{D}_\nu$ is denoted by

$$\chi_\nu(a).$$

The index ν will sometimes be omitted if a fixed representation is being considered.

In an absolutely irreducible representation $\mathfrak{D}_\nu$ of degree n_ν the center elements z are represented by diagonal matrices $E.\ \Theta_\nu(z)$ by Section 14.3, where Θ_ν is a homomorphism of the center into the field Ω. The trace of the matrix $E\cdot\Theta_\nu(z)$ is

$$\chi_\nu(z) = n_\nu\cdot\Theta_\nu(z). \tag{14.11}$$

In particular, the identity element of $\mathfrak{o}$ is represented by the identity matrix E whose trace is n_ν:

$$\chi_\nu(1) = n_\nu.$$

In the following we shall require that the degree n_ν of the irreducible representations not be divisible by the characteristic of the field Ω. We may then divide (14.11) by n_ν and obtain

$$\Theta_\nu(z) = \frac{\chi_\nu(z)}{n_\nu}. \tag{14.12}$$

In this manner the homomorphisms of the center are expressed in terms of the characters.

Theorem: *A completely reducible representation of an algebra $\mathfrak{o}$ in a field Ω of characteristic 0 is uniquely determined up to equivalence by the traces of the matrices of the representation.*

Proof: If $\mathfrak{R}$ is the radical of $\mathfrak{o}$, then every completely reducible representation of $\mathfrak{o}$ is also such a representation of $\mathfrak{o}/\mathfrak{R}$. The traces of the matrices representing the elements of $\mathfrak{o}/\mathfrak{R}$ are known by hypothesis. Suppose that

$$\mathfrak{o}/\mathfrak{R} = \mathfrak{a}_1 + \cdots + \mathfrak{a}_n;$$

let the identity elements of $\mathfrak{a}_1, \ldots, \mathfrak{a}_n$ be $e_1, \ldots e_n$. In the irreducible representation $\mathfrak{D}_\nu$ the elements e_ν is then represented by the n_ν-rowed identity matrix; the corresponding trace is thus

$$S_\nu(e_\nu) = n_\nu,$$

while

$$S_\nu(e_\mu) = 0 \qquad \text{for} \quad \mu \neq \nu.$$

Now a completely reducible representation is known if it is known how often each irreducible representation $\mathfrak{D}_\nu$ occurs in it. If the representation $\mathfrak{D}_\nu$ occurs q_ν times, then the representation consists of q_1 blocks $\mathfrak{D}_1$, q_2 blocks $\mathfrak{D}_2$, and so on. The trace of e_ν in this representation is then

$$S(e_\nu) = q_\nu n_\nu. \tag{14.13}$$

The q_ν can be computed from (14.13) as soon as the traces $S(e_\nu)$ are known. This completes the proof.

Remark: The traces of all the elements of $\mathfrak{o}$ are known if the traces of the basis elements of $\mathfrak{o}$ are known. For example, if $\mathfrak{o}$ is the group ring of a finite group, it is only necessary to know the traces of the group elements, and the representation is already determined. If $a_1, \ldots, a_n$ are the basis elements and $\chi_\nu(a_i)$ are their traces in the irreducible representations, then for an arbitrary representation

$$S(a_i) = \sum_{\nu=1}^{s} q_\nu \chi_\nu(a_i). \tag{14.14}$$

The numbers q_ν are uniquely determined by the equations according to the theorem above. Equations (14.14) afford a computational method of decomposing a given completely reducible representation into irreducible components by computing only the traces. However, the characters of the irreducible representations must first be known.

14.5 REPRESENTATIONS OF FINITE GROUPS

We begin with the following theorem.

Maschke's Theorem: *Every representation of a finite group $\mathfrak{G}$ in a field* P *whose characteristic does not divide the order h of the group is completely reducible.*

Proof: We suppose that the representation module $\mathfrak{M}$ is reducible and that $\mathfrak{N}$ is a minimal submodule. We shall show that $\mathfrak{M}$ can be represented as a direct sum $\mathfrak{N}+\mathfrak{N}''$, where $\mathfrak{N}''$ is again a representation module.

As a vector space, $\mathfrak{M}$ decomposes according to the scheme $\mathfrak{N}+\mathfrak{N}'$; however, $\mathfrak{N}'$ is not necessarily invariant under $\mathfrak{o}$. If y is an element of $\mathfrak{N}'$ and a an element of $\mathfrak{G}$, then ay can be uniquely represented as the sum of an element of $\mathfrak{N}$ and an element y' of $\mathfrak{N}'$; thus,

$$ay \equiv y' \qquad (\text{mod } \mathfrak{N}).$$

For fixed a the element y' is uniquely determined by y and depends linearly on y: $ay \equiv y'$ and $az \equiv z'$ imply $a(y+z) \equiv y'+z'$ and $ay\beta \equiv y'\beta$ for $\beta \in \mathrm{P}$. We may therefore write:

$$y' = A'y; \qquad A'y \equiv ay \qquad (\text{mod } \mathfrak{N}),$$

where A' is a linear transformation into $\mathfrak{N}'$ which depends on a. Indeed, the A' form a representation of the group $\mathfrak{G}$ since $a \to A'$ and $b \to B'$ imply $ab \to A'B'$.

We now put

$$\frac{1}{h}\sum_a a^{-1}A'y = Qy = y'';$$

y'' depends linearly on y, and the y'' therefore form a linear subspace $\mathfrak{N}'' = \mathfrak{Q}\mathfrak{N}'$. It also follows that modulo $\mathfrak{N}$

$$y'' \equiv \frac{1}{h}\sum_a a^{-1}ay = y.$$

Each element of $\mathfrak{M}$ is therefore congruent modulo $\mathfrak{N}$ not only to an element y'

of $\mathfrak{N}'$, but also to a uniquely determined element y'' of $\mathfrak{N}''$; that is, we have the direct-sum representation

$$\mathfrak{M} = \mathfrak{N}+\mathfrak{N}''.$$

Finally, for each element b of $\mathfrak{G}$,

$$\begin{aligned} by'' &= \frac{1}{h}\sum_a ba^{-1}A'y \\ &= \frac{1}{h}\sum_a (ab^{-1})^{-1}(A'B'^{-1})B'y \\ &= QB'y \in Q\mathfrak{N}' = \mathfrak{N}''; \end{aligned}$$

$\mathfrak{N}''$ is therefore transformed into itself by the operators b of $\mathfrak{G}$; that is, $\mathfrak{N}''$ is a representation module.

If $\mathfrak{N}''$ is reducible, then it can be treated in the same way by splitting off a minimal submodule, and so on. A complete decomposition of the module, and hence of the representation, finally results. This completes the proof of Maschke's theorem.

By section 14.1, every representation of $\mathfrak{G}$ can be extended to a representation of the group ring

$$\mathfrak{o} = a_1\mathrm{P}+\cdots+a_h\mathrm{P};$$

conversely, every representation of $\mathfrak{o}$ provides a representation of $\mathfrak{G}$ in a natural way. It now follows from Maschke's theorem that every representation of $\mathfrak{o}$ is completely reducible. This is true, in particular, of the regular representation provided by $\mathfrak{o}$ itself as the representation module. Thus $\mathfrak{o}$ is the direct sum of minimal left ideals and is therefore *semisimple* by Theorem 13, Section 13.7. By section 14.2, the minimal left ideals of $\mathfrak{o}$ give all the irreducible representations.

The number of absolutely irreducible representations is equal to the rank of the center by Section 14.3, and the center of the group ring consists of all those sums

$$\sum_\lambda a\lambda\beta\lambda \qquad (a\lambda \in \mathfrak{G},\ \beta\lambda \in \mathrm{P}), \tag{14.15}$$

in which conjugate group elements have the same coefficients, as is easily seen. The elements conjugate to an element a form a *class*. If k_a is the sum of the elements of this class, then (14.15) is a sum of such sums k_a with coefficients in P. We thus have the theorem: *The center of the group ring is generated by the class sums k_a.* The rank of the center is therefore equal to the number of classes, and from this there follows: *The number of inequivalent, absolutely irreducible representations of a group is equal to the number of classes of conjugate elements.*

By section 14.2, the relation

$$n_1^2+n_2^2+\cdots+n_s^2 = h$$

holds between the degrees $n_1, \ldots, n_s$ of the irreducible representations.

One representation of first degree which is always available is the "identity representation" in which every group element is mapped onto 1. If there are still other representations of first degree, then a proper normal subgroup with an Abelian factor group is present, for the matrices of a representation of first degree commute and form an Abelian group homomorphic to the group. Conversely, if a proper normal subgroup with Abelian factor group is present, then the characters of this factor group give representations of first degree. All other representations are of higher degree.

Example 1: *The symmetric group* $\mathfrak{S}_3$. There are three classes and thus three representations. The alternating group has two cosets $\mathfrak{K}_0$ and $\mathfrak{K}_1$, those of the even and odd substitutions. The two characters are

$$\chi(\mathfrak{K}_0) = 1, \qquad \chi(\mathfrak{K}_1) = \pm 1$$

and these determine the representations of first degree. Since

$$n_1{}^2 + n_2{}^2 + n_3{}^2 = 6,$$

the third representation must have degree 2. The permutations of three vectors e_1, e_2, e_3 in a plane whose sum is zero give a faithful representation of this permutation group; it is easy to show that the representation is irreducible. If e_1 and e_2 are taken as base vectors, then the representation is as follows:

$$\begin{cases}(1\,2)\,e_1 = e_2\\(1\,2)\,e_2 = e_1\end{cases} \quad \begin{cases}(1\,3)\,e_1 = -e_1 - e_2\\(1\,3)\,e_2 = e_2\end{cases} \quad \begin{cases}(2\,3)\,e_1 = e_1\\(2\,3)\,e_2 = -e_1 - e_2\end{cases}$$

$$\begin{cases}(1\,2\,3)\,e_1 = e_2\\(1\,2\,3)\,e_2 = -e_1 - e_2\end{cases} \quad \begin{cases}(1\,3\,2)\,e_1 = -e_1 - e_2\\(1\,3\,2)\,e_2 = e_1.\end{cases}$$

Example 2: *The quaternion group* $\mathfrak{Q}_8$ is the group of eight quaternions: ± 1, $\pm j$, $\pm k$, $\pm l$. It has two generators j and k which satisfy the relations

$$j^4 = 1, \qquad k^2 = j^2, \qquad kj = j^3 k.$$

There are five classes and thus five representations. The normal subgroup $\{1, j^2\}$ has as factor group the Klein four-group whose four characters give four linear representations. Since

$$n_1{}^2 + n_2{}^2 + n_3{}^2 + n_4{}^2 + n_5{}^2 = 8,$$

the remaining representations must have degree 2. If we assign to the group elements $1, j, j^2, j^3, k, jk, j^2k, j^3k$ the quaternions $1, j, -1, -j, k, l, -k, -l$, then we obtain a homomorphic mapping of the group ring $\mathfrak{o}$ onto the field of quaternions; the field of quaternions must therefore occur among the two-sided composition factors of $\mathfrak{o}$. The decomposition of $\mathfrak{o}$ in the rational base field $\mathbb{Q}$ is therefore

$$\mathfrak{o} = \mathfrak{a}_1 + \mathfrak{a}_2 + \mathfrak{a}_3 + \mathfrak{a}_4 + \mathfrak{a}_5,$$

where $\mathfrak{a}_1, \mathfrak{a}_2, \mathfrak{a}_3, \mathfrak{a}_4$ are isomorphic to $\mathbb{Q}$ and $\mathfrak{a}_5$ is isomorphic to the field of quaternions. If we go over to the algebraically closed base field (it suffices in this case to adjoin $i = \sqrt{-1}$), then the field of quaternions splits and we obtain the matrix representation

$$j \to \begin{pmatrix} i & 0 \\ 0 & -i \end{pmatrix}, \quad k \to \begin{pmatrix} 0 & 1 \\ -1 & 0 \end{pmatrix}, \quad l \to \begin{pmatrix} 0 & i \\ i & 0 \end{pmatrix}.$$

Example 3: *The alternating group* $\mathfrak{A}_4$ can be handled in precisely the same manner as the symmetric group $\mathfrak{S}_3$; this is left to the reader. There are four representations of degrees 1, 1, 1, 3.

Example 4: *The symmetric group* $\mathfrak{S}_4$. There are five classes and thus five representations. The Klein four-group {1, (12) (34), (13) (24), (14) (23)} has a factor group isomorphic to $\mathfrak{S}_3$ for which three irreducible representations of degrees 1, 1, and 2 have already been found; these also give representations of degrees 1, 1, and 2 of $\mathfrak{S}_4$. Denoting these degrees by n_1, n_2, and n_3, we have

$$n_1^2 + n_2^2 + n_3^2 + n_4^2 + n_5^2 = 24,$$

and hence

$$n_4^2 + n_5^2 = 18.$$

This holds only for $n_4 = 3$ and $n_5 = 3$. If we introduce four vectors e_1, e_2, e_3, e_4 with sum zero, then the permutations of these four vectors give a faithful representation of third degree of $\mathfrak{S}_4$. If e_1, e_2, e_3 are chosen as base vectors, then the representation is as follows:

$$\begin{cases} (1\,2)\, e_1 = e_2 \\ (1\,2)\, e_2 = e_1 \\ (1\,2)\, e_3 = e_3 \end{cases} \quad \begin{cases} (1\,3)\, e_1 = e_3 \\ (1\,3)\, e_2 = e_2 \\ (1\,3)\, e_3 = e_1 \end{cases} \quad \begin{cases} (1\,4)\, e_1 = -e_1 - e_2 - e_3 \\ (1\,4)\, e_2 = \quad e_2 \\ (1\,4)\, e_3 = \quad e_3 \end{cases}$$

$$\begin{cases} (1\,2\,3)\, e_1 = e_2 \\ (1\,2\,3)\, e_2 = e_3 \\ (1\,2\,3)\, e_3 = e_1 \end{cases} \quad \text{etc.}$$

Since the representation is faithful, it cannot be reduced to representations of first and second degrees; it is therefore irreducible. If the matrices of this representation corresponding to the odd substitutions are multiplied by -1, then another representation of third degree is obtained which is likewise faithful and therefore irreducible; it is not equivalent to the original representation, since its trace is different. We have thus found all the representations.

Exercises

14.3. The element $s = \sum_{a \in \mathfrak{G}} a$ of the group ring $\mathfrak{o}$ satisfies the equations

$$bs = s \quad \text{for} \quad b \in \mathfrak{G}.$$

What left ideal is generated by s? What representation belongs to this ideal? Which idempotent element is contained in this ideal?

14.4. If the number h of the group elements is divisible by the characteristic of the field, then the ideal Exercise 14.3 is nilpotent. This implies that the condition that the characteristic not divide h in Maschke's theorem is also necessary.

14.6 GROUP CHARACTERS[2]

THE KRONECKER PRODUCT TRANSFORMATION

Suppose that A' and A'' are linear transformations on the vector spaces $(u_1, \ldots, u_n)$ and $(v_1, \ldots, v_m)$, respectively:

$$A'u_k = \sum_i u_i \alpha'_{ik}$$

$$A''v_l = \sum_j v_j \alpha''_{jl}.$$

We form a product space from the two vector spaces according to Section 94 which is generated by the products $u_k v_l$ and define:

$$A(u_k v_l) = (A'u_k)(A''v_l) = \sum_i \sum_j u_i v_j \alpha'_{ik} \alpha''_{jl}. \tag{14.16}$$

The linear transformation A on the product space thus defined is called the *Kronecker product transformation* and is denoted by $A' \times A''$. It follows from (14.16) that the matrix elements of A are $\alpha'_{ik}\alpha''_{jl}$. The trace of A is

$$\sum_i \sum_j \alpha'_{ii}\alpha''_{jj} = \sum_i \alpha'_{ii} \cdot \sum_j \alpha''_{jj} = S(A') \cdot S(A'');$$

thus *the trace of the product transformation $A' \times A''$ is the product of the traces of the transformations A' and A''.*

If two transformations B' and A' are applied successively to the u and two transformations B'' and A'' to the v, then the products $u_k v_l$ undergo the transformations $B' \times B''$ and $A' \times A''$; that is,

$$(A' \times A'') \cdot (B' \times B'') = A'B' \times A''B''. \tag{14.17}$$

If the matrices $A', B', \ldots$ form one representation D' of a group $\mathfrak{G}$ and the matrices $A'', B'', \ldots$ form another representation D'' of the same group, then it follows from (14.17) that the product transformations $A = A' \times A''$, $B = B' \times B'', \ldots$ again form a representation. This *product representation* of the representations $\mathfrak{D}'$, $\mathfrak{D}''$ will be denoted by $\mathfrak{D}' \times \mathfrak{D}''$.

[2]*Literature*: A development of the representation theory of finite groups independent of the theory of algebras can be found in the paper by I. Schur, "Neue Begründung der Theorie der Gruppencharactere," *Sitzungsber. Berlin*, 1905, p. 406. A generalization of this theory to infinite groups has been given by J. v. Neumann, "Almost Periodic Functions in Groups," *Trans. Amer. Math. Soc.*, Vol. 36, 1934. For further literature see B. L. v. d. Waerden, "Gruppen von Linearen Transformationen," *Ergebn. Math.*, Vol. IV, No. 2, Berlin, 1935.

If we write $\mathfrak{D}'+\mathfrak{D}''$ for a reducible representation which decomposes into $\mathfrak{D}'$ and $\mathfrak{D}''$ and consider equivalent representations to be the same, then the following relations are easily verified:

$$\mathfrak{D}'+\mathfrak{D}'' = \mathfrak{D}''+\mathfrak{D}'$$

$$\mathfrak{D}'\times\mathfrak{D}'' = \mathfrak{D}''\times\mathfrak{D}'$$

$$D'+(\mathfrak{D}''+\mathfrak{D}''') = (\mathfrak{D}'+\mathfrak{D}'')+\mathfrak{D}'''$$

$$\mathfrak{D}'\times(\mathfrak{D}''\times\mathfrak{D}''') = (\mathfrak{D}'\times\mathfrak{D}'')\times D'''$$

$$\mathfrak{D}'\times(\mathfrak{D}''+\mathfrak{D}''') = \mathfrak{D}'\times\mathfrak{D}''+\mathfrak{D}'\times\mathfrak{D}'''$$

$$(\mathfrak{D}''+\mathfrak{D}''')\times\mathfrak{D}' = \mathfrak{D}''\times\mathfrak{D}'+\mathfrak{D}'''\times\mathfrak{D}'.$$

In particular, if $\mathfrak{G}$ is a finite group whose order is not divisible by the characteristic of the field P, then every representation decomposes completely into irreducible representations $\mathfrak{D}_\nu$, and we have:

$$\mathfrak{D}_\lambda\times\mathfrak{D}_\mu = \sum_v c_{\lambda\mu}{}^\nu\mathfrak{D}_\nu, \tag{14.18}$$

where the $c_{\lambda\mu}{}^\nu$ are nonnegative integers. In (14.18), ν is not an exponent but rather an index.

For the traces it follows from (14.18) that

$$S_\lambda(a)\cdot S_\mu(a) = \sum_v c_{\lambda\mu}{}^\nu S_\nu(a).$$

If the representations are absolutely irreducible, and the traces are thus characters, we may also write:

$$\chi_\lambda(a)\cdot\chi_\mu(a) = \sum_\nu c_{\lambda\mu}{}^\nu\chi_\nu(a) \qquad (\textit{first character relation}). \tag{14.19}$$

THE CHARACTERS AS CLASS FUNCTIONS

If a and a' are conjugate group elements,

$$a' = bab^{-1},$$

then for the representing matrices it follows that

$$A' = BAB^{-1}.$$

Thus A and A' have the same trace; that is,

$$S(bab^{-1}) = S(a)$$

and, in particular,

$$\chi(bab^{-1}) = \chi(a).$$

If we again combine all group elements conjugate to a into a class $\mathfrak{K}_a$, then each character has the same value for all elements of a class.

If h_a is the number of elements of the class $\mathfrak{K}_a$ and k_a is the sum of the elements of this class (in the group ring $\mathfrak{o}$), then the character of k_a is the sum of the characters of the elements of the class:

$$\chi(k_a) = h_a \cdot \chi(a).$$

We now assume that neither the order h of the group nor the degrees n_ν of the absolutely irreducible representations $\mathfrak{D}_\nu$ are divisible by the characteristic of the base field. As we have seen in Section 14.5, the quantities k_a generate the center $\mathfrak{Z}$ of the group ring $\mathfrak{o}$. By Section 14.4, the homomorphisms Θ_ν of $\mathfrak{Z}$ are related to the characters χ_ν by the relations

$$\Theta_\nu(z) = \frac{\chi_\nu(z)}{n_\nu};$$

in particular,

$$\Theta_\nu(k_a) = \frac{\chi_\nu(k_a)}{n_\nu} = \frac{h_a}{n_\nu} \chi_\nu(a). \tag{14.20}$$

The product $k_a k_b$ is a sum of group elements which again belongs to $\mathfrak{Z}$ and can therefore be expressed in terms of the class sums k_a with integer coefficients

$$k_a \cdot k_b = \sum_c g_{ab}{}^c k_c. \tag{14.21}$$

The homomorphism property of Θ_ν is now expressed in the equation

$$\Theta_\nu(k_a) \cdot \Theta_\nu(k_b) = \sum_c g_{ab}{}^c \Theta_\nu(k_c), \tag{14.22}$$

which may be rewritten

$$h_a h_b \chi_\nu(a) \chi_\nu(b) = n_\nu \sum_c g_{ab}{}^c h_c \chi_\nu(c) \qquad \textit{(second character relation)} \tag{14.23}$$

by (14.20).

In the sums (14.21), (14.22). and (14.23), c runs through a system of representatives of all classes. If we permit c to run over all group elements, then the factor h_c on the right side of (14.23) is to be omitted. Since the Θ_ν are the only homomorphisms of $\mathfrak{Z}$, the characters χ_ν are the only solutions of (14.23).

CONJUGATE CHARACTERS

For every representation $a \to A$ there is a *conjugate* (or *contragredient*) *representation* $a \to A'^{-1}$, where A' is the transposed matrix of A. Indeed, under this correspondence,

$$ab \to (AB)'^{-1} = (B'A')^{-1} = A'^{-1}B'^{-1}.$$

The conjugate of the conjugate representation is again the original representation. If the representation $a \to A$ is reducible, then the conjugate representation is also reducible, and conversely. The conjugate of an irreducible representation is therefore again irreducible.

If we go from A to an equivalent representation $P^{-1}AP$, then the conjugate representation goes over into

$$(P^{-1}AP)'^{-1} = P'A'^{-1}P'^{-1},$$

and therefore likewise into an equivalent representation.

If $\mathfrak{D}_{\nu'}$ denotes the irreducible representation conjugate to $\mathfrak{D}_\nu$ and if $\mathfrak{D}_\nu(a) = A$, then

$$\mathfrak{D}_{\nu'}(a^{-1}) = A',$$

and hence

$$\chi_{\nu'}(a^{-1}) = \chi_\nu(a),$$

since the trace of A' is equal to that of A. The *character* $\chi_\nu{}'$ *conjugate* to χ_ν is also denoted by $\bar{\chi}_\nu$.

Every character is a sum of roots of unity. For every element a of $\mathfrak{G}$ generates a cyclic subgroup $\mathfrak{C}$ whose order m divides the order h of $\mathfrak{G}$. Every irreducible representation $\mathfrak{D}_\nu$ of $\mathfrak{G}$ provides a representation of $\mathfrak{C}$, and this representation decomposes completely into representations of first degree whose matrix elements are mth roots of unity. The trace of the representing matrix is the sum of the diagonal elements and therefore a sum of mth roots of unity—for example,

$$\chi(a) = \zeta^{v_1} + \zeta^{v_2} + \cdots + \zeta^{v_n}, \tag{14.24}$$

where ζ is a primitive mth root of unity.

FURTHER CHARACTER RELATIONS

If $S(c)$ is the trace of the group element c in the regular representation, then

$$S(c) = \sum_\nu n_\nu \chi_\nu(c),$$

since the regular representation contains the irreducible representation $\mathfrak{D}_\nu$ precisely n_ν times. The trace $S(c)$ can also be computed directly. The group elements $a_1, \ldots, a_h$ form a basis for the vector space $\mathfrak{o}$ of the regular representation, and

$$ca_i = a_k.$$

Terms with $i = k$ occur only when c is equal to the identity element 1 of the group; in this case each i is equal to the corresponding k. Hence

$$S(1) = h; \qquad S(c) = 0 \qquad \text{for} \quad c \neq 1,$$

and therefore

$$\sum_\nu n_\nu \chi_\nu(c) = \begin{cases} h & \text{for } c = 1 \\ 0 & \text{for } c \neq 1. \end{cases} \tag{14.25}$$

Summing (14.23) over all $_\nu$ and using (14.25), we obtain

$$h_a h_b \sum_\nu \chi_\nu(a)\chi_\nu(b) = g_{ab}{}^1 \cdot h. \tag{14.26}$$

Now the number $g_{ab}{}^1$, whenever it occurs, indicates that a product $a'b'$ is equal to 1, where a' belongs to the class $\mathfrak{K}_a$ and b' to the class $\mathfrak{K}_b$. The number is therefore zero if $\mathfrak{K}_a$ and $\mathfrak{K}_b$ contain no two elements inverse to one another. If such a pair of elements is present, say $b = a^{-1}$, then for every element $a' = cac^{-1}$ of $\mathfrak{K}_a$ there is a corresponding inverse element $b' = a'^{-1} = cbc^{-1}$ in $\mathfrak{K}_b$; hence

$$g_{ab}{}^1 = h_a = h_b.$$

Then (14.26), after division by h_b, becomes

$$h_a \sum_v \chi_v(a)\chi_v(b) = \begin{cases} h & \text{for } \mathfrak{K}_b = \mathfrak{K}_{a^{-1}} \\ 0 & \text{for } \mathfrak{K}_b \neq \mathfrak{K}_{a^{-1}} \end{cases} \quad \textit{(third character relation)}. \quad (14.27)$$

In the special case $a = 1$ we again obtain (14.25).

Now let $a_1, \ldots, a_s$ be a system of representatives of all the classes. Putting

$$\chi_{\nu\mu} = \chi_\nu(a_\mu)$$

$$\eta_{\mu\nu} = \frac{h_\mu}{h}\bar{\chi}_\nu(a_\mu) = \frac{h_\mu}{h}\chi_\nu(a_\mu{}^{-1}),$$

relation (14.27) states that the matrices $X = (\chi_{\mu\nu})$ and $Y = (\eta_{\mu\nu})$ are inverse to one another:

$$YX = E \quad \text{or} \quad Y = X^{-1}. \quad (14.28)$$

From (14.28) it follows that

$$XY = E$$

or, written out explicitly,

$$\frac{1}{h}\sum_{\mathfrak{K}_a} h_a\chi_\nu(a)\bar{\chi}_\mu(a) = \begin{cases} 1 & \text{for } \nu = \mu \\ 0 & \text{for } \nu \neq \mu. \end{cases} \quad (14.29)$$

Here a runs through a system of representatives of all the classes. If a runs over all group elements, then the factors h_a are to be omitted. This now implies the *orthogonality of the characters*:

$$\sum_{a\in\mathfrak{G}} \bar{\chi}_\mu(a)\chi_\nu(a) = \begin{cases} h & \text{for } \nu = \mu \\ 0 & \text{for } \nu \neq \mu \end{cases} \quad \textit{(fourth character relation)}. \quad (14.30)$$

In particular, if $\mu = 0$, that is, if χ_μ is the character χ_0 of the identity representation, then (14.30) implies

$$\sum_a \chi_\nu(a) = \begin{cases} h & \text{for } \nu = 0 \\ 0 & \text{for } \nu \neq 0. \end{cases} \quad (14.31)$$

The fact that the matrices X and Y are inverse to one another can be used to find the idempotent elements of the center which generate the simple two-sided ideals in $\mathfrak{o}$, for by Section 14.5 we have the following expressions for the basis elements k_a of the center $\mathfrak{Z}$:

$$k_a = \sum_\nu e_\nu \Theta_\nu(k_a) = \sum_\nu e_\nu \frac{h_a}{n_\nu}\chi_\nu(a). \quad (14.32)$$

Multiplying by $\bar{\chi}_\mu(a)$ and summing over all classes $\mathfrak{K}_a$, we obtain

$$\sum_{\mathfrak{K}_a} k_a \bar{\chi}_\mu(a) = e_\mu \cdot \frac{h}{n_\mu}$$

or

$$e_\nu = \sum_{\mathfrak{K}_a} k_a \frac{n_\nu}{h} \chi_\nu(a^{-1}).$$

14.7 THE REPRESENTATIONS OF THE SYMMETRIC GROUPS[3]

We consider the group $\mathfrak{S}_n$ of permutations of n digits 1, 2, . . . , n and seek its absolutely irreducible representations in, say, the field Ω of all algebraic numbers. It will turn out that these representations are actually rational, that is, are found in the field $\mathfrak{Q}$ of rational numbers.

We start with the group ring $\mathfrak{o} = s_1\Omega + \cdots + s_{n!}\Omega$ and consider its left ideals. Every such left ideal is a direct sum of minimal left ideals; these ideals provide the irreducible representations. Since each left ideal is generated by an idempotent element, we first look for the idempotent elements.

We write the numerals 1, 2, . . . , n in any order in h successive rows (h arbitrary) so that in the νth row there are α_ν numbers and the conditions

$$\begin{cases} \alpha_1 \geqq \alpha_2 \geqq \cdots \geqq \alpha_h \\ \sum_{\nu=1}^{h} \alpha_\nu = n \end{cases} \tag{14.33}$$

are satisfied. We write the first elements of the h rows all under one another, likewise the second elements, and so on, for example, as in the following schema in which the points represent numerals:

$$\begin{matrix} \cdot\ \cdot\ \cdot \\ \cdot\ \cdot\ \\ \cdot\ \cdot\ \end{matrix} \qquad (\alpha_1, \alpha_2, \alpha_3) = (3, 2, 2); \qquad n = 7.$$

Such an arrangement of the numerals 1, 2, . . . , n we call a *schema* Σ_α. The index α denotes the sequence $(\alpha_1, \alpha_2, \ldots, \alpha_h)$. The possible indices are ordered by the following convention: $\alpha > \beta$ if the first nonvanishing difference $\alpha_\nu - \beta_\nu$ is positive. For example, in the case $n = 5$,

$$(5) > (4, 1) > (3, 2) > (3, 1, 1) > (2, 2, 1) > (2, 1, 1, 1) > (1, 1, 1, 1, 1).$$

Given a schema Σ_α, we denote by p all those permutations which permute the numerals within the rows of a schema but leave the rows themselves invariant; similarly, we denote by q those permutations which permute only the numerals

[3]I am indebted to a conversation with J. von Neumann for the simplified proofs in the Frobenius theory (*Sitzungsber. Berlin*, 1903, p. 328) which appear in this section.

within the columns of a schema. For each fixed q the symbol σ_q denotes the number $+1$ or -1 according to whether q is an even or odd permutation. If s is any permutation we denote by $s\,\Sigma_\alpha$ the schema into which Σ_α is transformed by the permutation s. It is easily seen that if q leaves the columns of Σ_α invariant, then sqs^{-1} leaves the columns of $s\,\Sigma_\alpha$ invariant, and conversely. Finally, we put (in the group ring $\mathfrak{o}$), for each fixed Σ_α,

$$S_\alpha = \sum_p p$$

$$A_\alpha = \sum_q q\sigma_q.$$

The following rules are easily verified:

$$pS_\alpha = S_\alpha p = S_\alpha \tag{14.34}$$

$$A_\alpha q\sigma_q = qA_\alpha\sigma_q = A_\alpha. \tag{14.35}$$

From (14.34) and (14.35) it now follows that S_α and A_α are idempotent up to a factor f_α. The additional algebraic properties of S_α and A_α follow from the following *combinatorial lemma.*

Lemma: *Let Σ_α and Σ_β be two schemata of the above type, and let $\alpha \geqq \beta$. If in Σ_α there are nowhere two numerals in a single row which occur in Σ_β in the same column, then $\alpha = \beta$ and the schema Σ_α can be transformed by a permutation of the form pq into the schema Σ_β:*

$$pq\Sigma_\alpha = \Sigma_\beta.$$

(Here p and q refer to Σ_α; that is, p leaves the rows and q the columns of Σ_α invariant.)

Proof: $\alpha \geqq$ implies $\alpha_1 \geqq \beta_1$. In the first row of Σ_α there are α_1 numerals. If these same numerals in Σ_β are all in distinct columns, then Σ_β must have at least α_1 columns; from this it follows that $\alpha_1 \leqq \beta_1$ and hence $\alpha_1 = \beta_1$. These numerals can all be brought into the first row of Σ_β by a permutation q_1' which leaves the columns of Σ_β invariant.

Further, $\alpha \geqq \beta$ implies $\alpha_2 \geqq \beta_2$. In the second row of Σ_α there are α_2 numerals. If these are all in distinct columns in $q_1'\,\Sigma_\beta$, then, apart from the first row, $q_1'\,\Sigma_\beta$ must still have at least α_2 columns. This implies that $\alpha_2 \leqq \beta_2$ and hence $\alpha_2 = \beta_2$. These numerals can all be brought into the second row of Σ_β by a permutation which leaves both the columns of $q_1'\,\Sigma_\beta$ and the first row invariant.

Continuing in this manner, we finally obtain a schema $q'\,\Sigma_\beta = q_h' \cdots q_2'q_1'\,\Sigma_\beta$ whose rows coincide with those of Σ_α. Therefore Σ_α can be transformed into $q'\,\Sigma_\beta$ by a permutation p:

$$q'\Sigma_\beta = p\Sigma_\alpha.$$

The permutation $q' = q_h' \cdots q_2'q_1'$ leaves the columns of Σ_β invariant; it therefore also leaves the columns of $q'\,\Sigma_\beta = p\,\Sigma_\alpha$ invariant. For appropriate q, then,

$$q' = pq^{-1}p^{-1}$$

and hence

$$pq^{-1}p^{-1}\,\Sigma_\beta = p\,\Sigma_\alpha$$

$$\Sigma_\beta = pq\,\Sigma_\alpha, \qquad \text{Q.E.D.}$$

The combinatorial lemma implies first of all that

$$A_\beta S_\alpha = 0 \qquad \text{for} \quad a > \beta. \tag{14.36}$$

For by the lemma if $\alpha > \beta$ there must exist a pair of numerals which occur in a single row of Σ_α and in a single column of Σ_β. If t is the transposition which interchanges the numerals of this pair, then, by (14.34) and (14.35),

$$A_\beta S_\alpha = A_\beta t t^{-1} S_\alpha = -A_\beta S_\alpha,$$

which gives (14.36).

Similarly,

$$S_\alpha A_\beta = 0 \qquad \text{for} \quad \alpha > \beta.$$

Now all transforms of A_β are also annihilated by S_α:

$$S_\alpha s A_\beta s^{-1} = 0 \qquad \text{for} \quad \alpha > \beta;$$

since $sA_\beta s^{-1}$ is again an A_β which belongs to the permuted schema $s\sum_\beta$. On multiplying by $s\Omega$ and summing over all s in $\mathfrak{G}$, this result implies

$$S_\alpha(\textstyle\sum s\Omega)A_\beta = (0)$$

or

$$S_\alpha \mathfrak{o} A_\beta = (0) \qquad (\alpha > \beta). \tag{14.37}$$

The left ideals $\mathfrak{o}A_\beta$ with $\beta < \alpha$ are therefore annihilated by S_α; this means that S_α is represented by zero in the representation provided by $\mathfrak{o}A_\beta$. On the other hand, $S_\alpha A_\alpha \neq 0$, since the coefficient of the identity element in the product $S_\alpha A_\alpha$ does not vanish. Therefore S_α is not represented by zero in the representation given by $\mathfrak{o}A_\alpha$; this representation thus contains at least one irreducible component which occurs in no $\mathfrak{o}A_\beta$ with $\beta < \alpha$. This irreducible component we shall now determine more explicitly.

The element $S_\alpha A_\alpha = \sum_p \sum_q pq\sigma_q$ has, by (14.34) and (14.35), the property

$$pS_\alpha A\alpha q\sigma_q = S\alpha A\alpha.$$

We now prove that up to a factor, $S_\alpha A_\alpha$ is the only element with this property. We show: *if an element a of* $\mathfrak{o}$ *has the property*

$$paq\sigma_q = a \tag{14.38}$$

for all p and q, then a must have the form $(S_\alpha A_\alpha)\cdot\gamma$.

Proof: We put

$$a = \sum_s s\gamma_s \qquad (\gamma_s \in \Omega). \tag{14.39}$$

Substituting (14.39) in (14.38) gives:

$$\sum_s s\gamma_s = \sum_s psq\sigma_q\gamma_s. \tag{14.40}$$

On the left side, only one term with pq occurs, namely $pq\gamma_{pq}$; on the right side there is also only one term containing pq, namely the term with $s = 1$. Equating coefficients gives

$$\gamma_{pq} = \sigma_q\gamma_1.$$

We now select an s which does not have the form pq. Then $s\,\Sigma_\alpha$ is distinct from all the $pq\,\Sigma_\alpha$ and thus by the combinatorial lemma there are two numerals i, j which occur in Σ_α in a single row and also in $s\,\Sigma_\alpha$ in a single column. If t is the transposition of these two numerals, $t = (jk)$, then $t' = s^{-1}ts$ interchanges only the numerals $s^{-1}j$ and $s^{-1}k$ which appear in the same column in $s^{-1}s\,\Sigma_\alpha = \Sigma_\alpha$. Therefore t is a permutation of type p, and t' a permutation of type q. In (14.40) we may therefore put $p = t$ and $q = t'$; for this special s, then,

$$psq = tss^{-1}ts = s$$

$$\sigma_q = -1;$$

comparison of the terms with s on the left and right in (14.40) gives

$$\gamma_s = -\gamma_s, \qquad \gamma_s = 0.$$

In (14.39), therefore, only terms with $s = pq$, $\gamma_s = \sigma_q\gamma_1$ occur, and hence

$$a = \sum_{p,q} pq\sigma_q\gamma_1 = (S_\alpha A_\alpha)\gamma_1, \qquad \text{Q.E.D.}$$

From what has just been proved it follows immediately that for every element b of $\mathfrak{o}$ the element $S_\alpha bA_\alpha$ has the form $(S_\alpha A_\alpha)\gamma$, since, for each p and each q,

$$pS_\alpha bA_\alpha q\sigma_q = S_\alpha bA_\alpha.$$

Thus

$$S_\alpha\mathfrak{o}A_\alpha \subseteq (S_\alpha A_\alpha)\Omega.$$

Putting $S_\alpha A_\alpha = I_\alpha$, it follows that

$$I_\alpha\mathfrak{o}I_\alpha \subseteq S_\alpha\mathfrak{o}A_\alpha \subseteq I_\alpha\Omega. \tag{14.41}$$

We now assert that $\mathfrak{o}I_\alpha$ is a minimal left ideal. Indeed, if $\mathfrak{l}$ is a subideal of $\mathfrak{o}I_\alpha$, then it follows from (14.41) that

$$I_\alpha\mathfrak{l} \subseteq I_\alpha\Omega,$$

and hence, since $I_\alpha\Omega$ is a minimal Ω-module, either

$$I_\alpha\mathfrak{l} = I_\alpha\Omega \qquad \text{or} \qquad I_\alpha\mathfrak{l} = (0).$$

In the first case it follows that $\mathfrak{o}I_\alpha = \mathfrak{o}I_\alpha\Omega \subseteq \mathfrak{o}I_\alpha\mathfrak{l} \subseteq \mathfrak{l}$, and hence $\mathfrak{l} = \mathfrak{o}I_\alpha$. In the second case it follows that $\mathfrak{l}^2 \subseteq \mathfrak{o}I_\alpha\mathfrak{l} = (0)$, and hence $\mathfrak{l} = (0)$, since there are no nilpotent ideals except (0).

The minimal left ideals $\mathfrak{o}I_\alpha$ and $\mathfrak{o}I_\beta$ are not operator isomorphic for $\alpha > \beta$. For from (14.37), for $\alpha > \beta$,

$$S_\alpha \mathfrak{o} I_\beta = S_\alpha \mathfrak{o} S_\beta A_\beta \subseteqq S_\alpha \mathfrak{o} A_\beta = (0),$$

and hence, for each a' of $\mathfrak{o}I_\beta$,

$$S_\alpha a' = 0.$$

If now $\mathfrak{o}I_\alpha \cong \mathfrak{o}I_\beta$, then it would follow that, for each a of $\mathfrak{o}I_\alpha$,

$$S_\alpha a = 0;$$

this, however, is not true for $a = I_\alpha = S_\alpha A_\alpha$, since $S_\alpha{}^2 A_\alpha = f_\alpha S_\alpha A_\alpha \neq 0$.

Each left ideal $\mathfrak{o}I_\alpha$ provides an irreducible representation $\mathfrak{D}_\alpha$, and these representations are inequivalent for distinct α by the above remarks.

The number of representations $\mathfrak{D}_\alpha$ thus found is equal to the number of solutions of (14.33). This number is at the same time the number of classes of conjugate permutations; for each such class consists of all elements which decompose into cycles of definite lengths $\alpha_1, \alpha_2, \ldots, \alpha_h$, and these lengths can be ordered in accordance with the conditions (14.33). However, since the number of *all* inequivalent irreducible representations is given by the number of classes of conjugate permutations, it follows that *up to equivalence the representations* $\mathfrak{D}_\alpha$ *exhaust all irreducible representations of the symmetric groups* $\mathfrak{S}_n$.

In the foregoing the minimal left ideals $\mathfrak{o}I_\alpha$ were rationally determined. This implies the *rationality of the irreducible representations* (as well as of the characters).

14.8 SEMIGROUPS OF LINEAR TRANSFORMATIONS

We begin with a base field P and consider sets of linear transformations whose matrix elements belong either to P itself or to a commutative extension field Λ of P. Such a set is called a *semigroup* if it contains the product of any two transformations of the set. The *linear hull* of a system of transformations over P consists of all linear combinations of transformations of the system with coefficients in P. In the following we shall consider only systems containing finitely many linearly independent transformations over P, whose linear hull thus has finite rank over P. Under these hypotheses the linear hull of a semigroup is an algebra $\mathfrak{A}$ of finite rank over P. Each element of this algebra is a linear transformation. We therefore have an algebra $\mathfrak{A}$ over P in a definite faithful representation $\mathfrak{D}$.

The principal question of interest here is: *How does an irreducible representation* $\mathfrak{D}$ *decompose when the field* Λ *is extended?*

We shall always assume that the representation $\mathfrak{D}$ does not contain the null representation as a component.

The following two theorems are basic for the theory.

Theorem 1: *If the representation* $\mathfrak{D}$ *is completely reducible, then the algebra* $\mathfrak{A}$ *is semisimple.*

Theorem 2: *If the representation $\mathfrak{D}$ is irreducible or decomposes into equivalent irreducible components, then $\mathfrak{A}$ is simple.*

Proof of 1: If $\mathfrak{R}$ is the radical of $\mathfrak{A}$, then in every irreducible representation the elements of $\mathfrak{R}$ are represented by zero. Since $\mathfrak{D}$ is a faithful representation, it follows that $\mathfrak{R} = \{0\}$.

Proof of 2: The algebra $\mathfrak{A}$ is semisimple in any case and is thus the direct sum of simple algebras: $\mathfrak{A} = \mathfrak{a}_1 + \cdots + \mathfrak{a}_s$. In an irreducible representation all the $\mathfrak{a}_\mu$ except for a single $\mathfrak{a}_\nu$ are represented by zero according to Section 14.2. This fact remains in force if the representation is repeated several times. If the representation is faithful, then there can be only one $\mathfrak{a}_\nu$, that is, the algebra $\mathfrak{A}$ is simple.

A theorem due to Burnside, which was generalized by Frobenius and Schur, follows immediately from Theorem 1.

Burnside's Theorem: *In an absolutely irreducible semigroup of matrices of degree n there are precisely n^2 linearly independent matrices.*

Generalization: *If a semigroup of matrices in the field Λ decomposes into absolutely irreducible components among which occur s inequivalent components of degrees $n_1, \ldots, n_s$, then the semigroup contains precisely*

$$n_1^2 + n_2^2 + \cdots + n_s^2$$

linearly independent matrices over Λ.

Proof of the Generalization: The linear hull over Λ of the given semigroup is the sum of s complete matrix rings of degrees $n_1, n_2, \ldots, n_s$ over Λ and therefore has rank $n_1^2 + n_2^2 + \cdots + n_s^2$.

In fields of characteristic zero we also have the following.

Trace Theorem: *If there exists a one-to-one, product-preserving correspondence between two semigroups (or, more generally, if both semigroups may be interpreted as representations of a single abstract semigroup) and if the traces of corresponding matrices are equal, then the two semigroups (or the two representations) are equivalent.*

Proof: Arranging corresponding matrices A and B of the two semigroups in the manner

$$\begin{pmatrix} A & 0 \\ 0 & B \end{pmatrix}, \tag{14.42}$$

we obtain a new, completely reducible semigroup $\mathfrak{g}$ whose linear hull is an algebra $\mathfrak{A}$. The elements of $\mathfrak{A}$ are linear combinations of the matrices (14.42) and thus decompose in the same manner into two components, each of which provides a representation of $\mathfrak{A}$. The traces of these two representations are certain linear combinations of the traces of the original matrices A and B and therefore are the same for both representations. Thus (Section 14.4), the two representations of $\mathfrak{A}$ are equivalent. This proves the assertion.

If $\Lambda = \mathsf{P}$, the converse of Theorems 1 and 2 follows immediately by Section 14.2. If, however, Λ is a proper extension field of P, then we must proceed somewhat more carefully.

Theorem 1a: *If $\mathfrak{A}$ is semisimple and Λ is separable over* P, *then every representation $\mathfrak{D}$ of $\mathfrak{A}$ in Λ is completely reducible.*

Theorem 2a: *If $\mathfrak{A}$ is simple and central over* P, *then every representation of $\mathfrak{A}$ in Λ decomposes into equivalent irreducible components.*

Proof: By section 14.1, every representation of $\mathfrak{A}$ in Λ is provided by a representation of $\mathfrak{A} \times \Lambda$. If now $\mathfrak{A}$ is semisimple and Λ is separable over P, then, by Section 13.12, $\mathfrak{A} \times \Lambda$ is also semisimple, and therefore every representation of $\mathfrak{A} \times \Lambda$ in Λ is completely reducible. If $\mathfrak{A}$ is central and simple over P, then $\mathfrak{A} \times \Lambda$ is likewise simple, again by Section 13.12, and hence every representation of $\mathfrak{A} \times \Lambda$ in Λ decomposes into equivalent irreducible components. Both assertions are herewith proved.

We call a semigroup *central* over P if its linear hull is central, that is, if the center of the linear hull is equal to the base field P.

Taking Theorems 1 and 2 into account, we may also formulate Theorems 1a and 2a as follows:

Theorem 1b: *A completely reducible semigroup of linear transformations in* P *remains completely reducible under every separable extension of the base field* P.

Theorem 2b: *A central irreducible semigroup of linear transformations in* P *remains irreducible or decomposes into equivalent irreducible components under every extension of the base field.*

The following assertion can be proved in the same way as Theorem 1b.

Theorem 1c: *A completely reducible semigroup remains completely reducible under every extension of the base field if the center of the linear hull is a direct sum of separable fields over* P.

14.9 DOUBLE MODULES AND PRODUCTS OF ALGEBRAS

We noted in Section 14.1 that every representation of a hypercomplex system $\mathfrak{S}$ in a commutative field K containing the base field P can be obtained from a representation of the extended system $\mathfrak{S}_K$. In the language of representation modules, this means that every module having $\mathfrak{S}$ as left and K as right multiplier domain may be interpreted as a left $\mathfrak{S}_K$-module. The proof was based on the fact that if $\mathfrak{S} = a_1\mathrm{P} + \cdots + a_n\mathrm{P}$, and thus $\mathfrak{S}_K = a_1\mathrm{K} + \cdots + a_n\mathrm{K}$, and if u is an element of the module, then left multiplication by an element of $\mathfrak{S}_K$ is defined by

$$(a_1\kappa_1 + \cdots + a_n\kappa_n)u = a_1u\kappa_1 + \cdots + a_nu\kappa_n.$$

Verification of the rules for the $\mathfrak{S}_K$-module presents no difficulties; essential use is made of commutativity only in the proof of the associative law

$$(bc)u = b(cu).$$

If $b = a_1\kappa_1$ and $c = a_2\kappa_2$ (it clearly suffices to consider this special case), then the associative law follows from the relations

$$(a_1\kappa_1\cdot a_2\kappa_2)u = (a_1a_2\kappa_1\kappa_2)u = (a_1a_2)u(\kappa_1\kappa_2)$$
$$a_1\kappa_1(a_2\kappa_2\cdot u) = a_1\kappa_1(a_2u\kappa_2) = a_1(a_2u\kappa_2)\kappa_1 = (a_1a_2)u(\kappa_2\kappa_1).$$

These two expressions are equal, since $\kappa_1\kappa_2 = \kappa_2\kappa_1$.

The situation can also be saved when K is a skew field or, more generally, an arbitrary ring, by constructing an inverse ring K′ to K, that is, a ring which is anti-isomorphic to K. If K is an algebra over P, then K′ is also an algebra over P. If K is a skew field, then K′ is also a skew field.

Every module having $\mathfrak{S}$ as left and K *as right multiplier domain may be interpreted as a left* ($\mathfrak{S}\times$ K′)*-module.*

Proof: Let $\mathfrak{S} = a_1\mathsf{P}+\cdots+a_n\mathsf{P}$ so that $\mathfrak{S}\times\mathsf{K}' = a_1\mathsf{K}'+\cdots+a_n\mathsf{K}'$; we then define

$$(a_1\kappa_1'+\cdots+a_n\kappa_n')u = a_1u\kappa_1+\cdots+a_nu\kappa_n. \tag{14.43}$$

All the rules are now easily verified. The associative law $(bc)u = b(cu)$ follows from

$$(a_1\kappa_1'\cdot a_2\kappa_2')u = (a_1a_2\kappa_1'\kappa_2')u = (a_1a_2)u(\kappa_2\kappa_1)$$
$$a_1\kappa_1'(a_2\kappa_2'\cdot u) = a_1\kappa_1'(a_2u\kappa_2) = a_1(a_2u\kappa_2)\kappa_1 = (a_1a_2)u(\kappa_2\kappa_1).$$

In the same way, a left ($\mathfrak{S}\times$ K′)-module may be interpreted as a left $\mathfrak{S}$- and right K-module by the definition $u\kappa = \kappa'u$. *Isomorphic* ($\mathfrak{S}\times$ K′)*-modules hereby produce isomorphic double modules, and conversely.*

These facts have many applications. Henceforth, K shall always be a division algebra, and $\mathfrak{S}$ shall be a simple algebra with identity over P. Let at least one of the two algebras, $\mathfrak{S}$ or K, be central over P. The product $\mathfrak{S}\times$ K′ is then simple by Section 13.12. By section 14.2, all simple left ($\mathfrak{S}\times$ K′)-modules are isomorphic to each other and to the simple left ideals of $\mathfrak{S}\times$ K′. Therefore all simple double modules ($\mathfrak{S}$ on the left and K on the right) are isomorphic. From this we have the following.

All irreducible representations of $\mathfrak{S}$ in K *are equivalent.*

Since $\mathfrak{S}$ is simple, all these representations are faithful. Each such representation maps $\mathfrak{S}$ isomorphically onto a subring Σ of the complete matrix ring K_r. Any two such representations $s\to S_1$ and $s\to S_2$, which map $\mathfrak{S}$ onto Σ_1 and Σ_2, are equivalent. By Section 12.4, this means that there exists a fixed matrix Q, independent of s, which takes S_1 into S_2:

$$S_2 = Q^{-1}S_1Q. \tag{14.44}$$

From this we easily obtain the following.

Automorphism Theorem: *If Σ_1 and Σ_2 are two isomorphic, simple subalgebras of the central simple algebra K_r, then any isomorphism between Σ_1 and Σ_2 which leaves the elements of the base field fixed is given by an inner automorphism of K_r according to* (14.44).

Indeed, any two such isomorphic algebras Σ_1 and Σ_2 may always be interpreted as representations of a single algebra $\mathfrak{S}$. If these representations are reducible, then they decompose into the same number of irreducible representations since their degrees are both equal to r. Since these representations are equivalent, the decomposable ones are also.

As a special case we obtain the following.

Every automorphism of K_r *which leaves the elements of the center* P *invariant is an inner automorphism.*

When speaking of isomorphisms and automorphisms of algebras with identity in the following, we shall always mean those which leave the elements of the base field P fixed. To these belong, in any case, the inner automorphisms.

Let $\mathfrak{S}$ again be a simple algebra, and let K be a division algebra over P. Let one of the two algebras, $\mathfrak{S}$ or K, be central. Then $\mathfrak{S} \times \mathsf{K}'$ is simple and is thus isomorphic to a complete matrix ring Δ_t over a skew field Δ. We now wish to see what can be said about this skew field Δ.

Quite generally, Δ is the right endomorphism ring of a simple $(\mathfrak{S} \times \mathsf{K}')$-module which, as was remarked at the outset, may be interpreted as a double module ($\mathfrak{S}$ on the left, K on the right). Each endomorphism of the $(\mathfrak{S} \times \mathsf{K}')$-module corresponds to an endomorphism of this double module $\mathfrak{M}$ in a one-to-one manner; Δ is therefore isomorphic to the right endomorphism ring of the double module $\mathfrak{M}$. The inverse skew field Δ' is thus isomorphic to the left endomorphism ring of the double module $\mathfrak{M}$. Thus Δ' may be identified with this endomorphism ring.

If the double module $\mathfrak{M}$ is interpreted as a vector space over K, then the elements a of $\mathfrak{S}$ induce linear transformations A of this vector space:

$$au = Au.$$

As we have seen, $\mathfrak{S}$ is isomorphic to a subring Σ of K_r under the representation $a \to A$. By Section 13.9, the left endomorphisms of $\mathfrak{M}$, and thus the elements of Δ', are those linear transformations L of this vector space which commute with the transformations A:

$$LA = AL \qquad \text{for all} \quad A \in \Sigma.$$

The ring Δ' is thus the *centralizer of* Σ *in* K_r, that is, the ring of those matrices L in K_r which commute with all matrices A of Σ.

We have thus obtained the following.

Structure Theorem for Products: *Let* $\mathfrak{S}$ *be a simple algebra (with identity), and let* K *be a division algebra over* P. *Let one of the two algebras be central over* P, *and let* K' *be anti-isomorphic to* K. *Then* $\mathfrak{S} \times \mathsf{K}'$ *is isomorphic to a complete matrix ring* Δ_t *over a skew field* Δ. *The only irreducible representation of* $\mathfrak{S}$ *in* K *maps* $\mathfrak{S}$ *faithfully onto a subring* Σ *of* K_r. *The centralizer* Δ' *of* Σ *in* K_r *is anti-isomorphic to* Δ.

The degree r of the representation $\mathfrak{S} \to \Sigma$ is the rank of the double module $\mathfrak{M}$ over K. If $\mathfrak{M}$ is interpreted as an $(\mathfrak{S} \times \mathsf{K}')$-module, the the rank of this module over K' is likewise r. A simple left ideal $\mathfrak{l}$ of $\mathfrak{S} \times \mathsf{K}'$ may be chosen for $\mathfrak{M}$; the rank of this left ideal is then

$$(\mathfrak{l} : \mathsf{K}') = r.$$

The simple ring $\mathfrak{S} \times \mathsf{K}' \cong \Delta_t$ is the direct sum of t such left ideals; its rank over K' is therefore tr. From this there follows the important rank relation

$$(\Sigma : \mathsf{P}) = (\mathfrak{S} : \mathsf{P}) = (\mathfrak{S} \times \mathsf{K}' : \mathsf{K}') = tr. \tag{14.45}$$

The formulation of the structure theorem is somewhat simpler if we start with Σ in place of $\mathfrak{S}$ and consider the isomorphic algebra $\Sigma \times \mathsf{K}'$ instead of $\mathfrak{S} \times \mathsf{K}'$. We thus take in the complete matrix ring K_r a subring Σ, of which it is assumed that the matrices form an irreducible system. Further, let K or Σ (or both) be central over P. The structure theorem then reads as follows.

Theorem: *$\Sigma \times \mathsf{K}'$ is isomorphic to a complete matrix ring over a skew field Δ. The centralizer Δ' of Σ in K_r is anti-isomorphic to Δ. The rank of Σ over P is r.*

The requirement that Σ be an irreducible system of linear transformations may also be relaxed. Since $\Sigma \times \mathsf{K}'$ is simple, each matrix representation of Σ in K is completely reducible, and the irreducible components are equivalent. By appropriate choice of basis, the matrices of the system Σ can therefore be brought to the form:

$$A = \begin{pmatrix} A_1 & & & \\ & \cdot & & \\ & & \cdot & \\ & & & \cdot \\ & & & & A_1 \end{pmatrix}, \tag{14.46}$$

with s equal blocks A_1 along the diagonal. The matrices A_1 form an irreducible system Σ_1 to which the structure theorem above may be applied. The centralizer of the system Σ_1, which consists of matrices L_1 that commute with all the matrices A_1 of Σ_1, is again a division algebra Δ' anti-isomorphic to Δ. The centralizer T of Σ consists of the matrices

$$L = \begin{pmatrix} L_{11} \dots L_{1s} \\ \cdot \qquad \cdot \\ \cdot \qquad \cdot \\ \cdot \qquad \cdot \\ L_{s1} \dots L_{ss} \end{pmatrix}, \tag{14.47}$$

where the L_{ik} are taken from Δ'. Thus $\mathsf{T} \cong \Delta'_s$.

The product relation

$$(\Sigma : \mathsf{P})(\mathsf{T} : \mathsf{P}) = (\mathsf{K}_r : \mathsf{P}) \tag{14.48}$$

holds between the ranks of the rings Σ and T which commute element-wise; this relation is easily verified.

It follows easily from (14.48) that the centralizer of T is again Σ.

This symmetric relation between the systems Σ and T is basic in the "Galois theory" which is developed in grea: generality in Jacobson, *Structure of Rings*, Chapters VI and VII.

We now pass to some applications of the structure theorem.

1. *The structure of* $\mathsf{K} \times \mathsf{K}'$. Let K be a central division algebra over P. We may then choose $\Sigma = \mathsf{K}$ and apply the structure theorem. The degree r of the matrices is equal to 1 in this case; the system Σ is trivially irreducible. The centralizer Δ' of K in K is the center P of K. Hence $\Delta = \mathsf{P}$ also. The rank relation (14.45) gives

$$(\mathsf{K} : \mathsf{P}) = t.$$

We therefore obtain the result: $\mathsf{K} \times \mathsf{K}'$ *is a complete matrix ring over the base field* P. *The degree* t *of the matrices is equal to the linear rank* $(\mathsf{K} : \mathsf{P})$.

2. *The maximal commutative subfields of a division algebra.* Let K be a division algebra over P. If K is not central over P, then we choose the center Z of K as the new base field P. Now suppose that Σ is a maximal commutative subfield of K. The centralizer of Σ in K is Σ itself. For if θ commutes with all the elements of Σ, then the skew field $\Sigma(\theta)$ is a field, and since Σ is maximal θ must be contained in Σ.

Thus $\Delta = \Sigma$, and $\mathsf{Z} \times \mathsf{K}'$ is therefore a complete matrix ring over Σ. The inverse ring to $\Sigma \times \mathsf{K}'$

$$\mathsf{K} \times \Sigma' = \mathsf{K} \times \Sigma = \mathsf{K}_{\Sigma}$$

is thus also a complete matrix ring over Σ; that is, Σ is a splitting field of K. The representation of K_{Σ} as a complete matrix ring Σ_t is absolutely irreducible. In Section 13.12 we have called the degree t of an absolutely irreducible matrix representation of K in a suitable extension field Σ of P the *index* m of the division algebra K. Hence $t = m$ and $r = 1$. The rank relation (14.45) now gives

$$(\Sigma : \mathsf{P}) = t = m,$$

and we thus obtain the following.

The maximal commutative subfields of a division algebra K *with center* P *are splitting fields of* K *and their field degree* $(\Sigma : \mathsf{P})$ *is equal to the index* m *of the division algebra.*

3. As an application of this theorem, we determine *all division algebras over the field* $\mathbb{R}$ *of real numbers.*

As commutative division algebras over P, we have P and $\mathsf{P}(i)$, the fields of real and complex numbers. We now assume that the algebra K is noncommutative. If Z is the center and Σ a maximal commutative subfield of K, then

$$\mathsf{P} \subseteqq \mathsf{Z} \subseteqq \Sigma \subset \mathsf{K}; \qquad (\Sigma : \mathsf{Z}) = m; \qquad (\mathsf{K} : \mathsf{Z}) = m^2.$$

Since K is noncommutative, we must have $m > 1$. For the fields Z and Σ, only P and $\mathsf{P}(i)$ are possibilities. Since $m > 1$, $\Sigma \neq \mathsf{Z}$ and hence

$$\Sigma = \mathsf{P}(i), \qquad \mathsf{Z} = \mathsf{P}, \qquad m = 2.$$

The algebra K sought can therefore only have rank $m^2 = 4$.

According to the automorphism theorem, the isomorphism of $\mathsf{P}(i)$ which carries i into $-i$ must be given by an inner automorphism of K; that is, there exists a k with the property

$$kik^{-1} = -i. \tag{14.49}$$

Since k is not contained in $\Sigma = \mathsf{P}(i)$, it follows that $\Sigma(k) = \mathsf{K}$; hence $\mathsf{K} = \mathsf{P}(i, k)$. It follows from (14.49) that

$$k^2ik^{-2} = i;$$

that is, k^2 commutes with i. Since k^2 also commutes with k, k^2 lies in the center: $k^2 = a \in \mathsf{P}$.

If it were the case that $a \geqq 0$, then $a = b^2$,

$$k^2 - b^2 = (k-b)(k+b) = 0$$

$$k-b = 0 \qquad \text{or} \quad k+b = 0,$$

and hence $k \in \mathsf{P}$, which is impossible. Therefore, $a<0$: $a = -b^2$ $(b \neq 0)$. After multiplying k by a real factor b^{-1}, we may assume that $k^2 = -1$ without destroying the other properties of k. For i and k we therefore have the relations

$$ki = -ik$$

$$i^2 = k^2 = -1.$$

But these are just the properties which characterize the algebra of quaternions. Hence *the algebra of quaternions is the only noncommutative division algebra over the field of real numbers.*

In the same way we can prove: *every central division algebra of index* 2 *over the field* $\mathbb{Q}$ *of rational numbers is a generalized quaternion algebra.*

4. *Determination of all finite skew fields* (skew fields with finitely many elements).

If K is a finite skew field, Z its center, and m the index of K over Z, then every element of K is contained in a maximal commutative subfield Σ of degree m over Z. Now all commutative extensions Σ of degree m of a Galois field Z of p^n elements are equivalent (they are obtained by the adjunction of all roots of the equation $x^q = x$, $q = p^{nm}$). These fields therefore all arise from a single one, Σ_0, by transformation with elements of K:

$$\Sigma = \kappa\Sigma_0\kappa^{-1}.$$

If the zero element of K is omitted, K becomes a group $\mathfrak{G}$, Σ_0 becomes a subgroup $\mathfrak{H}$, and Σ becomes a conjugate subgroup $\kappa\mathfrak{H}\kappa^{-1}$, and these conjugate subgroups together make up the entire group $\mathfrak{G}$ (since every element of K is contained in some Σ). We now need the following group theory lemma.

Lemma: *A proper subgroup $\mathfrak{H}$ of a finite group $\mathfrak{G}$ together with its conjugates $s\mathfrak{H}s^{-1}$ cannot exhaust the entire group $\mathfrak{G}$.*

Proof: Let n and N be the orders of $\mathfrak{H}$ and $\mathfrak{G}$, respectively, and let j be the index of $\mathfrak{H}$ so that $N = j \cdot n$. If s and s' belong to the same coset $s\mathfrak{H}$, so that $s' = sh$, then

$$s'\mathfrak{H}s'^{-1} = sh\mathfrak{H}h^{-1}s^{-1} = s\mathfrak{H}s^{-1}.$$

There are thus at most as many distinct $s\mathfrak{H}s^{-1}$ as there are cosets, that is, at most j. If these $s\mathfrak{H}s^{-1}$ (to which $\mathfrak{H}$ also belongs) exhaust the group $\mathfrak{G}$, then they must be disjoint, for otherwise they could not supply the necessary $N = j \cdot n$ elements. Since, however, two distinct $s\mathfrak{H}s^{-1}$ have the identity element in common, they can never be disjoint, and we have reached a contradiction.

For our case it follows from the lemma that $\mathfrak{H}$ cannot be a *proper* subgroup of $\mathfrak{G}$; therefore $\mathfrak{H} = \mathfrak{G}$ and hence $\mathsf{K} = \Sigma_0$. Thus K is commutative. We have now proved the following.

Every skew field with finitely many elements is commutative and thus a Galois field.

For an alternate proof of this theorem due to MacLagan-Wedderburn, see E. Witt, *Abh. Math. Sem. Hamburg*, **8**, 413 (1931).

14.10 THE SPLITTING FIELDS OF A SIMPLE ALGEBRA

It may be assumed that a simple algebra $\mathfrak{A}$ is a complete matrix ring over a division algebra K:

$$\mathfrak{A} = \mathsf{K}_r.$$

By Section 13.12, the splitting fields of K are at the same time splitting fields of $\mathfrak{A}$, and conversely. In studying the splitting fields we may therefore restrict our consideration to the division algebra K. Further, it may be assumed that the center of K is the base field P; K is then central over P.

By Section 14.9, the maximal commutative subfields of K are splitting fields of K. There thus exist splitting fields Σ of finite degree over P. We therefore restrict ourselves henceforth to finite extension fields Σ of P.

Each such field Σ can be irreducibly imbedded in K_r by Section 14.9. We may therefore interpret Σ as an irreducible system of matrices in K_r at the outset. If now Σ is a splitting field of K, then this means that $\Sigma \times \mathsf{K}'$ is a complete matrix ring over Σ:

$$\Sigma \times \mathsf{K}' = \Sigma_t, \qquad \text{hence} \quad \Delta = \Sigma.$$

The inverse ring Δ' is then likewise equal to Σ. The centralizer of Σ is therefore equal to Σ; that is, any element of K_r which commutes with all the elements of Σ lies in Σ. From this it follows that Σ is a maximal commutative subfield (even a maximal commutative subring) of K_r.

Conversely, let Σ be a maximal commutative subfield of the matrix ring K_r. If Σ is reducible, then the matrices A of the system Σ can be obtained by combining submatrices A_1 in accordance with (14.46). These submatrices form a system Σ_1 isomorphic to Σ which is likewise maximal. We may therefore, without loss of generality, assume at the beginning that the system Σ is irreducible.

The centralizer Δ' of Σ is a skew field whose elements θ commute with all the elements of Σ. If such an element θ were not contained in Σ, then $\Sigma(\theta)$ would be a proper extension of Σ in K_r, which contradicts the maximality of Σ. Hence $\Delta' = \Sigma$. But then also $\Delta = \Sigma$; that is, Σ is a splitting field of K.

We have thus obtained the following characterization of the splitting fields.

Every maximal commutative subfield of a complete matrix ring K_r *is a splitting field of* K; *conversely, every splitting field can be represented (even irreducibly) as a maximal subfield of* K_r.

For the case of the irreducible imbedding of Σ in K_r we have from (14.45) the rank relation

$$(\Sigma : \mathsf{P}) = tr.$$

Here t is again the degree of the absolutely irreducible representation of K in Σ; that is, t is equal to the index m of the division algebra K. Hence

$$(\Sigma : \mathsf{P}) = mr.$$

From this it follows that *the field degree of a splitting field* Σ *of* K *is always divisible by the index* m *of* K. *The maximal commutative subfields of* K *are splitting fields of smallest possible rank* m.

We prove finally the following theorem.

Theorem: *Every central division algebra* K *over* P *has at least one separable splitting field.*

For the proof we need the following lemma.

Lemma: *In a field of characteristic* p *every* p^f*-rowed matrix* A *which satisfies an equation of the form*

$$A^{p^e} = E\zeta \qquad (E = \textit{identity matrix}) \tag{14.50}$$

has a characteristic polynomial (cf. Section 12.6) *of the form*

$$\chi(x) = x^{p^f} - \beta$$

and hence, if $p^f > 1$, *has trace zero.*

Proof of the Lemma: We can adjoin the p^eth roots of ζ to the base field, and we may therefore assume that $\zeta = \eta^{pe}$. If the matrix A is interpreted as the matrix of a linear transformation of a vector space, then, for every vector v,

$$0 = (A^{p^e} - \zeta)v = (A^{p^e} - \eta^{p^e})v = (A - \eta)^{p^e} v.$$

The elementary divisors $f_\nu(x)$ of the matrix A are divisors of $(x-\eta)^{p^e}$ by definition (Section 12.5), and they are therefore powers of $(x-\eta)$. The characteristic polynomial $\chi(x)$ is a product of the elementary divisors and is thus likewise a power of $(x-\eta)$. Since $\chi(x)$ is a polynomial of degree p^f, it follows that

$$\chi(x) = (x-\eta)^{p^f} = x^{p^f} - \eta^{p^f} = x^{p^f} - \beta.$$

Proof of the Existence of Separable Splitting Fields: Let Z be a maximal separable subfield of K, and let Δ' be the centralizer of Z in K. By the structure theorem of Section 14.9, $Z \times K'$ is isomorphic to a complete matrix ring Δ_t, where Δ is anti-isomorphic to Δ'. The center of $Z \times K'$ is $Z \times P = P$, since P is the center of K'. Thus, the center of Δ_t is also Z. The center of the complete matrix ring Δ_t is equal to the center of Δ, and therefore the center of Δ' is equal to Z.

If now θ is an element of Δ not belonging to Z, then $Z(\theta)$ is inseparable and, indeed, is of reduced degree 1, since otherwise $Z(\theta)$ would contain a separable subfield containing Z. Therefore θ satisfies an irreducible equation of the form

$$\theta^{p^e} = \zeta, \qquad \zeta \text{ in } Z. \tag{14.51}$$

The same is true (with $p^e = 1$) if θ itself lies in Z.

If Σ is a maximal commutative subfield of Δ', then Σ over Z as base field has reduced degree 1 and thus has field degree p^f. And Σ is a splitting field of Δ'; that is, $\Delta' \times \Sigma$ is a complete matrix ring over Σ and has degree p^f. In this matrix representation all elements of Δ' have trace zero if $p^f > 1$ by the lemma; if A is the matrix representing θ, then the matrix equation (14.50) follows from (14.51). All the matrices of $\Delta' \times \Sigma$ are linear combinations of matrices of Δ' with coefficients in Σ, the base field of the matrix ring. All these matrices thus have trace zero for $p^f > 1$. This, however, is contradicted by the fact that we are here concerned with the *complete* matrix ring. Hence $p^f = 1$ and $Z = \Sigma$ are the only remaining possibilities. Now Z is itself a maximal subfield of K and is thus a splitting field.

14.11 THE BRAUER GROUP. FACTOR SYSTEMS

We partition the central simple algebras over a fixed base field P into classes by assigning to a class [K] all those algebras which are isomorphic to complete matrix rings over the same division algebra K.

If K and Λ are two such division algebras, then $K \times \Lambda$ is again central and simple (Section 13.12), and hence

$$K \times \Lambda \cong \Delta_t. \tag{14.52}$$

It follows from (14.52) that

$$\begin{aligned} K_r \times \Lambda_s &= K \times P_r \times \Lambda \times P_s \cong \Delta_t \times P_{rs} \\ &= \Delta \times P_t \times P_{rs} = \Delta \times P_{trs} = \Delta_{trs}\,; \end{aligned}$$

all products $K_r \times \Lambda_s$ of algebras of the classes [K] and [Λ] therefore belong to a class [Δ]. This class is called the *product* of the classes [K] and [Λ]. Since further,

$$K \times \Lambda \cong \Lambda \times K$$

$$K \times (\Lambda \times \Gamma) = (K \times \Lambda) \times \Gamma,$$

the product is commutative and associative. There is also an identity class: the class [P] of the base field. Finally, for every class [K] there is an inverse class: the class [K′] of the division algebra K′ anti-isomorphic to K. Thus *the classes of central simple algebras over* P *form an Abelian group*. This group was first studied by R. Brauer and is called the *Brauer group of algebra classes*.

Those algebra classes having a given commutative field Σ over P as splitting field always form a subgroup of the Brauer group. Indeed, a splitting field of K is, by Section 13.12, also a splitting field of the entire class [K] as well as a splitting field of the inverse class [K′], since K′ is anti-isomorphic to K and therefore $K' \times \Sigma$ is also anti-isomorphic to $K \times \Sigma$. If K and Λ both have the splitting field Σ, so that

$$K \times \Sigma \cong \Sigma_s, \qquad \Lambda \times \Sigma \cong \Sigma_t,$$

then it follows that

$$\begin{aligned}(K \times \Lambda) \times \Sigma &\cong K \times \Sigma_t \cong K \times \Sigma \times P_t\\ &\cong \Sigma_s \times P_t = \Sigma \times P_s \times P_t \cong \Sigma_{st},\end{aligned}$$

and thus Σ is also a splitting field of the product $K \times \Lambda$ and therefore of the entire product class $[K \times \Lambda]$.

By the last theorem of Section 14.10, each Brauer algebra class [K] has a separable splitting field, say the field $P(\theta)$. If all the conjugates of θ are also adjoined, a normal, separable splitting field Σ is obtained. This field can be irreducibly represented by Section 14.10 as a maximal commutative subfield of a simple algebra $\mathfrak{A} = K_r$ which belongs to the class [K].

We shall now prove: *the algebra* $\mathfrak{A}$ *is a crossed product of the field* Σ *with its Galois group* $\mathfrak{G}$ *in the sense of Section* 13.3.

It follows first of all from Section 13.3 that Σ is its own centralizer in $\mathfrak{A} = K_r$; that is, any element of $\mathfrak{A}$ which commutes with all the elements of Σ lies in Σ.

As in Section 13.3, we denote by $S, T, \ldots$ the elements of the Galois group $\mathfrak{G}$ and by β^S the element of Σ which is the image of β under the automorphism S. The product ST is again defined by

$$\beta^{ST} = (\beta^S)^T.$$

The automorphisms S are generated by inner automorphisms of $\mathfrak{A}$ according to the automorphism theorem of Section 14.9. Therefore there exists for each S an element u_S in $\mathfrak{A}$ having an inverse u_S^{-1} in $\mathfrak{A}$ so that, for all β of Σ,

$$u_S^{-1}\beta u_S = \beta^S$$

or

$$\beta u_S = u_S\beta^S. \tag{14.53}$$

The element $u_{ST}^{-1}u_Su_T$ commutes with all the elements of Σ by (14.53) and is therefore itself an element of Σ. Putting

$$u_{ST}^{-1}u_Su_T = \delta_{S,T},$$

we obtain the multiplication rule

$$u_Su_T = u_{ST}\delta_{S,T}. \tag{14.54}$$

Here $\delta_{S,T} \neq 0$, since $\delta_{S,T}$ has an inverse $u_T^{-1}u_S^{-1}u_{ST}$.

The composition rules (14.53) and (14.54) are precisely the same as formulas (13.36) and (13.37), by which the crossed product was defined. It follows from these rules, as was previously proved, that the u_S are linearly independent over Σ. The linear combinations of the u_S with coefficients in Σ,

$$a = \sum_S u_S \beta_S,$$

form a ring $\mathfrak{A}_1$ in $\mathfrak{A}$ having rank n over Σ and thus rank n^2 over P; here $n = (\Sigma:\mathrm{P})$ is the rank of Σ over P. By Section 14.10,

$$n = (\Sigma : \mathrm{P}) = rm.$$

The rank of $\mathfrak{A} = \mathrm{K}_r$ over P is

$$r^2(\mathrm{K} : \mathrm{P}) = r^2m^2 = n^2.$$

Since $\mathfrak{A}_1$ and $\mathfrak{A}$ thus have the same rank n^2 and $\mathfrak{A}_1$ is contained in $\mathfrak{A}$, it follows that $\mathfrak{A}_1 = \mathfrak{A}$; that is, $\mathfrak{A}$ is a crossed product of the field Σ with the group $\mathfrak{G}$.

The fact that algebras $\mathfrak{A} = \mathrm{K}_r$ could be represented as crossed products was first recognized by Emmy Noether. The system $\{\delta_{S,T}\}$ of elements $\delta_{S,T}$ is therefore called a *Noether factor set* of the algebra $\mathfrak{A}$ or of the algebra class [K]. The following assertion is clearly true.

The structure of the algebra $\mathfrak{A}$ is completely determined if the field Σ and the factor set $\{\delta_{S,T}\}$ are known.

The converse is not true. If $\mathfrak{A}$ and Σ are given, then the imbedding of Σ in $\mathfrak{A}$ is, to be sure, uniquely determined up to inner automorphisms of $\mathfrak{A}$, but the u_S are not uniquely determined by the imbedding; on the contrary, according to (13.39), they may be replaced by

$$v_S = u_S\gamma_S \qquad (\gamma_S \neq 0). \tag{14.55}$$

This is the only freedom present; for it the v_S as well as the u_S have property (14.53),

$$\beta v_S = v_S\beta^S,$$

then $v_S u_S^{-1}$ commutes with all elements β of Σ,

$$\beta v_S u_S^{-1} = v_S\beta^S u_S^{-1} = v_S u_S^{-1}\beta.$$

If we put $v_S u_S^{-1} = \gamma_S$, then the γ_S are elements of Σ and we have

$$v_S = \gamma_S u_S.$$

Replacement of the u_S by the v_S implies, as we have seen in Section 13.3, that the factor set $\{\delta_{S,T}\}$ is replaced by the *associated factor set* $\{\epsilon_{S,T}\}$:

$$\epsilon_{S,T} = \frac{\gamma_S^T\gamma_T}{\gamma_{ST}}\delta_{S,T}. \tag{14.56}$$

There is thus a one-to-one correspondence between Brauer algebra classes [K] *with a fixed splitting field which is normal and separable and the classes of associated factor systems* $\{\delta_{S,T}\}$ *in* Σ *which satisfy the associativity conditions* (13.38).

Thus far we have started with a normal splitting field Σ. However, R. Brauer has shown that a factor system can be defined relative to a splitting field of a simple algebra K_r which is not normal.

Let Δ be a finite splitting field which need not be normal. Let $\vartheta = \vartheta_1$ be a primitive element of Δ, such that $\Delta = \mathsf{P}(\vartheta)$, and let $\vartheta_\alpha (\alpha = 1, 2, \ldots, n)$ be the conjugates of ϑ in an appropriate normal extension field Σ.

Up to equivalence there is only one absolutely irreducible representation of K_r by matrices in Δ. Let $a \to A$ be this representation, and let $a \to A_\alpha$ be the representations arising from the first when the field isomorphisms $\vartheta \to \vartheta_\alpha$ are applied to the matrix elements of the representation. Since these representations are all equivalent (there is also in Σ, up to equivalence, only one irreducible representation), there exist matrices $P_{\alpha\beta}$ which transform the representation A_α into A_β:

$$A_\alpha = P_{\alpha\beta} A_\beta P_{\alpha\beta}^{-1}.$$

The matrix $P_{\alpha\beta}$ may be taken in the field $\mathsf{P}(\vartheta_\alpha, \vartheta_\beta)$, since the two representations are already equivalent in this field. The $P_{\alpha\beta}$ can further be chosen so that each isomorphism of $\mathsf{P}(\vartheta_\alpha, \vartheta_\beta)$ which takes $\vartheta_\alpha, \vartheta_\beta$ into a conjugate pair $\vartheta_\gamma, \vartheta_\delta$ also takes $P_{\alpha\beta}$ into $P_{\gamma\delta}$. For this purpose we need only select from each class of conjugate pairs a pair α, β, determine a $P_{\alpha\beta}$ for this pair, and derive the remaining $P_{\gamma\delta}$ from the relevant isomorphisms.

We now have

$$\begin{aligned} A_\alpha = P_{\alpha\beta} A_\beta P_{\alpha\beta}^{-1} &= P_{\alpha\beta} P_{\beta\gamma} A_\gamma P_{\beta\gamma}^{-1} P_{\alpha\beta}^{-1} \\ &= P_{\alpha\beta} P_{\beta\gamma} P_{\alpha\gamma}^{-1} A_\alpha P_{\alpha\gamma} P_{\beta\gamma}^{-1} P_{\alpha\beta}^{-1}. \end{aligned}$$

The matrix $P_{\alpha\beta} P_{\beta\gamma} P_{\alpha\gamma}^{-1}$ therefore commutes with all the matrices A_α of an absolutely irreducible representation, and it is therefore a multiple of the identity matrix

$$P_{\alpha\beta} P_{\beta\gamma} P_{\alpha\gamma}^{-1} = c_{\alpha\beta\gamma} E \qquad (14.57)$$

$$P_{\alpha\beta} P_{\beta\gamma} = c_{\alpha\beta\gamma} P_{\alpha\gamma}.$$

The *Brauer factor set* $\{c_{\alpha\beta\gamma}\}$ is defined by (14.57). It has the following properties.

1. $c_{\alpha\beta\gamma}$ belongs to the field $P(\vartheta_\alpha, \vartheta_\beta, \vartheta_\gamma)$.
2. $c_{\alpha\beta\gamma} c_{\alpha\gamma\delta} = c_{\alpha\beta\delta} c_{\beta\gamma\delta}$.
3. $c_{\alpha\beta\gamma}^S = c_{\alpha'\beta'\gamma'}$ if S is an isomorphism of the field $P((\vartheta_\alpha, \vartheta_\beta, \vartheta_\gamma)$ which takes $\vartheta_\alpha, \vartheta_\beta, \vartheta_\gamma$ into $\vartheta_{\alpha'}, \vartheta_{\beta'}, \vartheta_{\gamma'}$.

Property 1 follows immediately from the definition of the $c_{\alpha\beta\gamma}$, property 2 follows from the associative law for the matrices $P_{\alpha\beta}$, and property 3 is a consequence of the behavior of the $P_{\alpha\beta}$ under the isomorphism S.

If $P_{\alpha\beta}$ is replaced by $k_{\alpha\beta} P_{\alpha\beta}$, where the nonzero field elements $k_{\alpha\beta}$ are required to satisfy the same conjugation relations as the $P_{\alpha\beta}$, then the system of the $c_{\alpha\beta\gamma}$ goes over into an *associated factor set*

$$c'_{\alpha\beta\gamma} = \frac{k_{\alpha\beta}k_{\beta\gamma}}{k_{\alpha\gamma}} c_{\alpha\beta\gamma}. \tag{14.58}$$

On the other hand, if the representation $a \to A$ is replaced by an equivalent representation $a \to QAQ^{-1}$, then the P_α are to be replaced by $Q_\alpha P_\alpha Q_\alpha{}^{-1}$; it may be easily verified that the factor set $c_{\alpha\beta\gamma}$ hereby remains unchanged. The factor set is thus uniquely determined up to associated sets by K_r and Δ alone.

The entire theory can be constructed on the basis of either the Noether or the Brauer factor set. The proofs become simpler and more transparent if both types of sets are used and their equivalence is demonstrated. Certain properties are more easily proved for the Noether set, and others are more easily proved for the Brauer set. We begin with the basic properties of the Brauer factor set.

If K_r is a complete matrix ring over the base field P, and thus $\mathsf{K}_r = \mathsf{P}_r$, then all the $P_{\alpha\beta}$ may be chosen equal to the identity matrix E. All the $c_{\alpha\beta\gamma}$ are then equal to 1, and it follows that *the factor set of an algebra which splits in the base field is associated to the identity set* $c_{\alpha\beta\gamma} = 1$.

We now seek the factor set for a direct product $\mathsf{K}_r \times \Lambda_s$. If $a \to A$ is the irreducible representation of K_r in the field Λ and $b \to B$ is that of Λ_s in the same field, then a representation of the product system $\mathsf{K}_r \times \Lambda_s$ is obtained if ab is represented by the Kronecker product $A \times B$ (Section 14.6). It may easily be seen that this representation is irreducible by computing its degree. Indeed, if the irreducible representation of K_r has degree n and that of Λ_s has degree m, then K_r has rank n^2 and Λ_s has rank m^2 (by the Burnside theorem, for example), and thus $\mathsf{K}_r \times \Lambda_s$ has rank n^2m^2, and the degree of the product representation is mn and thus coincides with the degree of the absolutely irreducible representation of $\mathsf{K}_r \times \Lambda_s$.

We can compute the factor set of the product representation. Here $A_\alpha = P_{\alpha\beta}{}^{-1} A_\beta P_{\alpha\beta}$ and $B_\alpha = Q_{\alpha\beta}{}^{-1} B_\beta Q_{\alpha\beta}$ imply

$$A_\alpha \times B_\beta = (P_{\alpha\beta} \times Q_{\alpha\beta})^{-1}\,(A_\beta \times B_\beta)\,(P_{\alpha\beta} \times Q_{\alpha\beta}),$$

and thus $P_{\alpha\beta} \times Q_{\alpha\beta}$ are the transformation matrices of the product representation. It follows similarly from

$$P_{\alpha\beta}P_{\beta\gamma} = c_{\alpha\beta\gamma}P_{\alpha\gamma} \qquad \text{and} \quad Q_{\alpha\beta}Q_{\beta\gamma} = d_{\alpha\beta\gamma}Q_{\alpha\gamma}$$

that

$$(P_{\alpha\beta} \times Q_{\alpha\beta})\,(P_{\beta\gamma} \times Q_{\beta\gamma}) = c_{\alpha\beta\gamma}d_{\alpha\beta\gamma}(P_{\alpha\gamma} \times Q_{\alpha\gamma}).$$

Hence, $\{c_{\alpha\beta\gamma}d_{\alpha\beta\gamma}\}$ *is a factor system of the product algebra* $\mathsf{K}_r \times \Lambda_s$.

If we apply this result first to the case $\mathsf{K} \times \mathsf{P}_r = \mathsf{K}_r$, it follows, since the $d_{\alpha\beta\gamma}$ are in this case equal to 1, that *the matrix ring* K_r *has the same factor set as the skew field* K. Thus, to each Brauer algebra class there corresponds a single factor set up to associated sets.

Combining these results, we have this statement: *To each element of the Brauer group of algebra classes with the splitting field* Δ *there corresponds a factor set*

$\{c_{\alpha\beta\gamma}\}$ which is uniquely determined up to associated sets; to the identity element there hereby corresponds the identity set and to the product of two group elements there corresponds the product of the factor sets.

We now investigate how the Brauer factor set of an algebra behaves under extension of the splitting field. Let $\Delta' = \mathsf{P}(\vartheta')$ be a finite separable extension field of $\Delta = \mathsf{P}(\vartheta)$. Each isomorphism $\vartheta' \to \vartheta'_\alpha$ of the field Δ' induces an isomorphism $\vartheta \to \vartheta_\alpha$ of the field Δ; to each α' there thus corresponds an α. The representation $a \to A$ of K_r in Δ can be left unchanged when going over to Δ'. The conjugate representations A_α then also remain unchanged; that is, $A'_{\alpha'} = A_\alpha$ if the number α corresponds to the number α'. For the transformation matrices the corresponding rule reads: $P'_{\alpha'\beta'} = P_{\alpha\beta}$ if the numbers α, β correspond to the numbers α', β'. Finally, the same simple rule holds for the factor set: $c'_{\alpha'\beta'\gamma'} = c_{\alpha\beta\gamma}$ *if the numbers α', β', γ' correspond to the numbers α, β, γ, that is, if the isomorphisms $\vartheta' \to \vartheta'_{\alpha'}$, $\vartheta' \to \vartheta'_{\beta'}$, $\vartheta' \to \vartheta'_{\gamma'}$, of the field Δ' induce the isomorphisms $\vartheta \to \vartheta_\alpha$, $\vartheta \to \vartheta_\beta$, $\vartheta \to \vartheta_\gamma$ of the field Δ.*

On the basis of this rule we may always pass from an arbitrary separable splitting field Δ to an encompassing normal field Σ. The isomorphisms $\vartheta \to \vartheta_\alpha$ of Σ are then the elements $S, T, \ldots$ of the Galois group: $\vartheta_\alpha = \vartheta^S$, $\vartheta_\beta = \vartheta^T$, and so on. In this case the elements S, T, R may be used as indices in place of α, β, γ, and we can write $c_{S,T,R}$ in place of $c_{\alpha\beta\gamma}$. In this new notation rule 3 reads:

$$c_{S}{}^{Q}{}_{,T,R} = c_{SQ,TQ,RQ}. \tag{14.59}$$

The connection with the Noether factor set can now be established. We shall compute the Brauer factor set for the crossed product K_r originally defined and show that it is the same as the Noether factor set except for notation.

We obtain an irreducible representation of K_r in Σ by interpreting K_r itself as a representation module. The basis elements of K_r as a right Σ-module are precisely the u_S. The matrix representing an element $a = u_S\beta$ (it suffices to consider only these elements, since all others are sums of such elements) is obtained by multiplying this element by all the basis elements u_T and expanding the products in terms of the u_T:

$$(u_S\beta)u_T = u_S u_T \beta^T = u_{ST}\delta_{S,T}\beta^T.$$

The representing matrix A therefore has the element $\delta_{S,T}\beta^T$ in the STth row and Tth column and has zeros elsewhere in that column. The conjugate matrix A^R therefore has the element

$$(\delta_{S,T}\beta^T)^R = \delta^R_{S,T}\beta^{TR}$$

in the Tth column and STth row.

We now try to determine the matrix $P_{1,R}$ which takes A into A^R:

$$AP_{1,R} = P_{1,R}A^R. \tag{14.60}$$

We take for $P_{1,R}$ the matrix which has the element $\delta_{Y,R}$ in the Yth column and YRth row and has zeros elsewhere in that column. Relation (14.60) is then satisfied, for on the left side the element $\delta_{S,\,TR}\beta^{TR}\delta_{T,\,R}$ stands in the Tth column and STRth row and on the right side this entry is $\delta_{ST,R}\,\delta^R_{S,T}\beta^{TR}$, which is the same by (13.38). We have thus found $P_{1,\,R}$. The other $P_{S,\,T}$ are obtained (by the agreement established in defining the $P_{\alpha\beta}$) by applying the automorphisms S to $P_{1,\,R}$:

$$P^S_{1,\,R} = P_{S,\,RS}.$$

The relation $P_{S,\,T}P_{T,\,R} = c_{S,\,T,\,R}P_{S,\,R}$ must be established only for the case $S = 1$, since by applying the isomorphism S the index 1 can be converted to S [cf. (14.59)]. We are therefore concerned only with

$$P_{1,\,R}P_{R,\,TR} = c_{1,\,R,\,TR}P_{1,\,TR}$$

or

$$P_{1,\,R}P^R_{1,\,T} = c_{1,\,R,\,TR}P_{1,\,TR}.$$

The left side has the element

$$\delta_{ST,\,R}\delta^R_{S,\,T} = \delta_{S,\,TR}\delta_{T,\,R}$$

in the Sth column and STRth row; the right side has the element $c_{1,\,R,\,TR}\delta_{S,\,TR}$ at this place. We must therefore put

$$c_{1,\,R,\,TR} = \delta_{T,\,R}. \tag{14.61}$$

The Noether factor set is known from formula (14.61) as soon as the Brauer set is given. But the structure of the algebra K_r is determined by the Noether factor set. We thus have the following.

A Brauer algebra class is uniquely determined by the splitting field Δ *and the factor set* $\{c_{\alpha\beta\gamma}\}$.

From previous consideration of the factor set of product algebras we found a homomorphism of the group of Brauer algebra classes with given splitting field Δ to the group of classes of their associated factor sets. On the basis of the uniqueness proof just given, it follows that this homomorphism is an *isomorphism*.

It is easily checked that the associativity condition (13.38) is a consequence of properties 1, 2, and 3 of the $c_{\alpha\beta\gamma}$. *Thus, to each system of field elements* $c_{\alpha\beta\gamma}$ *with properties* 1, 2, *and* 3 *there belongs an algebra class which is represented by a crossed product with the factor system* $\delta_{S,\,T}$ *defined by* (14.61).

The basic properties of the Brauer factor set also hold for the Noether set by (14.61). In particular, there is also in this case an isomorphism of the group of algebra classes with a fixed normal splitting field to the group of classes of their associated (Noether) factor sets. We make special mention of the following fact.

The crossed product K_r *is a complete matrix ring over the base field* P *if and only if its factor set* $\delta_{S,\,T}$ *is associated to the identity set*

$$\delta_{S,\,T} = \frac{c_S{}^T c_T}{c_{ST}}.$$

Exercises

14.5. Under an extension of the base field P to an extension field Λ the skew field K goes over into the simple algebra K_Λ. Prove that the Brauer factor set is hereby "shortened" in the following manner. Let the fields Δ and Λ be imbedded in a common extension field and seek out among the elements ϑ_α conjugate to ϑ those which are still conjugate to ϑ with respect to the new base field Λ. The $c_{\alpha\beta\gamma}$ belonging to three of the ϑ_α are retained, and all others are omitted. In the language of the Noether factor sets this states that only those $\delta_{S,T}$ are retained for which S and T belong to a fixed subgroup (which?) of the Galois group.

14.6. Using Exercise 14.5, answer the following question: Which subfields of Σ are splitting fields of an algebra with the factor set $\delta_{S,T}$?

14.7. Two cyclic algebras (δ, Σ, S) and (ϵ, Σ, S) are isomorphic if and only if δ differs from ϵ only by a factor of a norm. In particular, (δ, Σ, S) is a complete matrix ring over P if and only if δ is the norm of an element of Σ.

Chapter 15

GENERAL IDEAL THEORY OF COMMUTATIVE RINGS

15.1 NOETHERIAN RINGS

In this chapter we shall study the divisibility properties of ideals in commutative rings and find to what extent the simple laws which hold in, say, the ring of integers can be carried over to more general rings. In order to avoid complicated situations, it is convenient to restrict attention to rings in which each ideal has a finite basis. As we shall see, this condition is actually satisfied in a great many important cases.

We say that the *basis condition* holds in a ring $\mathfrak{o}$ if every ideal in $\mathfrak{o}$ has a finite basis. Commutative rings in which the basis condition holds are called *Noetherian rings.*[1]

The basis condition holds in any field, since (0) and (1) are the only ideals. It is also satisfied in the ring of integers and, more generally, in any principal ideal ring. Of course, it also holds in any finite ring. We shall see later that the basis condition holds in any factor ring $\mathfrak{o}/\mathfrak{a}$ if it holds in $\mathfrak{o}$. Finally, we have the following theorem, which goes back essentially to Hilbert.

Theorem: *If the basis condition holds in a ring* $\mathfrak{o}$ *and if* $\mathfrak{o}$ *has an identity element, then the basis condition also holds in the polynomial ring* $\mathfrak{o}[x]$.

Proof: Let $\mathfrak{A}$ be an ideal in $\mathfrak{o}[x]$. The leading coefficients of x in the polynomials of $\mathfrak{A}$, together with zero, form an ideal. Indeed, if α and β are the leading coefficients of the polynomials a and b,

$$a = \alpha x^n + \cdots$$
$$b = \beta x^m + \cdots,$$

and if it is assumed, say, that $n \geqq m$, then

$$a - bx^{n-m} = (\alpha x^n + \cdots) - (\beta x^n + \cdots)$$
$$= (\alpha - \beta)x^n + \cdots$$

[1]The definition used by many authors includes, in addition, the requirement that the ring have an identity element (Trans.).

is again a polynomial of $\mathfrak{A}$ with leading coefficient $\alpha-\beta$ or zero; similarly, if α is the leading coefficient of a, then $\lambda\alpha$ is the leading coefficient of λa or zero.

Now by hypothesis this ideal $\mathfrak{a}$ of the leading coefficients has a basis $(\alpha_1, \ldots, \alpha_r)$. Suppose that α_i is the leading coefficient of the polynomial

$$a_i = \alpha_i x^{n_i} + \cdots$$

of degree n_i, and let n be the greatest of the finite number of integers n_i.

We include the polynomials a_i in the basis being constructed for $\mathfrak{A}$. We shall now determine what further polynomials are necessary for a basis.

If

$$f = \alpha x^N + \cdots$$

is a polynomial of $\mathfrak{A}$ of degree $N \geqq n$, then α must belong to the ideal $\mathfrak{a}$:

$$\alpha = \sum \lambda_i \alpha_i.$$

We form the polynomial

$$f_1 = f - \sum (\lambda_i x^{N-n_i}) a_i.$$

The coefficient of x^N in this polynomial is

$$\alpha - \sum \lambda_i \alpha_i = 0;$$

f_1 therefore has degree $< N$. The polynomial f can thus be replaced modulo $(a_1, \ldots, a_r)$ by a polynomial of lower degree. We can continue in this manner until the degree is less than n. It is thus sufficient henceforth to consider polynomials of degree $< n$.

The coefficients of x^{n-1} in the polynomials of $\mathfrak{A}$ of degree $\leqq n-1$, together with zero, form an ideal $\mathfrak{a}_{n-1}$; let

$$(\alpha_{r+1}, \ldots, \alpha_s)$$

be a basis for this ideal. Let α_{r+i} again be the leading coefficient of the polynomial

$$a_{r+i} = \alpha_{r+i} x^{n-1} + \cdots.$$

We now also include the polynomials $a_{r+1}, \ldots, a_s$ in the basis. Every polynomial of degree $\leqq n-1$ can now be replaced modulo $(a_{r+1}, \ldots, a_s)$ by a polynomial of degree $\leqq n-2$; we have, as before, only to subtract an appropriate linear combination

$$\sum \lambda_{r+i} a_{r+i}.$$

We continue in this manner. The coefficients of x^{n-2} in the polynomials of degree $\leqq n-2$ together with zero form an ideal $\mathfrak{a}_{n-2}$ whose basis elements $\alpha_{s+1}, \ldots, \alpha_t$ correspond to polynomials $a_{s+1}, \ldots, a_t$. These polynomials we again include in the basis. We finally arrive at the ideal $\mathfrak{a}_0$ of the constants in $\mathfrak{A}$; its basis elements $(\alpha_{v+1}, \ldots, \alpha_w)$ belong to the polynomials $a_{v+1}, \ldots, a_w$. Each polynomial of $\mathfrak{A}$ must now reduce to zero modulo

$$(a_1, \ldots, a_r, a_{r+1}, \ldots, a_s, \ldots, a_{v+1}, \ldots, a_w).$$

The polynomials $a_1, \ldots, a_w$ therefore form a basis for the ideal $\mathfrak{A}$, and this completes the proof of the theorem.

Applying this theorem n times, we obtain the following generalization.

If the basis condition holds for a ring $\mathfrak{o}$ *with identity, then it also holds for any polynomial ring* $\mathfrak{o}[x_1, \ldots, x_n]$ *in a finite number of indeterminates* $x_1, \ldots, x_n$.

The most important special cases are the ring $\mathbb{Z}[x_1, \ldots, x_n]$ of integral polynomials and the polynomial ring $\mathsf{K}[x_1, \ldots, x_n]$ with coefficients in a field K. All these rings are Noetherian.

Hilbert formulated his condition only for these cases and in a form which may appear to be more general, as follows.

In every subset $\mathfrak{M}$ *of* $\mathfrak{o}$ *(not only in every ideal) there exist a finite number of elements* $m_1, \ldots, m_r$ *such that every element* m *of* $\mathfrak{M}$ *can be expressed in the form*

$$\lambda_1 m_1 + \cdots + \lambda_r m_r \qquad (\lambda_i \text{ in } \mathfrak{o}).$$

This condition is an immediate consequence of the basis condition for ideals, since if $\mathfrak{A}$ is the ideal generated by $\mathfrak{M}$, then $\mathfrak{A}$ has first of all a basis

$$\mathfrak{A} = (a_1, \ldots, a_s).$$

Each element a_i (as an element of the ideal generated by $\mathfrak{M}$) depends on finitely many elements of $\mathfrak{M}$:

$$a_i = \sum_k \lambda_{ik} m_{ik}.$$

All elements of $\mathfrak{A}$ therefore depend linearly on the finitely many m_{ik}; this holds, in particular, for elements of $\mathfrak{M}$.

It is more important that the basis condition is equivalent to the following "ascending chain condition."

Ascending Chain Condition, First Formulation: *If a chain of ideals* $\mathfrak{a}_1, \mathfrak{a}_2, \mathfrak{a}_3, \ldots$ *is given in* $\mathfrak{o}$ *and if each* $\mathfrak{a}_{i+1}$ *is a proper divisor of* $\mathfrak{a}_i$,

$$\mathfrak{a}_i \subset \mathfrak{a}_{i+1},$$

then the chain breaks off after finitely many terms.

Or we have the following, which amounts to the same thing.

Ascending Chain Condition, Second Formulation: *Given an ascending chain* $\mathfrak{a}_1, \mathfrak{a}_2, \mathfrak{a}_3, \ldots,$

$$\mathfrak{a}_i \subseteq \mathfrak{a}_{i+1},$$

there exists an n *such that*

$$\mathfrak{a}_n = \mathfrak{a}_{n+1} = \cdots.$$

That the ascending chain condition follows from the basis condition may be seen as follows.

Let $\mathfrak{a}_1, \mathfrak{a}_2, \mathfrak{a}_3, \ldots$ be an infinite chain such that $\mathfrak{a}_i \subseteq \mathfrak{a}_{i+1}$. The union $\mathfrak{v}$ of all the ideals $\mathfrak{a}_i$ is an ideal. For if a and b are contained in $\mathfrak{v}$, then a is in some $\mathfrak{a}_n$ and b in some $\mathfrak{a}_m$; a and b therefore both lie in $\mathfrak{a}_N$, where N is the larger of the numbers n and m, and hence $a-b$ is also in $\mathfrak{a}_N$ and thus in $\mathfrak{v}$. If a is in $\mathfrak{v}$ it is in some $\mathfrak{a}_n$, and hence λa is in $\mathfrak{a}_n$ and so in $\mathfrak{v}$.

This ideal has a basis $(a_1, \ldots, a_r)$ by hypothesis. Each a_i is contained in some ideal $\mathfrak{a}_{n_i}$. If n is the largest of the numbers n_i, then $a_1, \ldots, a_r$ all lie in $\mathfrak{a}_n$. Since all the elements of $\mathfrak{v}$ depend linearly on $a_1, \ldots, a_r$, it follows that all elements of $\mathfrak{v}$ are in $\mathfrak{a}_n$ and hence

$$\mathfrak{v} = \mathfrak{a}_n = \mathfrak{a}_{n+1} = \mathfrak{a}_{n+2} = \cdots.$$

Conversely, the ascending chain condition implies the basis condition. Thus, suppose that $\mathfrak{a}$ is an ideal and a_1 is any element of $\mathfrak{a}$. If a_1 does not generate the entire ideal, then there exist elements in $\mathfrak{a}$ which are not contained in (a_1); let a_2 be such an element. Then

$$(a_1) \subset (a_1, a_2).$$

If a_1 and a_2 still do not generate the entire ideal $\mathfrak{a}$, then there is a third element a_3 in $\mathfrak{a}$ which is not contained in (a_1, a_2), and so on. We thus obtain an ascending chain

$$(a_1) \subset (a_1, a_2) \subset (a_1, a_2, a_3) \subset \cdots,$$

which must break off after a finite number (say r) of terms. This implies that

$$(a_1, a_2, \ldots, a_r) = \mathfrak{a},$$

and hence $\mathfrak{a}$ has a finite basis.

If the ascending chain condition holds in a ring $\mathfrak{o}$, *then it also holds in any residue class ring* $\mathfrak{o}/\mathfrak{a}$.

Proof: An ideal $\bar{\mathfrak{b}}$ in $\mathfrak{o}/\mathfrak{a}$ is a set of residue classes. If we form the union of all these residue classes, we obtain an ideal $\mathfrak{b}$ in $\mathfrak{o}$. Conversely, $\bar{\mathfrak{b}}$ is uniquely determined by $\mathfrak{b}$ from

$$\bar{\mathfrak{b}} = \mathfrak{b}/\mathfrak{a}.$$

A chain of ideals $\bar{\mathfrak{b}}_1 \subset \bar{\mathfrak{b}}_2 \subset \bar{\mathfrak{b}}_3 \subset \ldots$ in $\mathfrak{o}/\mathfrak{a}$ thus gives rise to a chain of ideals $\mathfrak{b}_1 \subset \mathfrak{b}_2 \subset \mathfrak{b}_3 \subset \ldots$ in $\mathfrak{o}$; the last chain breaks off after a finite number of terms and so the first must also.

This also proves the assertion made at the beginning of the section that the basis condition for $\mathfrak{o}$ implies the basis condition for $\mathfrak{o}/\mathfrak{a}$.

The ascending chain condition has two other formulations which are sometimes more convenient in applications.

Ascending Chain Condition, Third Formulation: The Maximum Condition: *If the ascending chain condition holds in* $\mathfrak{o}$, *then every nonempty set of ideals contains a maximal ideal, that is, an ideal which is contained in no other ideal of the set.*

Proof: Let one ideal be distinguished in every nonempty set of ideals. If now in a set $\mathfrak{M}$ of ideals there were no maximal ideal, then every ideal of the set would be contained in another ideal of the set. We now find the distinguished ideal $\mathfrak{a}_1$ of $\mathfrak{M}$; in the set of ideals of $\mathfrak{M}$ which contain $\mathfrak{a}_1$ and $\neq \mathfrak{a}_1$ we find the distinguished ideal $\mathfrak{a}_2$, and so on. We finally obtain an infinite chain

$$\mathfrak{a}_1 \subset \mathfrak{a}_2 \subset \mathfrak{a}_3 \subset \cdots,$$

which is impossible by hypothesis.

Ascending Chain Condition, Fourth Statement: The Principle of Divisor Induction: *If the ascending chain condition holds in* $\mathfrak{o}$ *and if a property E can be proved for any ideal* $\mathfrak{a}$ *(in particular, for the unit ideal) under the hypothesis that it is satisfied for all proper divisors of* $\mathfrak{a}$, *then all ideals have property E.*

Proof: Suppose that some ideal does not have property E. Then by the third statement of the ascending chain condition there is a maximal ideal $\mathfrak{a}$ which does not have property E. Because of the maximality, all proper divisors of $\mathfrak{a}$ must have property E. Therefore $\mathfrak{a}$ also has property E; this is a contradiction.

15.2 PRODUCTS AND QUOTIENTS OF IDEALS

As in Section 3.6, the *greatest common divisor* (g.c.d.) or the *sum* of the ideals $\mathfrak{a}, \mathfrak{b}, \ldots$ is the ideal $(\mathfrak{a}, \mathfrak{b}, \ldots)$ generated by their union, and the *least common multiple* (l.c.m.) is the intersection $[\mathfrak{a}, \mathfrak{b}, \ldots] = \mathfrak{a} \cap \mathfrak{b} \ldots$. The same notation as for the sum of ideals is used for ideals generated by elements and ideals, for example,

$$(\mathfrak{a}, b) = (\mathfrak{a}, (b)).$$

It is obvious that $(\mathfrak{a}, \mathfrak{b}) = (\mathfrak{b}, \mathfrak{a})$, $((\mathfrak{a}, \mathfrak{b}), \mathfrak{c}) = (\mathfrak{a}, (\mathfrak{b}, \mathfrak{c})) = (\mathfrak{a}, \mathfrak{b}, \mathfrak{c})$, and so on. Furthermore,

$$((a_1, a_2, \ldots), (b_1, b_2, \ldots)) = (a_1, a_2, \ldots, b_1, b_2, \ldots),$$

or in words: *a basis for the greatest common divisor is obtained by writing down the bases for the individual ideals one after the other.*

If the elements of an ideal $\mathfrak{a}$ are multiplied by the elements of an ideal $\mathfrak{b}$, then in general the products ab do not form an ideal. The ideal generated by these products ab is called the *product* of the ideals $\mathfrak{a}$ and $\mathfrak{b}$ and is denoted by $\mathfrak{a}\cdot\mathfrak{b}$ or $\mathfrak{a}\mathfrak{b}$. It consists of all sums $\sum a_i b_i$ (a_i in $\mathfrak{a}$, b_i in $\mathfrak{b}$).

Clearly

$$\mathfrak{a}\cdot\mathfrak{b} = \mathfrak{b}\cdot\mathfrak{a}$$

$$(\mathfrak{a}\cdot\mathfrak{b})\cdot\mathfrak{c} = \mathfrak{a}\cdot(\mathfrak{b}\cdot\mathfrak{c});$$

we may therefore compute with products of ideals as with ordinary products. In particular, it makes sense to speak of the powers $\mathfrak{a}^\varrho$ of an ideal; they are defined by

$$\mathfrak{a}^1 = \mathfrak{a}; \qquad \mathfrak{a}^{\varrho+1} = \mathfrak{a}\cdot\mathfrak{a}^\varrho.$$

If $\mathfrak{a} = (a_1, \ldots, a_n)$ and $\mathfrak{b} = (b_1, \ldots, b_m)$, then it is clear that the product $\mathfrak{a}\mathfrak{b}$ is generated by the products $a_i b_k$. *A basis for the product is therefore obtained by multiplying all the basis elements of one factor with all basis elements of the other.*

In particular, for principal ideals

$$(a)\cdot(b) = (ab),$$

and in this case the definition of product thus coincides with the usual one.

The product $\mathfrak{a}\cdot(b)$ of an arbitrary ideal and a principal ideal consists of all products ab with a in $\mathfrak{a}$. We write for this simply $\mathfrak{a}b$ or $b\mathfrak{a}$.

A further rule is the "distributive law for ideals":

$$\mathfrak{a}\cdot(\mathfrak{b}, \mathfrak{c}) = (\mathfrak{a}\cdot\mathfrak{b}, \mathfrak{a}\cdot\mathfrak{c}). \tag{15.1}$$

The ideal $\mathfrak{a}\cdot(\mathfrak{b}, \mathfrak{c})$ is generated by the products $a(b+c)$ which, since

$$a(b+c) = ab+ac,$$

all lie in $(\mathfrak{a}\cdot\mathfrak{b}, \mathfrak{a}\cdot\mathfrak{c})$; conversely, $(\mathfrak{a}\cdot\mathfrak{b}, \mathfrak{a}\cdot\mathfrak{c})$ is generated by the products ab and the products ac which are all contained in $\mathfrak{a}\cdot(\mathfrak{b}, \mathfrak{c})$.

Rule (15.1) continues to hold if in place of $\mathfrak{b}$, $\mathfrak{c}$ several ideals or even an infinite number occur inside the parentheses.

Since all products ab lie in $\mathfrak{a}$, it follows that

$$\mathfrak{a}\cdot\mathfrak{b} \subseteq \mathfrak{a}$$

and similarly

$$\mathfrak{a}\cdot\mathfrak{b} \subseteq \mathfrak{b}.$$

This implies that

$$\mathfrak{a}\cdot\mathfrak{b} \subseteq [\mathfrak{a}, \mathfrak{b}]$$

or: *the product is divisible by the least common multiple.*

In the ring of integers the product of the least common multiple and the greatest common divisor of two ideals $\mathfrak{a}$, $\mathfrak{b}$ is equal to the product $\mathfrak{a}\mathfrak{b}$. This is not true in arbitrary rings; however,

$$[\mathfrak{a} \cap \mathfrak{b}]\cdot(\mathfrak{a}, \mathfrak{b}) \subseteq \mathfrak{a}\mathfrak{b}. \tag{15.2}$$

Proof:

$$[\mathfrak{a} \cap \mathfrak{b}]\cdot(\mathfrak{a}, \mathfrak{b}) = ([\mathfrak{a} \cap \mathfrak{b}]\cdot\mathfrak{a}, [\mathfrak{a} \cap \mathfrak{b}]\cdot\mathfrak{b}) \subseteq (\mathfrak{b}\cdot\mathfrak{a}, \mathfrak{a}\cdot\mathfrak{b}) = \mathfrak{a}\cdot\mathfrak{b}.$$

The ideal $\mathfrak{o}$ which consists of all elements of the ring under consideration is called the *unit ideal.* Of course,

$$\mathfrak{a}\cdot\mathfrak{o} \subseteq \mathfrak{a}.$$

If $\mathfrak{o}$ has an identity element, then conversely

$$\mathfrak{a} = \mathfrak{a}\cdot e \subseteq \mathfrak{a}\cdot\mathfrak{o},$$

and hence

$$\mathfrak{a}\cdot\mathfrak{o} = \mathfrak{a}.$$

In this case the ideal $\mathfrak{o}$ plays the role of an identity element for multiplication. It is generated by the identity element.

It is always true that

$$(\mathfrak{a}, \mathfrak{o}) = \mathfrak{o}; \qquad \mathfrak{a} \cap \mathfrak{o} = \mathfrak{a}.$$

The *quotient ideal* $\mathfrak{a} : \mathfrak{b}$, where $\mathfrak{a}$ is an ideal, is by definition the set of all elements γ of $\mathfrak{o}$ such that

$$\gamma b \equiv 0(\mathfrak{a}) \qquad \text{for all } b \text{ in } \mathfrak{b}. \tag{15.3}$$

This set is an ideal: if γ and δ both have property (15.3), then $\gamma - \delta$ does also, and if γ has this property, then so does $r\gamma$. It is assumed here that $\mathfrak{a}$ is an ideal; $\mathfrak{b}$ need not be an ideal, but can rather be any set or even a single element.

From the definition

$$\mathfrak{b}\cdot(\mathfrak{a} : \mathfrak{b}) \subseteq \mathfrak{a}.$$

In the ring of integers the quotient of two principal ideals (a), $(b) \neq 0$ is formed by omitting the factors in the number a which also occur in the number b; for example,

$$(12) : (2) = (6)$$

$$(12) : (4) = (3)$$

$$(12) : (8) = (3)$$

$$(12) : (5) = (12).$$

Expressed in another way: a is divided in the usual sense by the greatest common divisor (a, b).

In general rings there is the corresponding rule

$$\mathfrak{a} : \mathfrak{b} = \mathfrak{a} : (\mathfrak{a}, \mathfrak{b}),$$

which is easily proved but is not very important.

It is clear that $\mathfrak{a} \subseteq \mathfrak{a} : \mathfrak{b}$, since every element of $\mathfrak{a}$ has property (15.3). There are thus two extreme cases:

$$\mathfrak{a} : \mathfrak{b} = \mathfrak{o} \quad \text{and} \quad \mathfrak{a} : \mathfrak{b} = \mathfrak{a}.$$

The first case occurs, in particular, if $\mathfrak{b} \subseteq \mathfrak{a}$, for then

$$\gamma\mathfrak{b} \equiv 0(\mathfrak{b}) \equiv 0(\mathfrak{a}).$$

for any γ. The second case means that $\gamma(\mathfrak{b}) \equiv 0(\mathfrak{a})$ implies $\gamma \equiv 0(\mathfrak{a})$. Therefore $\mathfrak{b}$ may be canceled in the congruence $\gamma\mathfrak{b} \equiv 0(\mathfrak{a})$. In this case $\mathfrak{b}$ is said to be *relatively prime* to $\mathfrak{a}$ or simply prime to $\mathfrak{a}$; however, this expression is easily misunderstood, and we shall seldom use it, preferring rather to write the equation $\mathfrak{a} : \mathfrak{b} = \mathfrak{a}$ explicitly. In the case of nonzero integers a and b the criterion

$$\gamma b \equiv 0(a) \quad \text{implies} \quad \gamma \equiv 0(a)$$

is satisfied only if a and b have no common prime factor. In general cases, however, the expression "relatively prime" is *not symmetric*; for example, if $\mathfrak{a}$ is a prime ideal and $\mathfrak{b}$ a proper divisor of $\mathfrak{a}$ distinct from $\mathfrak{o}$, then

$$\mathfrak{a} : \mathfrak{b} = \mathfrak{a} \quad \text{and hence} \quad \mathfrak{b} \text{ is relatively prime to } \mathfrak{a},$$

but

$$\mathfrak{b} : \mathfrak{a} = \mathfrak{o} \quad \text{and hence} \quad \mathfrak{a} \text{ is not relatively prime to } \mathfrak{b}.$$

For example,

$$(0) : (2) = (0)$$

$$(2) : (0) = (1).$$

The following rule is important:

$$[\mathfrak{a}_1, \ldots, \mathfrak{a}_r] : \mathfrak{b} = [\mathfrak{a}_1 : \mathfrak{b}, \ldots, \mathfrak{a}_r : \mathfrak{b}]. \tag{15.4}$$

Proof: From

$$\gamma\mathfrak{b} \subseteq [\mathfrak{a}_1, \ldots, \mathfrak{a}_r]$$

it follows that

$$\gamma\mathfrak{b} \subseteq \mathfrak{a}_i \qquad \text{for every } i,$$

and conversely.

Exercises

15.1. Prove the rules:

$$(\mathfrak{a} : \mathfrak{b}) : \mathfrak{c} = \mathfrak{a} : \mathfrak{b}\mathfrak{c} = (\mathfrak{a} : \mathfrak{c}) : \mathfrak{b}$$

$$\mathfrak{a} : (\mathfrak{b}, \mathfrak{c}) = (\mathfrak{a} : \mathfrak{b}) \cap (\mathfrak{a} : \mathfrak{c}).$$

15.2. Demonstrate the equivalence of the following three assertions:

(*a*) $\mathfrak{a} : \mathfrak{b}_1 = \mathfrak{a}$ and $\mathfrak{a} : \mathfrak{b}_2 = \mathfrak{a}$

(*b*) $\mathfrak{a} : [\mathfrak{b}_1 \cap \mathfrak{b}_2] = \mathfrak{a}$

(*c*) $\mathfrak{a} : \mathfrak{b}_1\mathfrak{b}_2 = \mathfrak{a}$.

15.3 PRIME IDEALS AND PRIMARY IDEALS

We have already defined a prime ideal as an ideal whose residue class ring has no zero divisors.

In the ring of integers every natural number a is a product of powers of distinct prime numbers:

$$a = p_1^{\sigma_1} \cdots p_r^{\sigma_r}, \tag{15.5}$$

and hence every ideal (a) is a product of powers of prime ideals:

$$(a) = (p_1)^{\sigma_1} \cdots (p_r)^{\sigma_r},$$

In general rings we cannot expect the decomposition of the ideals to be so simple. For example, in the ring of integral polynomials in one indeterminate x, the ideal $(4, x)$, which is not prime, has only one prime divisor $(2, x)$ in addition to $\mathfrak{o}$; however, the ideal $(4, x)$ cannot be expressed as a power of $(2, x)$. In general, we cannot therefore expect a product representation of the ideals but rather at most a representation as an l.c.m. (intersection) with simplest possible components[2] corresponding to the representation of (a) as an l.c.m.,

$$(a) = [(p_1^{\sigma_1}), \ldots, (p_r^{\sigma_r})],$$

which follows from (15.5).

The ideals (p^σ) occurring in this representation have the following characteristic property: if a product ab is divisible by p^σ and the factor a is not, then the other factor b must contain at least one factor of p^σ. This may also be expressed by saying that a power b^ϱ must be divisible by p^σ. Thus

$$ab \equiv 0(p^\sigma)$$

and

$$a \not\equiv 0(p^\sigma)$$

imply

$$b^\varrho \equiv 0(p^\sigma).$$

Ideals with this property are called *primary ideals.*

An ideal $\mathfrak{q}$ *is primary if*

$$ab \equiv 0(\mathfrak{q}) \qquad \text{and} \quad a \not\equiv 0(\mathfrak{q})$$

imply that there exists a ϱ *such that*

$$b^\varrho \equiv 0\ (\mathfrak{q}).$$

This definition may also be stated as follows.

If $\bar{a}\bar{b} = 0$ *and* $\bar{a} \neq 0$ *in the residue class ring module* $\mathfrak{q}$, *then a power* $\bar{b}^\varrho$ *must vanish.*

If $\bar{a}\bar{b} = 0$ and $\bar{a} \neq 0$, then this means that $\bar{b}$ is a zero divisor. A ring element b with the property that b^ϱ vanishes is called *nilpotent.* Hence we may also say: *An ideal is primary if in its residue class ring every zero divisor is nilpotent.*

We note that this definition is but a slight modification of the definition of a prime ideal: in the residue class ring modulo a prime ideal every zero divisor is not only nilpotent but is even zero.

We shall see that the primary ideals in general rings play the same role as prime powers in the ring of integers: that is, under very general conditions every ideal can be represented as the intersection of primary ideals, and in this representation the essential structure of the ideals is expressed.

The primary ideals are not necessarily powers of prime ideals; this is shown by the example of the ideal $(4, x)$ considered previously, which is easily seen to be primary. The converse is not true either: in the ring of those integral polynomials $a_0+a_1x+\cdots+a_nx^n$, in which a_1 is divisible by 3 $\mathfrak{p} = (3x, x^2, x^3)$, is a

[2]An l.c.m. representation is in certain cases more useful than a product representation, namely when it is a question of whether or not an element b is divisible by an ideal $\mathfrak{m}$, that is, whether it is contained in $\mathfrak{m}$. If $\mathfrak{m} = [\mathfrak{a}_1, \ldots, \mathfrak{a}_r]$ then b belongs to $\mathfrak{m}$ if and only if it belongs to all the $\mathfrak{a}_\nu$.

prime ideal, but $\mathfrak{p}^2 = (9x^2, 3x^3, x^4, x^5, x^6)$ is not primary, since

$$9\cdot x^2 \equiv 0(\mathfrak{p}^2)$$
$$x^2 \not\equiv 0(\mathfrak{p}^2)$$
$$9^\varrho \not\equiv 0(\mathfrak{p}^2)$$

for every ϱ.

PROPERTIES OF PRIMARY IDEALS INDEPENDENT OF THE ASCENDING CHAIN CONDITION

Theorem 1: *For every primary ideal* $\mathfrak{q}$ *there is a prime ideal divisor which is defined as follows:* $\mathfrak{p}$ *is the set of all elements b such that some power* b^ϱ *lies in* $\mathfrak{q}$.

Proof: First, $\mathfrak{p}$ is an ideal: $b^\varrho \equiv 0(\mathfrak{q})$ implies $(rb)^\varrho \equiv 0(\mathfrak{q})$, and $b^\varrho \equiv 0(\mathfrak{q})$ and $c^\sigma \equiv 0(\mathfrak{q})$ imply

$$(b-c)^{\varrho+\sigma-1} \equiv 0(\mathfrak{q}),$$

since either b^ϱ or c^σ occurs in each summand of the expansion of $(b-c)^{\varrho+\sigma-1}$.

Second, $\mathfrak{p}$ is prime, for

$$ab \equiv 0(\mathfrak{p})$$
$$a \not\equiv 0(\mathfrak{p})$$

imply that there exists a ϱ such that

$$a^\varrho b^\varrho \equiv 0(\mathfrak{q})$$

and further

$$a^\varrho \not\equiv 0(\mathfrak{q}).$$

Hence there exists a σ such that

$$b^{\varrho\sigma} \equiv 0(\mathfrak{q}),$$

and this implies that

$$b \equiv 0(\mathfrak{p}).$$

Third, $\mathfrak{p}$ is a divisor of $\mathfrak{q}$:

$$\mathfrak{q} \equiv 0(\mathfrak{p});$$

for the elements of $\mathfrak{q}$ certainly have the property that a power lies in $\mathfrak{q}$.

Here $\mathfrak{p}$ is called the *associated prime ideal* of $\mathfrak{q}$; $\mathfrak{q}$ is called a primary ideal belonging to $\mathfrak{p}$. From the definition of a primary ideal we obtain the next theorem.

Theorem 2: *If* $ab \equiv 0(\mathfrak{q})$ *and* $a \not\equiv 0(\mathfrak{q})$, *then* $b \equiv 0(\mathfrak{p})$.

In a sense the following is a converse of this theorem.

Theorem 3: *If* $\mathfrak{p}$ *and* $\mathfrak{q}$ *are ideals which have the properties:*

(*a*) $ab \equiv 0(\mathfrak{q})$ *and* $a \not\equiv 0(\mathfrak{q})$ *imply* $b \equiv 0(\mathfrak{p})$

(*b*) $\mathfrak{q} \equiv 0(\mathfrak{p})$

(*c*) $b \equiv 0(\mathfrak{p})$ *implies* $b^\varrho \equiv 0(\mathfrak{q})$,

then $\mathfrak{q}$ *is primary and* $\mathfrak{p}$ *is its associated prime ideal.*

Proof: Here $ab \equiv 0(\mathfrak{q})$ and $a \not\equiv 0(\mathfrak{q})$ imply [by (*a*) and (*b*)] that $b^\varrho \equiv 0(\mathfrak{q})$. Hence $\mathfrak{q}$ is primary. It remains to show that $\mathfrak{p}$ consists of elements b such that a power b^ϱ is contained in $\mathfrak{q}$. The one half of this assertion is just (*c*). It remains then to show that $b^\varrho \equiv 0(\mathfrak{q})$ implies $b \equiv 0(\mathfrak{p})$. Let ϱ be the smallest natural number for which $b^\varrho \equiv 0(\mathfrak{q})$. If $\varrho = 1$, we are done by (*b*). For $\varrho > 1$ we have $b \cdot b^{\varrho-1} \equiv 0(\mathfrak{q})$, but $b^{\varrho-1} \not\equiv 0(\mathfrak{q})$; hence $b \equiv 0(\mathfrak{p})$ by (*a*).

In particular instances this theorem facilitates establishing the primary property and finding the associated prime ideal. It furthermore shows which properties uniquely determine the associated prime ideal.

Theorem 2 also holds if a and b are replaced by ideals $\mathfrak{a}$ and $\mathfrak{b}$.

Theorem 4: $\mathfrak{a}\mathfrak{b} \equiv 0(\mathfrak{q})$ *and* $\mathfrak{a} \not\equiv 0(\mathfrak{q})$ *imply* $\mathfrak{b} \equiv 0(\mathfrak{p})$.

For if $\mathfrak{b} \not\equiv 0(\mathfrak{p})$, then there exists an element b in $\mathfrak{b}$ which is not contained in $\mathfrak{p}$; similarly, there exists an element a in $\mathfrak{a}$ which does not lie in $\mathfrak{q}$. However, the product ab must lie in $\mathfrak{a}\mathfrak{b}$ and therefore in $\mathfrak{q}$. This contradicts the earlier result.

The corresponding theorem for prime ideals is proved in the same manner: $\mathfrak{a}\mathfrak{b} \equiv 0(\mathfrak{p})$ *and* $\mathfrak{a} \not\equiv 0(\mathfrak{p})$ *imply* $\mathfrak{b} \equiv 0(\mathfrak{p})$.

A corollary of this (obtained by applying the result $h-1$ times) is: $\mathfrak{a}^h \equiv 0(\mathfrak{p})$ *implies* $\mathfrak{a} \equiv 0(\mathfrak{p})$.

Another formulation of Theorem 4 is the following:

Theorem 4′: $\mathfrak{b} \equiv 0(\mathfrak{p})$ *implies* $\mathfrak{q} : \mathfrak{b} = \mathfrak{q}$.

The residue class ring $\mathfrak{o}/\mathfrak{q}$ contains the ideal $\mathfrak{p}/\mathfrak{q}$ (since $\mathfrak{p} \supseteqq \mathfrak{q}$). This ideal consists of all nilpotent elements, and hence of all zero divisors in the case $\mathfrak{q} \neq \mathfrak{o}$.

PROPERTIES OF PRIMARY IDEALS ASSUMING THE ASCENDING CHAIN CONDITION

If $\mathfrak{p}$ is the associated prime ideal of $\mathfrak{q}$, then a power of every element of $\mathfrak{p}$ lies in $\mathfrak{q}$. The smallest exponent required depends on the particular element and may increase without bound. However, if the ascending chain condition is assumed to hold in the ring $\mathfrak{o}$, then the exponents can no longer increase without bound.

Theorem 5: *A power* $\mathfrak{p}$ *is divisible by* $\mathfrak{q}$:

$$\mathfrak{p}^\varrho \equiv 0(\mathfrak{q}).$$

Poof: Let $(p_1, \ldots, p_r)$ be a basis for $\mathfrak{p}$, and suppose that $p_1{}^{\varrho_1}, \ldots, p_r{}^{\varrho_r}$ liei rn $\mathfrak{q}$. If we put

$$\varrho = \sum_1^r (\varrho_i - 1) + 1,$$

then $\mathfrak{p}^\varrho$ is generated by all products of the p_i, ϱ at a time; in each such product at least one factor p_i must occur more than $\varrho_i - 1$ times, and hence at least ϱ times. All generators of $\mathfrak{p}^\varrho$ therefore lie in $\mathfrak{q}$, whence the assertion.

The following relations now hold between a primary ideal $\mathfrak{q}$ and its associated prime ideal $\mathfrak{p}$:

$$\mathfrak{q} \equiv 0(\mathfrak{p}) \qquad (15.6)$$
$$\mathfrak{p}^{\varrho} \equiv 0(\mathfrak{q}).$$

The smallest integer ϱ for which these relations hold is called the *exponent* of $\mathfrak{q}$. In particular, the exponent is an upper bound for the exponents of the powers to which elements of $\mathfrak{p}$ must be raised (at least) in order to obtain elements of $\mathfrak{q}$.

If $\mathfrak{q}$ is primary, then equations (15.6) characterize the associated prime ideal $\mathfrak{p}$. For if a second prime ideal $\mathfrak{p}'$ likewise satisfied (15.6) with an exponent ϱ', then

$$\mathfrak{p}^{\varrho} \subseteq \mathfrak{q} \subseteq \mathfrak{p}', \quad \text{so that} \quad \mathfrak{p} \subseteq \mathfrak{p}'$$
$$\mathfrak{p}'^{\varrho'} \subseteq \mathfrak{q} \subseteq \mathfrak{p}, \quad \text{so that} \quad \mathfrak{p}' \subseteq \mathfrak{p},$$

and hence $\mathfrak{p}' = \mathfrak{p}$.

Theorem 6: $\mathfrak{a}\mathfrak{b} \equiv 0(\mathfrak{q})$ *and* $\mathfrak{a} \not\equiv 0(\mathfrak{q})$ *imply a power* $\mathfrak{b}^{\sigma} \equiv 0(\mathfrak{q})$.

Proof: It suffices to choose $\sigma = \varrho$; $\mathfrak{a}\mathfrak{b} \equiv 0(\mathfrak{q})$ and $\mathfrak{a} \not\equiv 0(\mathfrak{q})$ imply as before that $\mathfrak{b} \equiv 0(\mathfrak{p})$, and hence

$$\mathfrak{b}^{\varrho} \equiv 0(\mathfrak{p}^{\varrho}) \equiv 0(\mathfrak{q}).$$

An ideal $\mathfrak{q}$ having the property just considered is said to be *strongly primary* in contrast to the *weakly primary* ideals, or simply primary ideals, first defined. If the ascending chain condition holds, then the two concepts coincide. We have already seen that in this case the primary ideals are strongly primary, and the converse follows by specializing $\mathfrak{a}$ and $\mathfrak{b}$ above to principal ideals (a) and (b). If the ascending chain condition does not hold, then every strongly primary ideal is, to be sure, also weakly primary, but the converse need not hold. See the review of a work by A. Walfisch, "Über Primäre Ideale," in *Math. Rev.*, **5**, 226 (1944).

Exercises

15.3. The ideal $\mathfrak{a} = (x^2, 2x)$ in the ring of integral polynomials in one variable x is not primary. Nevertheless, $(x)^2 \subset \mathfrak{a} \subset (x)$ and (x) is a prime ideal.

15.4. If $\mathfrak{o}$ has an identity element, then $\mathfrak{o}$ is itself the only primary ideal belonging to the prime ideal $\mathfrak{o}$.

15.4 THE GENERAL DECOMPOSITION THEOREM

Henceforth $\mathfrak{o}$ shall be a Noetherian ring. Thus the basis condition, the ascending chain condition, the maximum condition, and the principle of divisor induction all hold in $\mathfrak{o}$.

An ideal $\mathfrak{m}$ is said to be *reducible* if it can be represented as the intersection of two proper divisors:

$$\mathfrak{m} = \mathfrak{a} \cap \mathfrak{b}, \quad \mathfrak{a} \supset \mathfrak{m}, \quad \mathfrak{b} \supset \mathfrak{m}.$$

If no such representation is possible, the ideal is said to be *irreducible.*

Prime ideals are examples of irreducible ideals, for if a prime ideal $\mathfrak{p}$ had a representation

$$\mathfrak{p} = \mathfrak{a} \cap \mathfrak{b}, \quad \mathfrak{a} \supset \mathfrak{p}, \quad \mathfrak{b} \supset \mathfrak{p},$$

then

$$\mathfrak{a}\mathfrak{b} \equiv 0(\mathfrak{a} \cap \mathfrak{b}) \equiv 0(\mathfrak{p}), \qquad \mathfrak{a} \not\equiv 0(\mathfrak{p}), \qquad \mathfrak{b} \not\equiv 0(\mathfrak{p}),$$

which is contrary to the fact that $\mathfrak{p}$ is prime.

The ascending chain condition now implies the following theorem.

First Decomposition Theorem: *Every ideal is the intersection of finitely many irreducible ideals.*

Proof: The theorem is true for irreducible ideals. Suppose then that $\mathfrak{m}$ is reducible:

$$\mathfrak{m} = \mathfrak{a} \cap \mathfrak{b}, \quad \mathfrak{a} \supset \mathfrak{m}, \quad \mathfrak{b} \supset \mathfrak{m}.$$

If the theorem is assumed to be true for all proper divisors of $\mathfrak{m}$, then it is true in particular for $\mathfrak{a}$ and $\mathfrak{b}$; thus

$$\mathfrak{a} = [\mathfrak{i}_1, \ldots, \mathfrak{i}_s]$$

$$\mathfrak{b} = [\mathfrak{i}_{s+1}, \ldots, \mathfrak{i}_r].$$

From this it follows, however, that

$$\mathfrak{m} = [\mathfrak{i}_1, \ldots, \mathfrak{i}_s, \mathfrak{i}_{s+1}, \ldots, \mathfrak{i}_r],$$

and hence the theorem is also true for $\mathfrak{m}$. Since it is true for the unit ideal (which is always irreducible), it is true in general by the "principle of divisor induction."

We proceed from the representation in terms of irreducible ideals to a representation in terms of primary ideals by way of the following theorem.

Theorem: *Every irreducible ideal is primary.*

Proof: Suppose that $\mathfrak{m}$ is not primary; it will be shown that $\mathfrak{m}$ is reducible.

Since $\mathfrak{m}$ is not primary, there exist two elements a and b with the properties

$$ab \equiv 0(\mathfrak{m})$$

$$b \not\equiv 0(\mathfrak{m})$$

$$b^{\varrho} \not\equiv 0(\mathfrak{m}) \qquad \text{for any } \varrho.$$

By the ascending chain condition, the sequence of quotient ideals

$$\mathfrak{m} : b, \mathfrak{m} : b^2, \ldots$$

must eventually terminate; that is, there exists a k such that

$$\mathfrak{m} : b^k = \mathfrak{m} : b^{k+1}.$$

We now assert that

$$\mathfrak{m} = (\mathfrak{m},a) \cap (\mathfrak{m},\mathfrak{o}b^k). \tag{15.7}$$

Both ideals on the right-hand side are divisors of $\mathfrak{m}$; indeed, they are even proper divisors, since the first contains a and the second contains b^{k+1}. We have to show that any element common to both these ideals belongs to $\mathfrak{m}$. Such an element c, as an element of $(\mathfrak{m}, \mathfrak{o}b^k)$, has the form

$$c = m + rb^k;$$

as an element of $(\mathfrak{m}, a)$, it has the property

$$cb \equiv 0(\mathfrak{m}b, ab) \equiv 0(\mathfrak{m}).$$

This implies

$$mb + rb^{k+1} = cb \equiv 0(\mathfrak{m})$$
$$rb^{k+1} \equiv 0(\mathfrak{m})$$

and hence, since $\mathfrak{m} : b^{k+1} = \mathfrak{m} : b^k$,

$$rb^k \equiv 0(\mathfrak{m})$$
$$c = m + rb^k \equiv 0(\mathfrak{m}).$$

This completes the proof of (15.7); $\mathfrak{m}$ is therefore reducible.

Since every ideal can be represented as the intersection of finitely many irreducible ideals and since every irreducible ideal is primary, we have found the following.

Theorem: *Every ideal can be represented as the intersection of a finite number of primary ideals.*

This theorem can be made still sharper. First, all redundant ideals of $\mathfrak{q}_i$ of a representation

$$\mathfrak{m} = [\mathfrak{q}_1, \ldots, \mathfrak{q}_r],$$

meaning all those $\mathfrak{q}_i$ which contain the intersection of the other ideals, can be omitted. We thus arrive at an *irredundant* representation,[3] that is, one in which no component $\mathfrak{q}_i$ contains the intersection of the remaining ideals. In such a representation it is still possible that several of the primary components might be combined to form a primary ideal, that is, that their intersection is again a primary ideal. The following theorems indicate when this is the case.

Theorem 1: *The intersection of finitely many primary ideals belonging to the same prime ideal is again primary and has the same associated prime ideal.*

Theorem 2: *An irredundant intersection of finitely many primary ideals not all belonging to the same prime ideal is not primary.*

The validity of these theorems does not depend on the ascending chain condition.

Proof of Theorem 1: Let

$$\mathfrak{m} = [\mathfrak{q}_1, \ldots, \mathfrak{q}_r],$$

where $\mathfrak{q}_1, \ldots, \mathfrak{q}_r$ all belong to $\mathfrak{p}$. We use Theorem 3 (Section 15.3). From

$$ab \equiv 0(\mathfrak{m}), \qquad a \not\equiv 0(\mathfrak{m})$$

[3]Some authors call this a *reduced (primary)* representation and require, in addition, that the $\mathfrak{q}_i$ have distinct associated prime ideals (*Trans.*).

it follows that

$$ab \equiv 0(\mathfrak{q}_\nu)$$

for all ν and

$$a \not\equiv 0(\mathfrak{q}_\nu)$$

for at least one ν, and hence that $b \equiv 0(\mathfrak{p})$.

It is furthermore clear that

$$\mathfrak{m} \equiv 0(\mathfrak{q}_\nu) \equiv 0(\mathfrak{p}).$$

If finally $b \equiv 0(\mathfrak{p})$, then it follows that

$$b^{\varrho_\nu} \equiv 0(\mathfrak{q}_\nu) \qquad \text{for all } \nu,$$

and hence

$$b^{\varrho} \equiv 0(\mathfrak{q}_\nu) \qquad \text{for all } \nu$$

$$b^{\varrho} \equiv 0(\mathfrak{m}),$$

where $\varrho = \max \varrho_\nu$. All three properties listed in Theorem 3 (Section 15.3) have herewith been established. Hence $\mathfrak{m}$ is primary, and $\mathfrak{p}$ is its associated prime ideal.

Proof of Theorem 2: Let

$$\mathfrak{m} = [\mathfrak{q}_1, \ldots, \mathfrak{q}_r] \qquad (r \geqq 2)$$

be an irredundant representation in which at least two of the associated prime ideals $\mathfrak{p}_\nu$ are distinct. We suppose at the outset that any group of primary ideals with the same associated prime ideal has been combined to a single primary ideal. The representation remains irredundant.

Among the finitely many prime ideals $\mathfrak{p}_\nu$ there is a minimal one, that is, one which does not contain any of the others. Let this ideal be $\mathfrak{p}_1$. Since $\mathfrak{p}_1$ does not contain the ideals $\mathfrak{p}_2, \ldots, \mathfrak{p}_r$, there exist elements a such that

$$\left.\begin{aligned} a_\nu &\not\equiv 0(\mathfrak{p}_1) \\ a_\nu &\equiv 0(\mathfrak{p}_\nu) \end{aligned}\right\} \qquad (\nu = 2, 3, \ldots, r),$$

and hence, for sufficiently large ϱ,

$$a_\nu{}^{\varrho} \equiv 0(\mathfrak{q}_\nu).$$

If $\mathfrak{q}_1 = \mathfrak{m}$, then the representation $\mathfrak{m} = [\mathfrak{q}_1, \ldots, \mathfrak{q}_r]$ would be reducible (indeed, $\mathfrak{q}_2, \ldots, \mathfrak{q}_r$ would be redundant). Therefore $\mathfrak{q}_1$ contains an element q_1 such that

$$q_1 \not\equiv 0(\mathfrak{m}).$$

The product

$$q_1(a_2 \cdots a_r)^{\varrho}$$

is contained both in $\mathfrak{q}_1$ and in $\mathfrak{q}_2, \ldots, \mathfrak{q}_r$ and hence in $\mathfrak{m}$. However, q_1 is not in $\mathfrak{m}$. If $\mathfrak{m}$ were primary, this would imply that

$$(a_2 \cdots a_r)^{\varrho\sigma} \equiv 0(\mathfrak{m})$$

$$(a_2 \cdots a_r)^{\varrho\sigma} \equiv 0(\mathfrak{p}_1),$$

and hence, since $\mathfrak{p}_1$ is prime, that

$$a_\nu \equiv 0(\mathfrak{p}_1)$$

for at least one ν, contrary to the choice of the a_ν.

If in an irredundant representation

$$\mathfrak{m} = [\mathfrak{q}_1, \ldots, \mathfrak{q}_r]$$

all the associated prime ideals $\mathfrak{p}_\nu$ are distinct, so that it is not possible to combine two or more ideals of the representation to a single primary ideal, then the representation is called a *representation by greatest primary ideals.* These greatest primary ideals are also called the *primary components* of $\mathfrak{m}$.

Any irredundant representation $\mathfrak{m} = [\mathfrak{q}_1, \ldots, \mathfrak{q}_r]$ can be transformed into a representation by greatest primary ideals by combining those primary ideals belonging to the same prime ideal. This completes the proof of the following theorem.

Second Decomposition Theorem: *Every ideal admits an irredundant representation as the intersection of finitely many primary components. These primary components all have distinct associated prime ideals.*

This second decomposition theorem, proved for polynomial rings by E. Lasker and in general by E. Noether, is the most important result of general ideal theory. We shall learn some applications of this theorem in Chapter 16. In the sections immediately following we shall investigate what may be said about the uniqueness of the primary components.

Exercises

15.5. Decompose the ideal $(9, 3x+3)$ in the ring of integral polynomials in one indeterminate into primary components.

15.6. For every ideal $\mathfrak{a}$ there is a product of powers of prime ideals $\mathfrak{p}_1^{\varrho_1} \cdot \mathfrak{p}_2^{\varrho_2} \cdots \mathfrak{p}_h^{\varrho_h}$ which is divisible by $\mathfrak{a}$ and such that each $\mathfrak{p}_\nu$ is a divisor of $\mathfrak{a}$.

15.7. If the ring $\mathfrak{o}$ has an identity element, then every ideal $\mathfrak{a}$ distinct from $\mathfrak{o}$ is divisible by at least one prime ideal.

15.8. The ideal $(4, 2x, x^2)$ in the ring of integral polynomials in one indeterminate is primary but reducible. [Decomposition: $(4, 2x, x^2) = (4, x) \cap (2, x^2)$.]

15.5 THE FIRST UNIQUENESS THEOREM

The decomposition of an ideal into primary components is not unique.

Example: The ideal

$$\mathfrak{m} = (x^2, xy)$$

in the polynomial ring $K[x, y]$ consists of all polynomial which are divisible by x and in which the linear terms are absent. The set of all polynomials divisibly by (x) is the prime ideal

$$\mathfrak{q}_1 = (x);$$

the set of all polynomials in which the linear and constant terms are absent is the primary ideal

$$\mathfrak{q}_2 = (x^2, xy, y^2).$$

Hence

$$\mathfrak{m} = [\mathfrak{q}_1, \mathfrak{q}_2].$$

This is an irredundant representation, and the associated prime ideals, (x) and (x, y), of $\mathfrak{q}_1$ and $\mathfrak{q}_2$ are distinct; this is therefore also a representation by greatest primary ideals. But in addition to this representation there is still another:

$$\mathfrak{m} = (\mathfrak{q}_1, \mathfrak{q}_3],$$

where

$$\mathfrak{q}_3 = (x^2, y),$$

for in order that a polynomial lie in $\mathfrak{m}$, it is sufficient to require that the polynomial be divisible by x and that it contain no linear term. If the field K is infinite, then there are even an infinite number of representations of this type:

$$\mathfrak{m} = [\mathfrak{q}_1, \mathfrak{q}^{(\lambda)}], \qquad \mathfrak{q}^{(\lambda)} = (x^2, y+\lambda x).$$

All these decompositions of $\mathfrak{m}$ have the common feature that the *number* of primary components and the associated prime ideals,

$$(x), (x, y),$$

are the same. This is true in general.

First Uniqueness Theorem: *In two irredundant representations of an ideal* $\mathfrak{m}$ *by primary components the number of components and the associated prime ideals are the same* (*although the components themselves need not be*).

Proof: For a primary ideal the assertion is trivial. We can therefore begin an induction on the number of primary components which occur in at least one representation.

Let

$$\mathfrak{m} = [\mathfrak{q}_1, \ldots, \mathfrak{q}_l] = [\mathfrak{q}'_1, \ldots, \mathfrak{q}'_{l'}]. \tag{15.8}$$

From among all the associated prime ideals $\mathfrak{p}_1, \ldots, \mathfrak{p}_l, \mathfrak{p}'_1, \ldots, \mathfrak{p}'_{l'}$, we choose a maximal one, that is, one which is contained in none of the others. Suppose that this is $\mathfrak{p}_1$, so that it belongs to the left side of (15.8). We now assert that it also belongs to the right side. For otherwise we could form quotients with respect to $\mathfrak{q}_1$ in (15.8):

$$[\mathfrak{q}_1 : \mathfrak{q}_1, \ldots, \mathfrak{q}_l : \mathfrak{q}_1] = [\mathfrak{q}'_1 : \mathfrak{q}_1, \ldots, \mathfrak{q}'_{l'} : \mathfrak{q}_1].$$

Now (for all $\nu > 1$) $\mathfrak{q}_1 \not\equiv 0(\mathfrak{p}_\nu)$, since otherwise $\mathfrak{p}_1 \equiv 0(\mathfrak{p}_\nu)$, contrary to the assumed maximality of $\mathfrak{p}_1$. Similarly, $\mathfrak{q}_1 \not\equiv 0(\mathfrak{p}'_\nu)$ for all ν. By Theorem 4′ (Section 15.3), therefore,

$$\mathfrak{q}_\nu : \mathfrak{q}_1 = \mathfrak{q}_\nu \qquad (\nu = 2, \dots, l)$$

$$\mathfrak{q}'_\nu : \mathfrak{q}_1 = \mathfrak{q}'_\nu \qquad (\nu = 1, \dots, l').$$

Since, furthermore, $\mathfrak{q}_1 : \mathfrak{q}_1 = \mathfrak{o}$, it follows that

$$[\mathfrak{o}, \mathfrak{q}_2, \dots, \mathfrak{q}_l] = [\mathfrak{q}'_1, \dots, \mathfrak{q}'_{l'}].$$

The right side is equal to $\mathfrak{m}$ and so the left side is also. The $\mathfrak{o}$ may be omitted, and hence

$$\mathfrak{m} = [\mathfrak{q}_2, \dots, \mathfrak{q}_l].$$

The first of the two representations of (15.8) would thus be reducible, contrary to hypothesis.

Every maximal prime ideal therefore belongs to *both* sides.

Suppose now that $l \leqq l'$; we wish to show that $l = l'$ and that (for an appropriate ordering) $\mathfrak{p}'_\nu = \mathfrak{p}_\nu$. Suppose that this has all been proved for ideals which can be represented by fewer than l primary components. We order the $\mathfrak{q}$ and $\mathfrak{q}'$ so that $\mathfrak{p}_1 = \mathfrak{p}'_1$ is a maximal associated prime ideal (belonging to both $\mathfrak{q}_1$ and $\mathfrak{q}'_1$).

If we form quotients with respect to the product $\mathfrak{q}_1\mathfrak{q}'_1$ on both sides of (15.8),

$$[\mathfrak{q}_1 : \mathfrak{q}_1\mathfrak{q}'_1, \dots, \mathfrak{q}_l : \mathfrak{q}_1\mathfrak{q}'_1] = [\mathfrak{q}'_1 : \mathfrak{q}_1\mathfrak{q}'_1, \dots, \mathfrak{q}'_{l'} : \mathfrak{q}_1\mathfrak{q}'_1],$$

then, by the same argument as before

$$\left.\begin{aligned} \mathfrak{q}_\nu : \mathfrak{q}_1\mathfrak{q}'_1 &= \mathfrak{q}_\nu \\ \mathfrak{q}'_\nu : \mathfrak{q}_1\mathfrak{q}'_1 &= \mathfrak{q}'_\nu \end{aligned}\right\} \qquad (\nu > 1).$$

Furthermore, since $\mathfrak{q}_1\mathfrak{q}'_1$ is divisible by both $\mathfrak{q}_1$ and $\mathfrak{q}'_1$,

$$\mathfrak{q}_1 : \mathfrak{q}_1\mathfrak{q}'_1 = \mathfrak{o}$$

$$\mathfrak{q}'_1 : \mathfrak{q}_1\mathfrak{q}'_1 = \mathfrak{o};$$

and hence

$$[\mathfrak{q}_2, \dots, \mathfrak{q}_l] = [\mathfrak{q}'_2, \dots, \mathfrak{q}'_{l'}].$$

Since now an irredundant representation by primary components occurs on the right and left, it follows by the induction hypothesis that $l' - 1 = l - 1$ and hence $l = l'$. In addition, $\mathfrak{p}_\nu = \mathfrak{p}'_\nu$ in an appropriate ordering for all $\nu > 1$. Since $\mathfrak{p}_1 = \mathfrak{p}'_1$, the proof has herewith been completed.

The uniquely determined ideals $\mathfrak{p}_1, \dots, \mathfrak{p}_l$ which occur as the associated prime ideals for an irredundant representation $\mathfrak{a} = [\mathfrak{q}_1, \dots, \mathfrak{q}_l]$ are called the *associated prime ideals of the ideal* $\mathfrak{a}$. Their most important property is the following.

If an ideal $\mathfrak{a}$ *is not divisible by any associated prime ideal of an ideal* $\mathfrak{b}$, *then* $\mathfrak{b} : \mathfrak{a} = \mathfrak{b}$, *and conversely.*

Proof: Let $\mathfrak{b} = [\mathfrak{q}_1, \ldots, \mathfrak{q}_l]$ be an irredundant representation. First suppose that $\mathfrak{a} \not\equiv 0(\mathfrak{p}_i)$ for $i = 1, \ldots, l$, where $\mathfrak{p}_i$ belongs to $\mathfrak{q}_i$. This implies

$$\mathfrak{q}_i : \mathfrak{a} = \mathfrak{q}_i$$

$$\begin{aligned} \mathfrak{b} : \mathfrak{a} &= [\mathfrak{q}_1, \ldots, \mathfrak{q}_l] : \mathfrak{a} \\ &= [\mathfrak{q}_1 : \mathfrak{a}, \ldots, \mathfrak{q}_l : \mathfrak{a}] \\ &= [\mathfrak{q}_1, \ldots, \mathfrak{q}_l] = b. \end{aligned}$$

Conversely, suppose that $\mathfrak{b} : \mathfrak{a} = \mathfrak{b}$. If $\mathfrak{a} \equiv 0(\mathfrak{p}_i)$ for some i, say $\mathfrak{a} \equiv 0(\mathfrak{p}_1)$, then it would follow that $\mathfrak{a}^\varrho \equiv 0(\mathfrak{q}_1)$, and hence

$$\mathfrak{a}^\varrho \cdot [\mathfrak{q}_2, \ldots, \mathfrak{q}_l] \equiv 0([\mathfrak{q}_1, \mathfrak{q}_2, \ldots, \mathfrak{q}_l]) \equiv 0(\mathfrak{b}).$$

Therefore, since $\mathfrak{a}$ and hence $\mathfrak{a}^\varrho$ may be canceled in any congruence (mod $\mathfrak{b}$),

$$[\mathfrak{q}_2, \ldots, \mathfrak{q}_l] \equiv 0(\mathfrak{b}),$$

contrary to the assumption that the representation is irredundant.

An important special case arises if $\mathfrak{a}$ is specialized to a principal ideal (a).

If an element a is divisible by no associated prime ideal of an ideal $\mathfrak{b}$, then $\mathfrak{b} : \mathfrak{a} = \mathfrak{b}$; that is, $ac \equiv 0(\mathfrak{b})$ always implies $c \equiv 0(\mathfrak{b})$.

The general theorem can be formulated in still another manner if $\mathfrak{a}$ is represented as the intersection of primary ideals $[\mathfrak{q}_1', \ldots, \mathfrak{q}_{l'}']$. Here $\mathfrak{a}$ is divisible by $\mathfrak{p}_i$ if and only if some $\mathfrak{q}_j'$ is, or, what amounts to the same thing, if some $\mathfrak{p}_j'$ is divisible by $\mathfrak{p}_i$. We thus have the following.

If no associated prime ideal of $\mathfrak{a}$ is divisible by an associated prime ideal of $\mathfrak{b}$, then $\mathfrak{b} : \mathfrak{a} = \mathfrak{b}$, and conversely.

15.6 ISOLATED COMPONENTS AND SYMBOLIC POWERS

In a commutative ring $\mathfrak{o}$ let $\mathfrak{S}$ be a nonempty set which with any two elements s and t also contains their product st. Such a set $\mathfrak{S}$ is said to be *multiplicatively closed.*

Let now $\mathfrak{m}$ be an ideal in $\mathfrak{o}$. Now $\mathfrak{m}_S$ is defined to be the set of all elements x of $\mathfrak{o}$ such that sx lies in $\mathfrak{m}$ for some s of S.

Here $\mathfrak{m}_S$ is an ideal (and is, of course, a divisor of $\mathfrak{m}$). Indeed, if x and y belong to $\mathfrak{m}_S$, then sx and $s'y$ belong to $\mathfrak{m}$; therefore

$$ss'(x-y) = s'(sx) - s(s'y)$$

also belongs to $\mathfrak{m}$, and hence $x-y$ belongs to $\mathfrak{m}_S$. If x belongs to $\mathfrak{m}_S$, then rx also belongs to $\mathfrak{m}_S$. It is clear that all elements of $\mathfrak{m}$ belong to $\mathfrak{m}_S$.

Then $\mathfrak{m}_S$ is called *the S-component of* $\mathfrak{m}$ or the *isolated component of* $\mathfrak{m}$ *determined by S.*

It will henceforth be assumed that $\mathfrak{o}$ is a Noetherian ring. If the ideal $\mathfrak{m}$ is represented as the intersection of primary ideals,

$$\mathfrak{m} = [\mathfrak{q}_1, \ldots, \mathfrak{q}_r], \tag{15.9}$$

then the primary ideals $\mathfrak{q}_i$ can be separated into those which intersect S, that is, those which have at least one element in common with S, and those which do not intersect S. If a $\mathfrak{q}_i$ contains an element s of S, then the associated prime ideal $\mathfrak{p}_i$ contains the same element s of S. Conversely, if $\mathfrak{p}_i$ has an element s in common with S, then $\mathfrak{q}_i$ has a power s^ϱ in common with S.

We now enumerate the $\mathfrak{q}_1$ so that $\mathfrak{q}_1, \ldots, \mathfrak{q}_h$ do not intersect the set S, while $\mathfrak{q}_{h+1}, \ldots, \mathfrak{q}_r$ do intersect S. We assert that

$$\mathfrak{m}_S = [\mathfrak{q}_1, \ldots, \mathfrak{q}_h]. \tag{15.10}$$

If $h = 0$, then (15.10) means simply that $\mathfrak{m}_S = \mathfrak{o}$.

Proof: If x is an element of $\mathfrak{m}_S$, and thus sx belongs to $\mathfrak{m}$, then, for $1 \leqq i \leqq h$,

$$sx \equiv 0(\mathfrak{q}_i), \quad s \not\equiv 0(\mathfrak{p}_i), \qquad \text{hence} \quad x \equiv 0(\mathfrak{q}_i),$$

that is, x belongs to $[\mathfrak{q}_1, \ldots, \mathfrak{q}_h]$. Conversely, if x belongs to $[\mathfrak{q}_1, \ldots, \mathfrak{q}_h]$ then if $r > h$ we can select an s_i in S for each i between $h+1$ and r which is contained in $\mathfrak{q}_i$. We now put

$$s = s_{h+1} \cdots s_r.$$

In the case $r = h$ we choose an arbitrary s in S. In both cases sx lies in all the $\mathfrak{q}_i$, that is, sx lies in $\mathfrak{m}$ and hence x belongs to $\mathfrak{m}_S$.

A primary component $\mathfrak{q}_i$ of $\mathfrak{m}$ is called *imbedded* if the associated prime ideal $\mathfrak{p}_i$ is a divisor of another associated prime ideal $\mathfrak{p}_j$ of $\mathfrak{m}$; a primary component is said to be *isolated* if this is not the case. In the first case the associated prime ideal $\mathfrak{p}_i$ is also said to be *imbedded* (imbedded in $\mathfrak{p}_j$); in the second case it is called *isolated.* Similarly, a subset $\{\mathfrak{q}_a, \mathfrak{q}_b, \ldots\}$ or $\{\mathfrak{p}_a, \mathfrak{p}_b, \ldots\}$ of the set of all $\mathfrak{q}_i$ or $\mathfrak{p}_i$ is said to be *isolated* if no $\mathfrak{p}_i$ of the subset is a divisor of a $\mathfrak{p}_j$ not belonging to the subset.

For given $\mathfrak{m} = [\mathfrak{q}_1, \ldots, \mathfrak{q}_r]$ there is associated with every set S closed under multiplication an isolated subset $\{\mathfrak{p}_1, \ldots, \mathfrak{p}_h\}$ consisting of those $\mathfrak{p}_i$ which contain no element of S. This subset is isolated, for if $\mathfrak{p}_i$ belongs to the subset and is a divisor of $\mathfrak{p}_j$, then $\mathfrak{p}_j$ also belongs to the subset. The intersection of the primary ideals $\mathfrak{q}_1, \ldots, \mathfrak{q}_h$ belonging to $\mathfrak{p}_1, \ldots, \mathfrak{p}_h$ is then the isolated component $\mathfrak{a}_S$.

An important special case arises if an isolated $\mathfrak{p}_i$ is selected, and S is taken to be the set of elements of $\mathfrak{o}$ not contained in $\mathfrak{p}_i$. This set is nonempty except in the trivial case $\mathfrak{m} = \mathfrak{o}$. Every other $\mathfrak{p}_j$ contains an element not divisible by $\mathfrak{p}_i$, that is, an element of S. Hence, it follows from (15.10) that

$$\mathfrak{m}_S = \mathfrak{q}_i.$$

Now $\mathfrak{m}_S$ is uniquely determined by $\mathfrak{m}$ and $\mathfrak{S}$ and hence by $\mathfrak{m}$ and $\mathfrak{p}_i$. The isolated $\mathfrak{p}_i$ are also uniquely determined by $\mathfrak{m}$. We thus have the following.

The isolated primary components $\mathfrak{q}_i$ *in* (15.9) *are uniquely determined.*

Exercise

15.9. Using the same method, prove the *Second Uniqueness Theorem*: the intersection $[\mathfrak{q}_a, \mathfrak{q}_b, \ldots]$ of an isolated set of primary components of an ideal $\mathfrak{m}$ is uniquely determined by the associated prime ideals $\mathfrak{p}_a, \mathfrak{p}_b, \ldots$.

SYMBOLIC POWERS

We saw in Section 15.3 that the powers $\mathfrak{p}^r$ of a prime ideal $\mathfrak{p}$ are not necessarily primary. If $\mathfrak{p}^r$ is represented as an intersection of primary components,

$$\mathfrak{p}^r = [\mathfrak{q}_1, \ldots, \mathfrak{q}_s],$$

then all the associated prime ideals $\mathfrak{p}_1, \ldots, \mathfrak{p}_s$ are divisors of $\mathfrak{p}^r$ and hence of $\mathfrak{p}$. If we form the product $\mathfrak{p}_1, \ldots \mathfrak{p}_s$, then a power of this product is divisible by all the $\mathfrak{q}_i$; it is therefore divisibly by $\mathfrak{p}^r$ and hence by $\mathfrak{p}$. Thus, one of the factors, say $\mathfrak{p}_1$, must be divisible by $\mathfrak{p}$. On the other hand, $\mathfrak{p}_1$ is a divisor of $\mathfrak{p}$; therefore, $\mathfrak{p}_1 = \mathfrak{p}$.

The other $\mathfrak{p}_i$ $(i \neq 1)$ are proper divisors of $\mathfrak{p}$. From this it follows that $\mathfrak{q}_1$ is an isolated primary component of $\mathfrak{p}^r$ and as such is uniquely determined. More precisely, $\mathfrak{q}_1$ is the isolated component $\mathfrak{p}_S{}^r$ of $\mathfrak{p}^r$ determined by S, where S is the set of elements of $\mathfrak{o}$ not divisible by $\mathfrak{p}$.

This uniquely defined primary component of $\mathfrak{p}^r$ belonging to the prime ideal $\mathfrak{p}_1 = \mathfrak{p}$ is called, following Krull, the *symbolic rth power* of $\mathfrak{p}$ and is denoted by $\mathfrak{p}^{(r)}$.

15.7 THEORY OF RELATIVELY PRIME IDEALS

In the following, the ring $\mathfrak{o}$ is assumed to have an identity element. This identity element then generates the unit ideal $\mathfrak{o}$:

$$\mathfrak{o} = (1).$$

Two ideals $\mathfrak{a}$ and $\mathfrak{b}$ are said to be *relatively prime* if they have no common divisor except $\mathfrak{o}$; their greatest common divisor is then $\mathfrak{o}$:

$$(\mathfrak{a}, \mathfrak{b}) = \mathfrak{o}.$$

This means that every element of $\mathfrak{o}$ can be represented as the sum of an element of $\mathfrak{a}$ and an element of $\mathfrak{b}$.

For this it is necessary and sufficient that the identity element (the generator of $\mathfrak{o}$) admit a representation as a sum

$$1 = a+b \tag{15.11}$$

(a in $\mathfrak{a}$, b in $\mathfrak{b}$). It follows then that

$$a \equiv 1(\mathfrak{b}), \qquad b \equiv 0(\mathfrak{b}) \tag{15.12}$$

$$a \equiv 0(\mathfrak{a}), \qquad b \equiv 1(\mathfrak{a}).$$

If two primary ideals $\mathfrak{q}_1$ and $\mathfrak{q}_2$ are relatively prime, then their associated prime ideals $\mathfrak{p}_1$ and $\mathfrak{p}_2$ are also relatively prime (for any common divisor of $\mathfrak{p}_1$ and $\mathfrak{p}_2$ is also a common divisor of $\mathfrak{q}_1$ and $\mathfrak{q}_2$). The converse is also true: *if* $\mathfrak{p}_1$ *and* $\mathfrak{p}_2$ *are relatively prime, then* $\mathfrak{q}_1$ *and* $\mathfrak{q}_2$ *are also.* For

$$1 = p_1+p_2$$

implies, on raising both sides to the $(\varrho+\sigma-1)$th power,

$$1 = p_1^{\varrho+\sigma-1}+\cdots+p_2^{\varrho+\sigma-1};$$

choosing now ϱ and σ so large that $p_1{}^\varrho$ lies in $\mathfrak{q}_1$ and $p_2\sigma$ in $\mathfrak{q}_2$, it follows that each term of the sum on the right side is in either $\mathfrak{q}_1$ or $\mathfrak{q}_2$, and hence

$$1 = q_1+q_2.$$

Theorem: *If two ideals* $\mathfrak{a}$ *and* $\mathfrak{b}$ *are relatively prime, then* $\mathfrak{a} : \mathfrak{b} = \mathfrak{a}$ *and* $\mathfrak{b} : \mathfrak{a} = \mathfrak{b}$.
Proof: $(\mathfrak{a}, \mathfrak{b}) = \mathfrak{o}$ and $a+b = 1$. It suffices to show that $\mathfrak{a} : \mathfrak{b} \subseteq \mathfrak{a}$. If x belongs to $\mathfrak{a} : \mathfrak{b}$, then $x\mathfrak{b} \subseteq \mathfrak{a}$. Hence $xb \equiv 0(\mathfrak{a})$, so that

$$x(a+b) \equiv 0(\mathfrak{a})$$

$$x \cdot 1 \equiv 0(\mathfrak{a});$$

x therefore belongs to $\mathfrak{a}$, Q.E.D.

The converse is not true as may be seen by the following example in the polynomial ring $K[x, y]$: the ideals (x) and (y) are prime to each other, but they are not relatively prime:

$$(x, y) \neq \mathfrak{o}$$

$$(x) : (y) = (x)$$

$$(y) : (x) = (y).$$

If $\mathfrak{a}$ and $\mathfrak{b}$ are relatively prime, then congruences can be solved as in number theory. Let two congruences

$$f(\xi) \equiv 0(\mathfrak{a})$$

$$g(\xi) \equiv 0(\mathfrak{b}) \qquad (f(x), g(x) \in \mathfrak{o}[x])$$

be given. We assume that each individual congruence is solvable. Let $\xi \equiv \alpha$ b ae solution of the first and $\xi \equiv \beta$ be a solution of the second congruence. We obtain an element ξ which solves both congruences in the following manner. Using the elements a and b which satisfy both (15.11) and (15.12), we form

$$\xi \doteq b\alpha + a\beta.$$

Then $\xi \equiv \alpha(\mathfrak{a})$ and $\xi \equiv \beta(\mathfrak{b})$; ξ is therefore a solution of both congruences.

Theorem: *In the case of two relatively prime ideals the least common multiple is equal to the product.*

Proof: In Section 15.2 it was shown that

$$\mathfrak{a}\mathfrak{b} \subseteq \mathfrak{a} \cap \mathfrak{b}$$
$$[\mathfrak{a} \cap \mathfrak{b}] \cdot (\mathfrak{a}, \mathfrak{b}) \subseteq \mathfrak{a}\mathfrak{b}.$$

If now $(\mathfrak{a}, \mathfrak{b}) = \mathfrak{o}$ and an identity element is present, then the second equation simplifies to

$$\mathfrak{a} \cap \mathfrak{b} \subseteq \mathfrak{a}\mathfrak{b};$$

therefore

$$\mathfrak{a} \cap \mathfrak{b} = \mathfrak{a}\mathfrak{b}, \qquad \text{Q.E.D.}$$

In order to formulate this theorem for more than two pairwise relatively prime ideals, we must first prove a lemma.

Lemma: *If* $\mathfrak{a}$ *is relatively prime to* $\mathfrak{b}$ *and to* $\mathfrak{c}$, *then* $\mathfrak{a}$ *is also relatively prime to both the product* $\mathfrak{b}\mathfrak{c}$ *and the intersection* $\mathfrak{b} \cap \mathfrak{c}$.

Proof: From

$$a + b = 1$$
$$a' + c = 1$$

it follows that

$$(a+b)(a'+c) = 1$$
$$aa' + ac + a'b + bc = 1$$
$$a'' + bc = 1,$$

where $a'' = aa' + ac + a'b$ is again an element of $\mathfrak{a}$. This implies

$$(\mathfrak{a}, \mathfrak{b}\,\mathfrak{c}) = \mathfrak{o}$$

and, *a fortiori*,

$$(\mathfrak{a}, \mathfrak{b} \cap \mathfrak{c}) = \mathfrak{o}.$$

Both assertions are herewith proved.

If now, $\mathfrak{b}_1, \mathfrak{b}_2, \ldots, \mathfrak{b}_n$ are pairwise relatively prime and if it has already been shown that

$$[\mathfrak{b}_1, \ldots, \mathfrak{b}_{n-1}] = \mathfrak{b}_1 \cdots \mathfrak{b}_{n-1},$$

then

$$[\mathfrak{b}_1, \ldots, \mathfrak{b}_n] = [\mathfrak{b}_1, \ldots, \mathfrak{b}_{n-1}] \cap \mathfrak{b}_n$$
$$= (\mathfrak{b}_1 \cdots \mathfrak{b}_{n-1}) \cap \mathfrak{b}_n$$
$$= \mathfrak{b}_1 \cdots \mathfrak{b}_{n-1} \cdot \mathfrak{b}_n,$$

and hence by induction we obtain the following.

Theorem: *The least common multiple of a finite number of pairwise relatively prime ideals is equal to their product.*

The previous remark on solution of congruences with respect to relatively prime ideals also holds for several pairwise relatively prime ideals.

If $\mathfrak{b}_1, \mathfrak{b}_2, \ldots, \mathfrak{b}_r$ *are pairwise relatively prime ideals, then* ξ *can always be determined from the congruences*

$$\xi \equiv \beta_i(\mathfrak{b}_i) \qquad (i = 1, 2, \ldots, r).$$

Proof: We proceed by induction. If η has already been found such that

$$\eta \equiv \beta_i(\mathfrak{b}_i) \qquad (i = 1, 2, \ldots, r-1),$$

then ξ is determined by

$$\xi \equiv \eta([\mathfrak{b}_1, \ldots, \mathfrak{b}_{r-1}])$$
$$\xi \equiv \beta_r(\mathfrak{b}_r),$$

which is always possible, since $\mathfrak{b}_r$ is relatively prime to $[\mathfrak{b}_1, \ldots, \mathfrak{b}_{r-1}]$.

Theorem: *If the ascending chain condition holds in* $\mathfrak{o}$, *then every ideal can be represented as the intersection of pairwise relatively prime ideals which themselves cannot be represented as the intersection of pairwise relatively prime proper divisors.*

Let

$$\mathfrak{m} = [\mathfrak{q}_1, \ldots, \mathfrak{q}_r]$$

be an irredundant representation of the given ideal $\mathfrak{m}$ by primary ideals. Let $\mathfrak{b}_1$ be the intersection of all the primary ideals which are connected with some fixed one of them by a chain of primary ideals which are not pairwise relatively prime. From the remaining ideals we form the ideals $\mathfrak{b}_2, \ldots, \mathfrak{b}_s$ successively in the same manner. The representation

$$\mathfrak{m} = [\mathfrak{b}_1, \ldots, \mathfrak{b}_s] \tag{15.13}$$

then has the desired properties. First, $\mathfrak{b}_i$ and $\mathfrak{b}_k$ for $i \neq k$ are indeed relatively prime, since the components of $\mathfrak{b}_i$ are relatively prime to those of $\mathfrak{b}_k$. Second, it is not possible to represent $\mathfrak{b}_1$, say, as the intersection of two pairwise relatively prime proper divisors. For if such a representation,

$$\mathfrak{b}_1 = \mathfrak{b} \cap \mathfrak{c} = \mathfrak{b}\mathfrak{c}$$
$$(\mathfrak{b}, \mathfrak{c}) = \mathfrak{o},$$

were possible, then every associated prime ideal of $\mathfrak{b}_1$ would have to be a divisor of $\mathfrak{b}\mathfrak{c}$ and hence either of $\mathfrak{b}$ or $\mathfrak{c}$; since now all these prime ideals are connected with a single one among them by a chain of prime ideals which are pairwise not relatively prime, it follows that if one of these prime ideals divides $\mathfrak{b}$, say, then they all divide $\mathfrak{b}$ and none divide $\mathfrak{c}$. But the primary components belonging to these prime ideals divide $\mathfrak{b}\mathfrak{c}$; therefore they must divide $\mathfrak{b}$ (since their prime ideals do not divide $\mathfrak{c}$). From this it follows that the intersection $\mathfrak{b}_1$ is a divisor of $\mathfrak{b}$:

$$\mathfrak{b} \subseteqq \mathfrak{b}_1,$$

contrary to the assumption that $\mathfrak{b}$ should be a proper divisor of $\mathfrak{b}_1$.

Our theorems imply that in place of the representation (15.13) we may also write a product representation

$$\mathfrak{m} = \mathfrak{b}_1\mathfrak{b}_2\cdots\mathfrak{b}_s.$$

Exercise

15.10. The intersection (15.13) is a direct intersection in the sense of Section 13.1. The residue class ring $\mathfrak{o}/\mathfrak{m} = \bar{\mathfrak{o}}$ is a direct sum of rings $\mathfrak{a}_i/\mathfrak{m} = \bar{\mathfrak{a}}_i$, each of which is isomorphic to a residue class ring $\mathfrak{o}/\mathfrak{b}_i$. (Put $\mathfrak{a}_i = [\mathfrak{b}_1, \ldots, \mathfrak{b}_{i-1}, \mathfrak{b}_{i+1}, \ldots, \mathfrak{b}_s]$ and apply the theorems of Section 13.1.)

15.8 SINGLE-PRIMED IDEALS

Let $\mathfrak{o}$ again be a Noetherian ring with identity.

The unit ideal $\mathfrak{o}$ is always a prime ideal. What primary ideals belong to it? The answer is, only $\mathfrak{o}$ itself; for if $\mathfrak{q}$ is a primary ideal belonging to $\mathfrak{o}$, then $1 \in \mathfrak{o}$ implies $1^\varrho \in \mathfrak{q}$ and hence $\mathfrak{q} = \mathfrak{o}$.

If now in a representation of an ideal $\mathfrak{a} \neq \mathfrak{o}$ as the intersection of primary ideals $[\mathfrak{q}_1, \ldots, \mathfrak{q}_r]$ the unit ideal occurs among the associated prime ideals, then the $\mathfrak{q}_i$ belonging to it is likewise equal to $\mathfrak{o}$ and is therefore redundant. *If the representation $\mathfrak{a} = [\mathfrak{q}_1, \ldots, \mathfrak{q}_r]$ is irredundant and $\mathfrak{a} \neq \mathfrak{o}$, then the unit ideal does not occur among the associated prime ideals.*

From this there follows immediately the next statement.

Every ideal $\mathfrak{a} \neq \mathfrak{o}$ has at least one prime ideal divisor $\mathfrak{p} \neq \mathfrak{o}$. If the ideal $\mathfrak{a}$ is not primary, then it has at least two prime ideal divisors $\neq \mathfrak{o}$.

An ideal which has no more than one prime ideal divisor in addition to $\mathfrak{o}$ is said to be *single-primed.*[4] By the preceding theorem, every single-primed ideal $\mathfrak{q}$ is primary. Moreover, the associated prime ideal $\mathfrak{p}$ is maximal, since if $\mathfrak{a}' \neq \mathfrak{o}$ were a proper divisor of $\mathfrak{p}$, then $\mathfrak{a}'$ would again have a prime divisor $\mathfrak{p}' \neq \mathfrak{o}$ which would also be a proper divisor of $\mathfrak{p}$; $\mathfrak{q}$ would then have two distinct prime divisors $\mathfrak{p}$ and $\mathfrak{p}'$ different from $\mathfrak{o}$, contrary to the assumption that $\mathfrak{q}$ is single-primed.

Now

$$\mathfrak{p}^\varrho \equiv 0(\mathfrak{q}). \tag{15.14}$$

If $\mathfrak{p}$ is maximal, relation (15.14) implies, conversely, that $\mathfrak{q}$ is single-primed. Indeed, if $\mathfrak{p}'$ is any prime ideal divisor of $\mathfrak{q}$, then (15.14) implies that

$$\mathfrak{p}^\varrho \equiv 0(\mathfrak{p}'),$$

and hence

$$\mathfrak{p} \equiv 0(\mathfrak{p}');$$

therefore, either $\mathfrak{p}' = \mathfrak{p}$ or $\mathfrak{p}' = \mathfrak{o}$. Thus $\mathfrak{q}$ has only $\mathfrak{p}$ and $\mathfrak{o}$ as prime divisors.

[4]The German term *einartig* for such an ideal is due to Dedekind.

The following concepts are therefore equivalent:

1. Single-primed ideal;
2. Primary ideal belonging to a maximal prime ideal $\mathfrak{p}$;
3. Divisor of a power $\mathfrak{p}^{\varrho}$ of a maximal prime ideal.

Furthermore, *if the ideal* $\mathfrak{m}$ *has an isolated single-primed primary component* $\mathfrak{q}$ *with associated prime ideal* $\mathfrak{p}$ *and exponent* ϱ, *then, for every integer* $\sigma \geqq \varrho$,

$$\mathfrak{q} = (\mathfrak{m}, \mathfrak{p}^{\sigma}). \tag{15.15}$$

Proof: From

$$\mathfrak{m} \equiv 0(\mathfrak{q})$$

and

$$\mathfrak{p}^{\sigma} \equiv 0(\mathfrak{q})$$

we conclude that

$$(\mathfrak{m}, \mathfrak{p}^{\sigma}) \equiv 0(\mathfrak{q}). \tag{15.16}$$

Let now

$$\mathfrak{m} = [\mathfrak{q}, \mathfrak{q}_2, \ldots, \mathfrak{q}_s]$$

be a representation of $\mathfrak{m}$ by primary components. The ideal $(\mathfrak{m}, \mathfrak{p}^{\sigma})$ is single-primed and therefore primary; $\mathfrak{p}$ is the associated prime ideal. The product $\mathfrak{q}\mathfrak{q}_2 \ldots \mathfrak{q}_s$ is divisible by $(\mathfrak{m}, \mathfrak{p}^{\sigma})$; since $\mathfrak{q}$ is *isolated*, $\mathfrak{q}_2 \ldots \mathfrak{q}_s$ is not divisible by $\mathfrak{p}$; therefore, $\mathfrak{q}$ must be divisible by $(\mathfrak{m}, \mathfrak{p}^{\sigma})$:

$$\mathfrak{q} \equiv 0(\mathfrak{m}, \mathfrak{p}^{\sigma}). \tag{15.17}$$

Then (15.16) and (15.17) imply (15.15).
Corollary: For

$$\mathfrak{p}^{\sigma} \equiv 0(\mathfrak{q}) \equiv 0(\mathfrak{m}, \mathfrak{p}^{\sigma+1}),$$

and hence

$$\mathfrak{p}^{\sigma} \equiv 0(\mathfrak{m}, \mathfrak{p}^{\sigma+1}). \tag{15.18}$$

For $\sigma < \varrho$, relation (15.18) no longer holds, for if

$$\mathfrak{p}^{\sigma} \equiv (0\mathfrak{m}, \mathfrak{p}^{\sigma+1})$$

for some $\sigma < \varrho$ then on multiplying by $\mathfrak{p}^{\varrho-\sigma-1}$ we could obtain

$$\mathfrak{p}^{\varrho-1} \equiv 0(\mathfrak{m}\mathfrak{p}^{\varrho-\sigma-1}, \mathfrak{p}^{\varrho}) \equiv 0(\mathfrak{m}, \mathfrak{q}) \equiv 0(\mathfrak{q}),$$

contrary to the definition of the exponent ϱ.

The exponent ϱ *of* $\mathfrak{q}$ *is therefore the least integer* σ *so that* (15.18) *holds.*

There are integral domains $\mathfrak{o}$ with identity in which (the ascending chain condition holds and) every nonzero prime ideal is maximal. Examples are the principal ideal rings (cf. Section 3.8) and certain "orders" in number and function fields (to be defined later) of which the ring $\mathbb{Z}[\sqrt{-3}]$ is a typical example. The ideal theory of these rings is especially simple. All primary ideals except the null ideal are single-primed, and any two distinct prime ideals $\neq (0)$ are relatively prime. From this it follows that any two primary ideals $\neq (0)$ belonging to distinct prime ideals are also relatively prime. Finally, all the primary components of an ideal are isolated and are thus uniquely determined. Thus *every ideal distinct from the null ideal can be uniquely represented as the intersection of pairwise relatively prime, single-primed primary ideals.* By Section 15.7 this intersection is also a product:

$$\mathfrak{a} = [\mathfrak{q}_1, \ldots, \mathfrak{q}_r] = \mathfrak{q}_1 \cdot \mathfrak{q}_2 \cdots \mathfrak{q}_r,$$

In principal ideal rings the primary ideals $\mathfrak{q}_i$ are powers of prime ideals. Whether or not this is the case in more general rings depends on a condition which we shall learn later, namely the condition of "integral closure."

15.9 QUOTIENT RINGS

In Section 3.3 we constructed a quotient field for any commutative ring without zero divisors. This construction can be immediately extended to commutative rings with zero divisors if the rings contain regular elements (elements which are not zero divisors). We admit only the regular elements as denominators and then form the ring of all quotients a/b, where a is any ring element and b any regular element.

The set of admissible denominators may be still further restricted. In a commutative ring R let S be a nonempty set of regular elements which with any two elements s and t also contains their product st. The quotients a/s (a in R, s in S) then form an extension ring of R: the *quotient ring* $R' = R/S$. This concept is due to H. Grell (*Math. Ann.* Vol. 97, p. 499).

If R' is any commutative extension ring of R, then any ideal $\mathfrak{a}$ of R generates an ideal $\mathfrak{a}'$ of R': the *extended ideal* of $\mathfrak{a}$ in R'. Conversely, the intersection of R with an ideal $\mathfrak{c}'$ of R' is always an ideal in R: the *contracted ideal* of $\mathfrak{c}'$ in R. The contracted ideals $\mathfrak{c}' \cap R$ are also called *distinguished ideals* in R (with respect to R').

A general investigation of the concepts of extended and contracted ideals may be found in the work of H. Grell just mentioned. Only quotient rings will be treated here; in this case the situation is very simple.

If $\mathfrak{a}$ is an ideal in R, then the extended ideal $\mathfrak{a}'$ in the quotient ring R' consists of all quotients a/s (a in $\mathfrak{a}$, s in S). If we form the contracted ideal $\mathfrak{a}' \cap R$ of this $\mathfrak{a}'$, then we obtain precisely the S-component $\mathfrak{a}_S$ defined in Section 15.6, that is, the set of all x such that sx lies in $\mathfrak{a}$ for some s of S.

If we start with an arbitrary ideal $\mathfrak{a}'$ of the quotient ring R' and form the contracted ideal

$$\mathfrak{a} = \mathfrak{a}' \cap R,$$

then the extended ideal of $\mathfrak{a}$ is again $\mathfrak{a}'$. The intersection of this extended ideal with R is $\mathfrak{a}$, and therefore in this case $\mathfrak{a}_s = \mathfrak{a}$. Conversely, if $\mathfrak{a}_S = \mathfrak{a}$, then $\mathfrak{a}$ is a contracted ideal, namely the contracted ideal of its extended ideal $\mathfrak{a}'$. *The distinguished ideals* $\mathfrak{a}$ *in* R *are thus characterized by the property* $\mathfrak{a}_S = \mathfrak{a}$.

From what has been said it follows immediately that there is a one-to-one correspondence between ideals $\mathfrak{a}'$ of R' and distinguished ideals $\mathfrak{a}$ of R: $\mathfrak{a}$ is the contracted ideal of $\mathfrak{a}'$, and $\mathfrak{a}'$ is the extended ideal of $\mathfrak{a}$. The intersection $\mathfrak{a}' \cap \mathfrak{c}'$ hereby obviously corresponds to the intersection $\mathfrak{a} \cap \mathfrak{c}$.

If the ascending chain condition for ideals holds in R, then it holds, in particular, for distinguished ideals and thus also for the ideals of R'. If in an intersection representation

$$\mathfrak{a} = [\mathfrak{q}_1, \ldots, \mathfrak{q}_r] \tag{15.19}$$

we order the primary ideals $\mathfrak{q}_i$ such that only $\mathfrak{q}_{h+1}, \ldots, \mathfrak{q}_r$ (or the associated prime ideals $\mathfrak{p}_{h+1}, \ldots, \mathfrak{p}_r$) contain elements of S, and thus go over into the identity ideal of R' in the extension, then we obtain, as in Section 15.6,

$$\mathfrak{a}_S = [\mathfrak{q}_1, \ldots, \mathfrak{q}_h]. \tag{15.20}$$

The $\mathfrak{q}_i$ on the right of (15.20) has the property that $\mathfrak{q}_S = \mathfrak{q}$, and they are therefore distinguished. Hence $\mathfrak{a}_S$ is distinguished. From the one-to-one correspondence between distinguished and extended ideals we obtain from (15.19) the representation

$$\mathfrak{a}' = [\mathfrak{q}_1', \ldots, \mathfrak{q}_h'] \tag{15.21}$$

for the extended ideal.

Comparison of (15.19) and (15.21) shows that in the transition from R to R' a reduction in the ideal balance takes place. All the ideals which contain elements of S, the primary ideals $\mathfrak{q}_{h+1}, \ldots, \mathfrak{q}_r$ in this case, have the unit ideal as extended ideal. Only distinguished ideals $\mathfrak{a}$ (with the property $\mathfrak{a}_S = \mathfrak{a}$) remain undamaged in the extension in the sense that the original $\mathfrak{a} = \mathfrak{a}_S$ may be regained as a contracted ideal from $\mathfrak{a}'$.

Exercises

15.11. If $\mathfrak{q}$ is primary and $\mathfrak{p}$ is its associated prime ideal, then the extended ideal $\mathfrak{q}'$ in the quotient ring R' is also primary and $\mathfrak{p}'$ is its associated prime ideal.

15.12. If $\mathfrak{q}'$ is a primary ideal belonging to a prime ideal $\mathfrak{p}'$ in an arbitrary ring R', then in any subring R the contracted ideal $\mathfrak{q} = \mathfrak{q}' \cap R$ is a primary ideal belonging to the prime ideal $\mathfrak{p} = \mathfrak{p}' \cap R$.

GENERALIZED QUOTIENT RINGS

If S is a multiplicatively closed set of R which contains zero divisors but does not contain zero, then, following Chevalley, a generalized quotient ring can be defined as follows. Let $\mathfrak{n} = (0)_S$ be the S-component of the null ideal. We first form the residue class ring $R^* = R/\mathfrak{n}$. The residue classes modulo $\mathfrak{n}$ of the elements of S form a multiplicatively closed set S^* in R^* which contains no zero divisors. We may now form the usual quotient ring $R' = R^*/S^*$. Its properties are similar to those of the ordinary quotient ring. The extended ideal of an ideal $\mathfrak{a}$ of R is formed by first forming the image $\mathfrak{a}^*$ of $\mathfrak{a}$ under the homomorphism $R \to R^*$ and then forming the ideal generated by $\mathfrak{a}^*$ in R'. The contracted ideal of an ideal $\mathfrak{c}'$ of R' is similarly formed by first taking the intersection with R^* and then forming the set of elements whose residue classes modulo $\mathfrak{n}$ belong to this intersection.

For further development of this theory the reader is referred to D. G. Northcott, *Ideal Theory*, Cambridge Tracts, in *Math.*, Vol. 42, Section 2.7.

15.10 THE INTERSECTION OF ALL POWERS OF AN IDEAL

In the following $\mathfrak{o}$ shall be a Noetherian ring with identity. The ring is called *null primary* if the null ideal is primary, that is, if $ab = 0$ implies $a = 0$ or $b^r = 0$.

W. Krull has shown in a fundamental paper[5] that in a null primary ring $\mathfrak{o}$, and thus in an integral domain in particular, the intersection of all powers of an ideal $\mathfrak{a}$ distinct from $\mathfrak{o}$ is the null ideal. For a prime ideal $\mathfrak{p} \neq \mathfrak{o}$ it is even the case that the intersection of all symbolic powers $\mathfrak{p}^{(r)}$ is the null ideal. Statements for arbitrary rings can also be obtained from these theorems. The basic ideas of this investigation will be presented here.

Theorem 1: *If* $\mathfrak{a}$ *and* $\mathfrak{d}$ *are ideals in a null primary ring* $\mathfrak{o}$ *and if*

$$\mathfrak{d} \subseteqq \mathfrak{a}\mathfrak{d}, \tag{15.22}$$

then $\mathfrak{a} = \mathfrak{o}$ *or* $\mathfrak{d} = (0)$.

Proof: Let $\mathfrak{d} = (d_1, \ldots, d_n)$. It then follows from (15.22) that

$$d_i = \sum a_{ik} d_k. \tag{15.23}$$

If we put as usual $\delta_{ik} = 0$ for $i \neq k$ and $\delta_{ii} = 1$, then (15.23) may be written

$$\sum (\delta_{ik} - a_{ik}) d_k = 0. \tag{15.24}$$

The determinant of this linear system of equations is

$$D = 1 - a,$$

where a belongs to the ideal $\mathfrak{a}$. If we multiply equation (15.24) by the subdeterminants of the kth column of the determinant D and add, we obtain

$$Dd_k = 0;$$

from this it follows that, for any element d of the ideal $\mathfrak{d}$,

$$(1 - a)d = Dd = 0.$$

[5]W. Krull, "Primidealketten in Allgemeinen Ringbereichen," *Sitzungsber. Heidelberger Akad.*, 1298, 7 Abh.

Hence either $(1-a)^r = 0$ or, if no power of $(1-a)$ is zero, $d = 0$ for all d of $\mathfrak{d}$. In the first case, $1 \equiv 0(\mathfrak{a})$ and hence $\mathfrak{a} = \mathfrak{o}$. In the second case, $\mathfrak{d} = (0)$.

Theorem 2: *If $\mathfrak{o}$ is a null primary ring and $\mathfrak{a} \neq \mathfrak{o}$, then the intersection of all powers of $\mathfrak{a}$ is the null ideal:*

$$\mathfrak{d} = [\mathfrak{a}, \mathfrak{a}^2, \ldots] = (0). \tag{15.25}$$

Proof: We first show that $\mathfrak{d} \subseteq \mathfrak{a}\mathfrak{d}$. To this end we represent $\mathfrak{a}\mathfrak{d}$ as an intersection of primary ideals:

$$\mathfrak{a}\mathfrak{d} = [\mathfrak{q}_1, \ldots, \mathfrak{q}_r].$$

For each i $\mathfrak{a}\mathfrak{d}$ is divisible by $\mathfrak{q}_i$, and hence either $\mathfrak{d}$ or a power $\mathfrak{a}^n$ is divisible by $\mathfrak{q}_i$. But $\mathfrak{d}$ is divisible by every power $\mathfrak{a}^n$. In both cases, therefore, $\mathfrak{d} \subseteq \mathfrak{q}_i$. This holds for every i, and it thus follows that

$$\mathfrak{d} = \mathfrak{a}\mathfrak{d}.$$

Theorem 1 now implies that $\mathfrak{d} = (0)$.

For prime ideals $\mathfrak{p} \neq \mathfrak{o}$ we obtain a somewhat stronger theorem.

Theorem 3: *In a null primary ring the intersection of all symbolic powers $\mathfrak{p}^{(r)}$ of a prime ideal distinct from $\mathfrak{o}$ is the null ideal:*

$$[\mathfrak{p}, \mathfrak{p}^{(2)}, \mathfrak{p}^{(3)}, \ldots] = 0). \tag{15.26}$$

Proof: Let S be the set of elements of $\mathfrak{o}$ not contained in $\mathfrak{p}$. We form the quotient ring $\mathfrak{o}_S$. Let the extended ideal of $\mathfrak{p}$ in $\mathfrak{o}_S$ be $\mathfrak{P}$. The extended ideal of $\mathfrak{p}^r$ is then evidently $\mathfrak{P}^r$. The contracted ideal of $\mathfrak{P}^r$ is

$$(\mathfrak{p}^r)_S = \mathfrak{p}^{(r)}.$$

The intersection of all the $\mathfrak{p}^{(r)}$ is equal to the intersection of all the $\mathfrak{P}^{(r)}$ with $\mathfrak{o}$. The intersection of all the $\mathfrak{P}^{(r)}$ is the null ideal by Theorem 2. Hence, the intersection of all the $\mathfrak{p}^{(r)}$ is the null ideal.

Theorems 1 and 2 can be extended to arbitrary rings of the types here considered. Let S be the set of all elements $s = 1-a$, where a runs through all elements of the ideal $\mathfrak{a}$. The set S is multiplicatively closed; the S-component $(0)_S$ of the null ideal may therefore be defined as the set of x for which an equation of the form

$$(1-a)x = 0 \qquad \text{with } a \text{ in } \mathfrak{a}$$

holds.

Theorem 1a: $\mathfrak{d} \subseteq \mathfrak{a}\mathfrak{d}$ *implies* $\mathfrak{d} \subseteq (0)_S$.

Theorem 2a: *The intersection of all powers of $\mathfrak{a}$ is $(0)_S$.*

The proof of Theorem 1a is the same as that of Theorem 1 up to the equation

$$(1-a)d = 0.$$

From this equation it follows immediately that

$$d \in (0)_S \qquad \text{for all } d \text{ in } \mathfrak{d}.$$

One half of Theorem 2a, namely

$$[\mathfrak{a}, \mathfrak{a}^2, \ldots] \subseteq (0)_S,$$

is proved exactly as in Theorem 2. The other half,

$$(0)_S \subseteq [\mathfrak{a}, \mathfrak{a}^2, \ldots],$$

is easy to prove. Indeed, if x is in $(0)_S$, then

$$(1-a)\,x = 0,$$

so that $x = ax$ and hence

$$x = ax = a^2x = a^3x = \cdots .$$

Every power of a therefore divides x.

Applying Theorems 1 and 2 to the residue class ring $\mathfrak{o}/\mathfrak{q}$ by a primary ideal $\mathfrak{q}$, we obtain the following.

Theorem 1b: *If* $\mathfrak{q}$ *is a primary ideal and*

$$\mathfrak{b} \equiv 0(\mathfrak{a}\mathfrak{b}, \mathfrak{q}), \tag{15.27}$$

then either $(\mathfrak{a}, \mathfrak{q}) = \mathfrak{o}$ *or* $\mathfrak{b} \equiv 0(\mathfrak{q})$.

Theorem 2b: *If an element* y *of* $\mathfrak{o}$ *satisfies a congruence*

$$y \equiv 0(\mathfrak{a}^n, \mathfrak{q}) \tag{15.28}$$

for every natural number n, *then either* $(\mathfrak{a}, \mathfrak{q}) = \mathfrak{o}$ *or* $y \equiv 0(\mathfrak{q})$.

Exercises

15.13. In a Noetherian ring with identity, the intersection of the symbolic powers of a prime ideal $\mathfrak{p} \neq \mathfrak{o}$ is equal to $(0)_S$.

15.14. How do Theorems 1b and 2b read if the primary ideal $\mathfrak{q}$ is replaced by an arbitrary ideal $\mathfrak{m}$? (Apply Theorems 1a and 2a to the residue class ring $\mathfrak{o}/\mathfrak{m}$.)

15.11 THE LENGTH OF A PRIMARY IDEAL. CHAINS OF PRIMARY IDEALS IN NOETHERIAN RINGS

Theorems 1 and 2 (Section 15.10) and their variations were used by Krull in the paper previously mentioned to derive theorems on the termination of chains of prime ideals:

$$\mathfrak{p}_1 \supset \mathfrak{p}_2 \supset \cdots .$$

Before we present these theorems we must first clarify the notion of the length of a primary ideal.

Let $\mathfrak{q}$ be a primary ideal with associated prime ideal $\mathfrak{p}$ in a Noetherian ring $\mathfrak{o}$. A sequence of primary ideals for the same $\mathfrak{p}$ which terminates with $\mathfrak{q}$,

$$\mathfrak{q}_1 \supset \mathfrak{q}_2 \supset \cdots \supset \mathfrak{q}_l = \mathfrak{q},$$

is called a *proper normal series for the primary ideal.* The word "proper" emphasizes the fact that each successive ideal is a proper multiple of the preceding one. The number l is called the *length* of the normal series. If the series cannot be further refined by inserting other primary ideals, then it is called a *composition series for the primary ideal* $\mathfrak{q}$.

We shall show that every proper normal series for a primary ideal $\mathfrak{q}$ can be refined to a composition series and that all composition series have the same length. This length is called the *length of the primary ideal* $\mathfrak{q}$.

In the proof we need consider only the case in which $\mathfrak{q}$ is the null ideal. The general case can be reduced to this case by forming the residue class ring with respect to $\mathfrak{q}$. After forming the residue classes, all ideals are divisors of the null ideal $\mathfrak{q}$, and hence all prime ideals are divisors of $\mathfrak{p}$.

The situation is still further simplified by going over to the quotient ring $\mathfrak{o}' = \mathfrak{o}/S$, where S is the set of all elements of $\mathfrak{o}$ not contained in $\mathfrak{p}$. In the extension from $\mathfrak{o}$ to $\mathfrak{o}'$ all proper divisors of $\mathfrak{p}$ go into the unit ideal $\mathfrak{o}'$, and only $\mathfrak{p}$ gives rise to a prime ideal $\mathfrak{p}'$ distinct from $\mathfrak{o}$. Since every prime ideal in $\mathfrak{o}'$ is the extended ideal of a prime ideal in $\mathfrak{o}$ (namely, of its contracted ideal), there is only one prime ideal $\mathfrak{p}'$ in $\mathfrak{o}'$ with the exception of $\mathfrak{o}'$ itself. Thus, only a single primary ideal (with associated prime ideal $\mathfrak{p}'$) occurs in the intersection representation of an ideal $\mathfrak{m}' \neq \mathfrak{o}'$, that is to say: *In $\mathfrak{o}'$ every ideal with the exception of $\mathfrak{o}'$ itself is primary with associated prime ideal $\mathfrak{p}'$.*

Let $\mathfrak{o}'$ and $\mathfrak{p}'$ henceforth be called $\mathfrak{o}$ and $\mathfrak{p}$ for simplicity. We consider $\mathfrak{o}$ as an additive group with $\mathfrak{o}$ itself as operator domain. The admissible subgroups are the ideals of $\mathfrak{o}$, that is, $\mathfrak{o}$ itself and the primary ideals with associated prime ideal $\mathfrak{p}$. A proper normal series in the group theory sense,

$$\mathfrak{o} \supset \mathfrak{q}_1 \supset \mathfrak{q}_2 \supset \cdots \supset \mathfrak{q}_l = (0),$$

therefore gives a proper normal series for the primary ideal $\mathfrak{q}_l = (0)$ if the initial term $\mathfrak{o}$ is omitted.

It was shown in Chapter 6 that if there exists a composition series in a group with operators, then every proper normal series can be refined to a composition series and all composition series have the same length l. Therefore we need only prove that a composition series exists.

To this end we form the normal series

$$\mathfrak{p} \supset \mathfrak{p}^2 \supset \cdots \supset \mathfrak{p}^{\varrho} = (0).$$

We may interpret $\mathfrak{p}^k/\mathfrak{p}^{k+1}$ as a vector space with $\mathfrak{o}/\mathfrak{p}$ as operator domain. Since $\mathfrak{p}$ is maximal, $\mathfrak{o}/\mathfrak{p}$ is a field. Since $\mathfrak{p}^k$ has a finite ideal basis, the vector space is finite-dimensional; therefore there exists a finite composition series from $\mathfrak{p}^k$

to $\mathfrak{p}^{k+1}$. Arranging these composition series in sequence for $k = 1, 2, \ldots, \varrho-1$, we obtain a composition series from $\mathfrak{p}$ to (0). This completes the proof.

Krull's theorems on chains of prime ideals all rest on the following theorem.

Principal Ideal Theorem: *If $(b) \neq \mathfrak{o}$ is a principal ideal and $\mathfrak{p}$ is an isolated associated prime ideal of (b), then every proper chain of prime ideals*

$$\mathfrak{p} \supset \mathfrak{p}_1 \supset \cdots$$

terminates at $\mathfrak{p}_1$.

Proof: Suppose that there exists a chain

$$\mathfrak{p} \supset \mathfrak{p}_1 \supset \mathfrak{p}_2. \tag{15.29}$$

By forming residue classes mod $\mathfrak{p}_2$, $\mathfrak{p}_2$ may be taken to be the null ideal. It is hereby achieved that the ring has no zero divisors. If we go over to the quotient ring $\mathfrak{o}/S$, where S is the set of elements of $\mathfrak{o}$ not belonging to $\mathfrak{p}$, then all ideals not divisible by $\mathfrak{p}$ go over into the unit ideal, whereas the ideals divisible by $\mathfrak{p}$ of the chain (15.29) remain distinct and prime. The quotient ring, which we again denote by $\mathfrak{o}$, has an identity element and no zero divisors. Since all the associated prime ideals of (b), with the exception of $\mathfrak{p}$, have gone over into the unit ideal, (b) is now a primary ideal with associated prime ideal $\mathfrak{p}$. Thus all divisors of (b) with the exception of $\mathfrak{o}$ are now primary ideals with associated prime ideal $\mathfrak{p}$. The ideal theory of $\mathfrak{o}$ has become much simpler by going over to the quotient ring; this makes the following proof considerably easier.

We denote the rth symbolic power of $\mathfrak{p}_1$ by $\mathfrak{p}_1^{(r)}$. The ideals of the chain

$$(\mathfrak{p}_1^{(1)}, b) \supseteqq (\mathfrak{p}_1^{(2)}, b) \supseteqq \cdots$$

are divisors of b, and they are thus primary with associated prime ideal $\mathfrak{p}$ by what has just been said. The number of distinct ideals in this chain cannot be greater than the length of the primary ideal (b), and all the ideals of the chain therefore eventually become equal:

$$(\mathfrak{p}_1^{(s)}, b) = (\mathfrak{p}_1^{(s+1)}, b) = \cdots$$

Now let $m \geqq s$. We first show that

$$\mathfrak{p}_1^{(m)} \subseteqq (b\mathfrak{p}_1^{(m)}, \mathfrak{p}_1^{(m+1)}). \tag{15.30}$$

Let x be an element of $\mathfrak{p}_1^{(m)}$. Then

$$x \in (\mathfrak{p}_1^{(m)}, b) = (\mathfrak{p}_1^{(m+1)}, b),$$

so that

$$x = y + br \qquad \text{with} \quad y \in \mathfrak{p}_1^{(m+1)},$$

and hence

$$br = x - y \equiv 0(\mathfrak{p}_1^{(m)}).$$

Now $\mathfrak{p}_1^{(m)}$ is primary by definition, and b is not divisible by the associated prime ideal $\mathfrak{p}_1$; hence r must be divisible by $\mathfrak{p}_1^{(m)}$. From this it follows that

$$x = y+br \equiv 0(\mathfrak{p}_1^{(m+1)}, b\mathfrak{p}_1^{(m)}),$$

wherewith (15.30) is proved.

By Theorem 1b (Section 15.10), it follows from (15.30) that

$$\mathfrak{p}_1^{(m)} \subseteqq \mathfrak{p}_1^{(m+1)},$$

and hence $\mathfrak{p}_1^{(m)} = \mathfrak{p}_1^{(m+1)}$ for $m \geqq s$, that is,

$$\mathfrak{p}_1^{(s)} = \mathfrak{p}_1^{(s+1)} = \mathfrak{p}_1^{(s+2)} = \cdots. \tag{15.31}$$

The ring $\mathfrak{o}$ has no zero divisors. The intersection of the symbolic powers of $\mathfrak{p}_1$ is therefore the null ideal by Theorem 3 (Section 15.10). It therefore follows from (15.31) that

$$\mathfrak{p}_1^{(s)} = (0). \tag{15.32}$$

However, $\mathfrak{p}_1^{(s)}$ is a primary ideal with associated prime ideal $\mathfrak{p}_1$, whereas (0) is the prime ideal $\mathfrak{p}_2$. This is a contradiction. A chain such as (15.29) is therefore impossible.

By repeated application of the principal ideal theorem, Krull proves the following generalization.

If $\mathfrak{p}$ *is an isolated associated prime ideal of* $\mathfrak{m} = (b_1, \ldots, b_r)$ *and if* $\mathfrak{m} \neq \mathfrak{o}$, *then every proper chain of prime ideals*

$$\mathfrak{p} \supset \mathfrak{p}_1 \supset \mathfrak{p}_2 \supset \cdots \tag{15.33}$$

terminates at $\mathfrak{p}_r$ *at latest.*

This theorem applies, in particular, if

$$\mathfrak{m} = \mathfrak{q} = (b_1, \ldots, b_r)$$

is a primary ideal and $\mathfrak{p}$ is the associated prime ideal. Since every ideal has a finite basis, the following theorem is obtained.

Theorem: *Every proper chain of prime ideals* (15.33) *breaks off after finitely many terms.*

For the proofs and applications to the theory of local rings, the reader is referred to the book *Ideal Theory* by Northcott.

Chapter 16

THEORY OF POLYNOMIAL IDEALS

In this chapter general ideal theory will be applied to polynomial domains $\mathfrak{o} = \mathsf{K}[x_1, \ldots, x_n]$, where K is an arbitrary commutative field. In addition to general ideal theory, only the knowledge of Chapters 1 through 6 and 10 is assumed.

16.1 ALGEBRAIC MANIFOLDS

Let Ω be an arbitrary extension field of the base field K. A sequence of n elements $\xi_1, \ldots, \xi_n$ of Ω is called a *point* ξ of the *affine space* $A_n(\Omega)$. A point ξ is called a *zero* of a polynomial f of $\mathfrak{o} = \mathsf{K}[x_1, \ldots, x_n]$ if $f(\xi_1, \ldots, \xi_n) = 0$.

An *algebraic manifold* or, as is today more commonly said, a *variety* M in $A_n(\Omega)$ is defined to be the set of common zeros of a finite number of polynomials $f_1, \ldots, f_r$; it is thus the set of all solutions of the equations

$$f_1(\xi) = 0, \ldots, f_r(\xi) = 0.$$

If we form the ideal $\mathfrak{a} = (f_1, \ldots, f_r)$ from the polynomials $f_1, \ldots, f_r$, we see that the common zeros of $f_1, \ldots, f_r$ are zeros of all the polynomials

$$f = g_1 f_1 + \cdots + g_r f_r$$

of the ideal $\mathfrak{a}$. Thus M may also be characterized as the set of common zeros of all polynomials of the ideal or, as we shall also say, the *zeros of the ideal* $\mathfrak{o}$. The fact that $\mathfrak{a}$ has a finite basis is no restriction, because of the Hilbert basis theorem (Section 15.1). Hence *a variety M consists of the zeros in $A_n(\Omega)$ of an ideal $\mathfrak{a}$ in* $\mathfrak{o} = \mathsf{K}[x_1, \ldots, x_n]$. Here M is called *the variety* (or the zero variety) *of the ideal* $\mathfrak{a}$.

A divisor of $\mathfrak{a}$, that is, an ideal $\mathfrak{c}$ containing $\mathfrak{a}$, defines a subvariety of M. It may happen, however, that distinct ideals define the same variety M. Among all such ideals one is singled out, namely, the set of *all* polynomials f which assume the value zero at all points of M. This set is clearly an ideal $\mathfrak{m}$; it is called the *associated ideal* of M. The variety of $\mathfrak{m}$ is again M; M is thus uniquely determined by $\mathfrak{m}$, and conversely.

The ascending chain condition holds in the ring $\mathfrak{o} = K[x_1, \ldots, x_n]$ and therefore the maximum condition also holds (Section 15.1). From this we obtain the following theorem.

Minimum Principal for Varieties: *In every nonempty set of varieties M there exists a minimal variety M^*, that is, one which contains no other variety of the set.*

Proof: Each variety M has an associated ideal $\mathfrak{m}$, and to distinct M belong distinct $\mathfrak{m}$. In the set of these ideals $\mathfrak{m}$ there exists a maximal ideal $\mathfrak{m}^*$ which belongs to a variety M^*. This variety M^* is minimal in the set.

If a polynomial f assumes the value zero at all points of a variety M we say that f *contains* M (since, in fact, the variety $f = 0$ then contains the variety M). The associated ideal $\mathfrak{m}$ of M consists of all polynomials which contain M.

The *intersection* $M \cap N$ of two varieties M and N is again a variety. Indeed, if M consists of the zeros of $\mathfrak{a} = (f_1, \ldots, f_r)$ and N of the zeros of $\mathfrak{b} = (g_1, \ldots, g_s)$, then $M \cap N$ consists of the zeros of the ideal

$$(\mathfrak{a}, \mathfrak{b}) = (f_1, \ldots, f_r, g_1, \ldots, g_s).$$

The union $M \vee N$ is also a variety. It is, in fact, defined by the intersection $\mathfrak{a} \cap \mathfrak{b}$ (or equivalently by the product $\mathfrak{a} \cdot \mathfrak{b}$). To begin with, each point of the union is either a zero of every polynomial of $\mathfrak{a}$ or a zero of every polynomial of $\mathfrak{b}$ and is thus in any case a zero of all the polynomials of $\mathfrak{a} \cap \mathfrak{b}$ (and, in particular, of those of $\mathfrak{a} \cdot \mathfrak{b}$). If a point ξ does not belong to the union $M \vee N$, then there exists a polynomial f in $\mathfrak{a}$ and a polynomial g in $\mathfrak{b}$ which are not zero at the point ξ; but then the product fg which belongs to $\mathfrak{a} \cap \mathfrak{b}$ (and also to $\mathfrak{a} \cdot \mathfrak{b}$) is not zero at the point ξ, and therefore ξ is not a zero of $\mathfrak{a} \cap \mathfrak{b}$ (or $\mathfrak{a} \cdot \mathfrak{b}$). The zeros of $\mathfrak{a} \cap \mathfrak{b}$ (and of $\mathfrak{a} \cdot \mathfrak{b}$) are thus the points of $M \vee N$, and only these.

We now restrict our consideration, as is usual in algebraic geometry, to *nonempty* varieties.

A variety M which can be represented as the union of two (nonempty) proper subvarieties is called *composite* or *reducible*. If we wish to emphasize that both subvarieties can be defined by equations with coefficients in the same base field K, then we say that *M is reducible over the base field* K. A variety which is not reducible is said to be *irreducible* or *indecomposable* (over the base field K).

Criterion: *A variety M is irreducible over K if and only if the associated ideal is prime, that is, if "fg contains M" implies that f or g contains M.*

Proof: Suppose first that M is reducible: $M = M_1 \vee M_2$, where M_1 and M_2 are proper subvarieties of M. In the associated ideal of M_1 there exists a polynomial f which does not contain M, since otherwise we would have $M_1 \supseteqq M$. Similarly, in the associated ideal of M_2 there exists a polynomial g which does not contain M. The product fg contains M_1 and M_2 and therefore M. Thus the associated ideal of M is not prime.

Suppose now that M is irreducible. If now fg were a product which contained M without f or g containing M, then M could be written as the union of two proper subvarieties M_1 and M_2 defined as follows: M_1 consists of all points of M which satisfy the equation $f = 0$, and M_2 of all points of M which satisfy the equation $g = 0$. Each point ξ of M then belongs to either M_1 or M_2, since $f(\xi)g(\xi) = 0$ implies that $f(\xi) = 0$ or $g(\xi) = 0$. However, this contradicts the irreducibility of M.

In the same manner we can prove the following.

If an irreducible variety M is contained in the union of two varieties M_1 and M_2, then M is contained in either M_1 or M_2.

A corresponding statement holds if M is contained in a union of $M_1, \ldots, M_r$.

Decomposition Theorem: *Every variety M defined over K can be represented as the union of finitely many irreducible varieties over K.*

Proof: Suppose that there exist varieties M which cannot be represented as a union of irreducible varieties; then in the set of these M there is a minimal such variety M^*. This variety must be reducible and can thus be represented as the union of two proper subvarieties M_1 and M_2. Since M^* is minimal, M_1 and M_2 can be represented as unions of irreducible varieties, but then M^* can also be so represented, contrary to hypothesis. This completes the proof of the decomposition theorem.

If the redundant terms are omitted from the decomposition

$$M = I_1 \vee I_2 \vee \cdots \vee I_r, \tag{16.1}$$

then the decomposition is *unique* up to order. Indeed, if

$$M = J_1 \vee J_2 \vee \cdots \vee J_s \tag{16.2}$$

is a second decomposition, then I_1 is contained in the union of the J_i. It is therefore contained in a single J_i which by appropriate enumeration may be taken to be J_1. Similarly, J_1 is contained in some I_k:

$$I_1 \subseteqq J_1 \subseteqq I_k.$$

If now $k \neq 1$, then I_1 would be redundant in (16.1); therefore $k = 1$ and $I_1 = J_1$. In the same manner we find that $I_2 = J_2, \ldots, I_r = J_r$ and $r = s$. This completes the proof of uniqueness.

The same theorems continue to hold if we consider only the points which belong to a fixed subset of the affine space $A_n(\Omega)$.[1]

16.2 THE UNIVERSAL FIELD

In classical algebraic geometry the field Ω from which the coordinates of the point ξ are taken is always the field of complex numbers. The more modern algebraic geometry starts with an arbitrary base field K. It is convenient to assume, following André Weil, that the extension field Ω containing the co-

[1]See W. Habicht, "Topologische Eigenschaften Algebraischer Mannigfaltigkeiten," *Math. Ann.*, Vol. 122, p. 181. On the decomposition of an irreducible variety over K under extension of the base field, see my paper "Über A. Weils Neubegründung der Algebraischen Geometrie," *Abh. Math. Sem. Hamburg*, Vol 22, p. 158.

ordinates of the point ξ is a *universal field* over K; that is, it is assumed first that Ω is *algebraically closed* and second that Ω has *infinite transcendency degree* over K. If K is given, such a universal field can be constructed by first adjoining infinitely many indeterminates $u_1, u_2, \ldots$ to K and then forming the algebraic closure according to Section 10.1.

The usefulness of the universal field rests on the following theorem.

Theorem: *Any field extension* $K(\alpha_1, \ldots, \alpha_n)$ *obtained by adjoining finitely many field elements* $\alpha_1, \ldots, \alpha_n$ *to* K *can be isomorphically imbedded in* Ω. This means that if any n elements $\alpha_1, \ldots, \alpha_n$ are given in any extension field Λ of K, then there exists an isomorphism

$$K(\alpha_1, \ldots, \alpha_n) \cong K(\alpha'_1, \ldots, \alpha'_n),$$

which leaves the elements of K fixed and takes $\alpha_1, \ldots, \alpha_n$ into the elements $\alpha'_1, \ldots, \alpha'_n$ of Ω.

Proof: The $\alpha_1, \ldots, \alpha_n$ can be enumerated so that $\alpha_1, \ldots, \alpha_r$ are algebraically independent over K, while the other α_i are algebraic over $K(\alpha_1, \ldots, \alpha_r)$. We now choose $\alpha'_1, \ldots \alpha'$ algebraically independent over K in Ω. Then there exists an isomorphism

$$K(\alpha_1, \ldots, \alpha_r) \cong K(\alpha'_1, \ldots, \alpha'_r), \tag{16.3}$$

which leaves the elements of K fixed and takes $\alpha_1, \ldots, \alpha_r$ into $\alpha'_1, \ldots, \alpha'_r$. If $r = n$, we are finished. If $r < n$, then α_{r+1} is a zero of an irreducible polynomial $\varphi(x)$ with coefficients in $K(\alpha_1, \ldots, \alpha_r)$. To this polynomials there corresponds a polynomial $\varphi'(x)$ with coefficients in $K(\alpha'_1, \ldots, \alpha'_r)$ which has a zero α'_{r+1} in Ω. By Section 18.1, the isomorphism (16.3) can be extended to an isomorphism

$$K(\alpha_1, \ldots, \alpha_{r+1}) \cong K(\alpha'_1, \ldots, \alpha'_{r+1}), \tag{16.4}$$

which takes α_{r+1} into α'_{r+1}. Continuing in this manner, we finally obtain the desired isomorphism

$$K(\alpha_1, \ldots, \alpha_n) \cong K(\alpha'_1, \ldots, \alpha'_n). \tag{16.5}$$

16.3 THE ZEROS OF A PRIME IDEAL

Let Ω again be a universal field over the base field K, and let $\mathfrak{o}$ be the polynomial domain $K[x_1, \ldots, x_n]$. If $\xi_1, \ldots, \xi_n$ are elements of an arbitrary extension field of K, then by Section 16.2 we can always find a field isomorphism which takes $\xi_1, \ldots, \xi_n$ into elements of Ω. For the following theorems it is therefore of no importance whether $\xi_1, \ldots, \xi_n$ are assumed to be elements of Ω or elements of an arbitrary extension field Λ of K. If the ξ_i are taken to be elements of Ω, then ξ is a point of the affine space $A_n(\Omega)$.

Such a point ξ is called a *generic zero* of an ideal $\mathfrak{p}$ if $f \in \mathfrak{p}$ implies $f(\xi) = 0$ and conversely. The ideal $\mathfrak{p}$ consists precisely of the polynomials $f(x)$ with the property $f(\xi) = 0$. We shall soon see that such an ideal $\mathfrak{p}$ is necessarily prime. We shall further show that each point ξ is a generic zero of a uniquely determined prime ideal $\mathfrak{p} \neq \mathfrak{o}$ and that, conversely, each prime ideal $\mathfrak{p} \neq \mathfrak{o}$ has a generic zero ξ which is uniquely determined up to isomorphism.

Theorem 1: *If $\xi_1, \ldots, \xi_n$ are elements of an arbitrary extension field of* K, *then the polynomials f in $\mathfrak{o} = \mathsf{K}[x_1, \ldots, x_n]$ for which $f(\xi) = 0$ form a prime ideal in $\mathfrak{o}$ which is distinct from $\mathfrak{o}$ itself.*

Proof: The equations $f(\xi) = 0$ and $g(\xi) = 0$ imply $f(\xi)-g(\xi) = 0$; also, $f(\xi) = 0$ implies $f(\xi)h(\xi) = 0$. The polynomials in question therefore form an ideal.

The statements $f(\xi)g(\xi) = 0$ and $g(\xi) \neq 0$ imply $f(\xi) = 0$, since a field has no zero divisors. The ideal is therefore prime. Since it does not contain the identity, it is distinct from $\mathfrak{o}$.

Example: Let $\xi_1, \ldots, \xi_n$ be linear functions of an indeterminate t with coefficients in the field K:

$$\xi_i = \alpha_i + \beta_i t. \tag{16.6}$$

The prime ideal then consists of all polynomials $f(x_1, \ldots, x_n)$ with the property that $f(\alpha_1+\beta_1 t, \ldots, \alpha_n+\beta_n t)$ vanishes identically in t, or (in geometric terms) of all polynomials which vanish at all points of the line which is given parametrically in n-dimensional space by (16.6). This example may help the reader to visualize all the theorems of this and the next section.

Theorem 2: *If $\mathfrak{p}$ is the prime ideal of Theorem 1, then $\Lambda = \mathsf{K}(\xi_1, \ldots, \xi_n)$ is isomorphic to the residue class field Π of $\mathfrak{o}$ with respect to $\mathfrak{p}$ and in such a way that the residue classes of $x_1, \ldots, x_n$ correspond to the elements $\xi_1, \ldots, \xi_n$.*

Proof: Let $\mathfrak{L}$ be the ring of those elements of Λ which can be written as polynomials in $\xi_1, \ldots, \xi_n$. Then $\Lambda = \mathsf{K}(\xi_1, \ldots, \xi_n)$ is the quotient field of $\mathfrak{L}$. We assign to each element $f(\xi_1, \ldots, \xi_n)$ of $\mathfrak{L}$ the element of the residue class ring $\mathfrak{o}/\mathfrak{p}$ represented by $f(x_1, \ldots, x_n)$. Since $f(\xi)-g(\xi) = 0$ implies $f-g \equiv 0(\mathfrak{p})$ or $f \equiv g(\mathfrak{p})$ and conversely, this correspondence is one-to-one. It is clear that sums and products go into sums and products. The rings $\mathfrak{L}$ and $\mathfrak{o}/\mathfrak{p}$ are therefore isomorphic. The quotient fields Λ and Π must then also be isomorphic.

Theorem 1 states that each point ξ is a generic zero of a unique prime ideal $\mathfrak{p}$. Theorem 2 states that ξ is uniquely determined by $\mathfrak{p}$ up to isomorphism.

Theorem 3: *Every prime ideal distinct from $\mathfrak{o}$ has a generic zero ξ in the universal field Ω.*

Proof: We assign to the polynomials of $\mathfrak{o}$ the elements of a new set $\mathfrak{o}'$ which contains the coefficient field K: to two polynomials congruent modulo $\mathfrak{p}$ there corresponds the same element, to polynomials which are not congruent there correspond distinct elements, and the elements of K correspond to themselves. This is always possible, since two elements of K are congruent with respect to $\mathfrak{p}$ if and only if they are equal, because $\mathfrak{p} \neq \mathfrak{o}$. The elements corresponding to $x_1, \ldots, x_n$ we shall call $\xi_1, \ldots, \xi_n$.

The set $\mathfrak{o}'$ is mapped onto the residue class ring of $\mathfrak{o}$ with respect to $\mathfrak{p}$ in a one-to-one fashion. If we define addition and multiplication in $\mathfrak{o}'$ corresponding to addition and multiplication in the residue class ring, then $\mathfrak{o}'$ is isomorphic to the residue class ring; it therefore has no zero divisors and thus admits the formation of a quotient field Λ.

Each element of $\mathfrak{o}'$ corresponds to at least one polynomial f of $\mathfrak{o}$ and can therefore be written as $f(\xi_1, \ldots, \xi_n)$. Thus, $\mathfrak{o} = \mathsf{K}[\xi_1, \ldots, \xi_n]$ and $\Lambda = \mathsf{K}(\xi_1, \ldots, \xi_n)$. Then Λ can be isomorphically imbedded in the universal field Ω by Section 16.2; we may therefore assume that $\Lambda \subseteqq \Omega$. The element $f(\xi_1, \ldots, \xi_n)$ is zero if and only if the polynomial f belongs to the zero residue class mod $\mathfrak{p}$. Hence ξ is a generic zero of $\mathfrak{p}$, and this completes the proof of Theorem 3.

By Theorem 3 every prime ideal $\mathfrak{p} \neq \mathfrak{o}$ has a generic zero in the universal field Ω which is uniquely determined up to isomorphism by Theorem 2. This point ξ is a zero of $\mathfrak{p}$ and therefore lies on the zero variety M of $\mathfrak{p}$. The associated ideal of M is again $\mathfrak{p}$, for if a polynomial f vanishes at all points of M, then, in particular, $f(\xi) = 0$ and hence $f \in \mathfrak{p}$. Since the associated ideal is prime, it follows that M is irreducible. We thus have the following.

Theorem 4: *Each prime ideal $\mathfrak{p} \neq \mathfrak{o}$ has an irreducible variety of zeros and is itself the associated ideal of this variety.*

If we begin with an irreducible variety M, then the associated ideal $\mathfrak{p}$ is prime by Section 16.1. The zeros of $\mathfrak{p}$ are precisely the points of M. If ξ is a generic zero of $\mathfrak{p}$, then ξ is called a *generic point of M over* K. Going back to the definitions, we see that this means the following.

A point ξ of M is a generic point of M over K *if every equation $f(\xi) = 0$ with coefficients in* K *which holds for ξ also holds for all points of M.*

By Theorem 3 every irreducible variety M has a generic point. Conversely, if a variety M has a generic point ξ, then the associated ideal of M is prime by Theorem 1 and thus M is irreducible. We now have the following.

Theorem 5: *M has a generic point over* K *if and only if M is irreducible over* K.

Exercise

16.1. The ideal

$$(x_1x_3 - x_2^2,\ x_2x_3 - x_1^3,\ x_3^2 - x_1^2x_2)$$

in $\mathsf{K}[x_1, x_2, x_3]$ is prime, since it has the generic zero (t^3, t^4, t^5).

16.4 THE DIMENSION

Let ξ be a generic point over K of an irreducible variety M, that is, a generic zero of the associated prime ideal $\mathfrak{p}$. If r is the degree of transcendency of the system $\{\xi_1, \ldots, \xi_n\}$, then there are exactly r algebraically independent elements among the ξ_i, say $\xi_1, \ldots, \xi_r$; the others are algebraically dependent on these. We may take indeterminates $t_1, \ldots, t_r$ for the $\xi_1, \ldots, \xi_r$; all ξ_i are then algebraic functions of these r indeterminates. The degree of transcendency remains unchanged if the generic point is taken by a field isomorphism into another generic point ξ; r therefore depends only on $\mathfrak{p}$ and is called the *dimension* of the prime ideal $\mathfrak{p}$ or of the variety M.

The dimension of the prime ideal $\mathfrak{p} \neq \mathfrak{o}$ may clearly be any number from 0 to n. We assign dimension -1 to the unit ideal $\mathfrak{o}$ which has no zeros.

If ξ is a generic zero of a prime ideal $\mathfrak{p}$ and ξ' an arbitrary zero of the same ideal, then to each polynomial $f(\xi)$ of $K[\xi]$ there corresponds the polynomial $f(\xi')$ of $K[\xi']$. Since $f(\xi) = g(\xi)$ implies $f(x) = g(x)\,(\mathfrak{p})$ and hence $f(\xi') = g(\xi')$, it follows that the correspondence $f(\xi) \to f(\xi')$ is single-valued. Since this correspondence obviously takes sums to sums and products to products, it is thus a *homomorphism*:

$$K[\xi] \sim K[\xi']. \tag{16.7}$$

If this correspondence is an isomorphism, then of course, ξ' is also a generic zero of $\mathfrak{p}$, and conversely.

In the case of a zero-dimensional ideal $\mathfrak{p}$, all the ξ are algebraic over K; thus all rational functions of ξ are polynomials and $K(\xi) = K[\xi]$. Hence $K[\xi]$ is a *field*. If now ξ' is again an arbitrary zero, then the homomorphism (16.7) is necessarily an isomorphism; indeed, a field has no other homomorphisms except those which are one-to-one and those which map the entire field onto the null ring. This implies the following theorem.

Theorem: *If a prime ideal is zero-dimensional, all its zeros are generic and equivalent.*[2]

The coordinates $\xi_1, \ldots, \xi_n$ or $\xi'_1, \ldots, \xi'_n$ are in this case algebraic over K. If we consider only zeros ξ or ξ' in a fixed universal field Ω, then all these zeros are conjugate over K. The number of these conjugate points in Ω is at most equal to the field degree of $K(\xi)$ over K (exactly equal if $K(\xi)$ is separable). Hence we have the following.

A zero-dimensional irreducible variety consists of finitely many points which are conjugate over K.

In particular, if the field K is already algebraically closed, then there exists only one zero ξ in the field K itself, and the associated ideal is

$$\mathfrak{p} = (x_1 - \xi_1, \ldots, x_n - \xi_n).$$

Theorem: *The distinct zeros of an r-dimensional prime ideal have transcendency degree $\leqq r$, and if the transcendency degree is equal to r the zero is generic.*

Proof: If ξ' is a zero of transcendency degree s, the homomorphism (16.7) holds. If $\xi'_1, \ldots, \xi'_s$ are algebraically independent, then $\xi_1, \ldots, \xi_s$ are also; for any algebraic relation among the ξ would imply the same relation among the ξ'. Hence $r \geqq s$. If $r = s$, all the ξ are algebraically dependent on $\xi_1, \ldots, \xi_s$. If a polynomial $f(\xi)$ not identically zero were to go over into zero under the homo-

[2]This means that they can be obtained from one another by isomorphisms which leave the elements of the base field K fixed.

morphism (16.7), then in the field $K(\xi)$ the element $1/f$ could be written in the following special form:

$$\frac{1}{f(\xi_1, \ldots, \xi_n)} = \frac{g(\xi_1, \ldots, \xi_n)}{h(\xi_1, \ldots, \xi_s)}.$$

This would imply

$$h(\xi_1, \ldots, \xi_s) = g(\xi_1, \ldots, \xi_n)f(\xi_1, \ldots, \xi_n).$$

If, under the homomorphism (16.7), f were to go into 0, then $h(\xi_1, \ldots, \xi_s)$ would also go into 0, that is,

$$h(\xi'_1, \ldots, \xi'_s) = 0,$$

contrary to the assumed algebraic independence of $\xi'_1, \ldots, \xi'_s$. Thus no nonzero polynomial goes into zero under the homomorphism (16.7); the homomorphism is therefore an isomorphism in the case $r = s$. This implies that ξ' is a generic zero.

Any zero ξ' of $\mathfrak{p}$ can be interpreted as a generic zero of an ideal $\mathfrak{p}'$. Then $f \equiv 0(\mathfrak{p})$ implies $f(\xi') = 0$ or $f \equiv 0(\mathfrak{p}')$; $\mathfrak{p}'$ is therefore a divisor of $\mathfrak{p}$. Conversely, every prime divisor $\mathfrak{p}'$ of $\mathfrak{p}$ distinct from $\mathfrak{o}$ can be obtained in this way, since any ideal $\mathfrak{p}' \neq \mathfrak{o}$ has a generic zero ξ'. From the preceding theorem we now obtain the following.

Every divisor $\mathfrak{p}'$ *of* $\mathfrak{p}$ *has dimension* $r' \leqq r$; *if* $r' = r$ *then* $\mathfrak{p}' = \mathfrak{p}$.

The dimension of an arbitrary variety is defined to be the highest of the dimensions of its irreducible components. The purely one-dimensional varieties are called *curves*, the two-dimensional are called *surfaces*, and the $(n-1)$-dimensional varieties are called *hypersurfaces*.

Exercises

16.2. A principal ideal (p), where p is an indecomposable, nonconstant polynomial, is an $(n-1)$-dimensional prime ideal.

16.3. Conversely, every $(n-1)$-dimensional prime ideal is a principal ideal.

16.4. The only n-dimensional variety in $A_n(\Omega)$ is $A_n(\Omega)$ itself; the associated ideal is the null ideal.

16.5 HILBERT'S NULLSTELLENSATZ. RESULTANT SYSTEMS FOR HOMOGENEOUS EQUATIONS

Every prime ideal distinct from $\mathfrak{o}$ has a generic zero in the universal field Ω. The unit ideal $\mathfrak{o}$ is a prime ideal without zeros.

We now prove more generally the following.

Theorem: *Any ideal* $\mathfrak{a} = (f_1, \ldots, f_r)$ *which has no zeros in* Ω *is the unit ideal.*

Proof: Suppose that there exists an ideal $\mathfrak{a} \neq \mathfrak{o}$ without zeros. By the maximum principle there then also exists a maximal ideal $\mathfrak{m} \neq \mathfrak{o}$ without zeros. This is a maximal ideal and therefore prime by Section 3.6. But a prime ideal $\mathfrak{m} \neq \mathfrak{o}$ has zeros.

This theorem may also be formulated as follows.

Theorem: *If the polynomials $f_1, \ldots, f_r$ have no common zero in $A_n(\Omega)$, then*

$$1 = g_1 f_1 + \cdots + g_r f_r. \tag{16.8}$$

This theorem is a special case of the Hilbert's Nullstellensatz, which states: *If f is a polynomial of $K[x_1, \ldots, x_n]$ which vanishes at all the common zeros of $f_1, \ldots, f_r$ in $A_n(\Omega)$, then*

$$f^q = h_1 f_1 + \cdots + h_r f_r \tag{16.9}$$

for some natural number q.

Proof: The general case can be reduced to the special case just proved by an aritifice due to A. Rabinowitsch (*Math. Ann.*, Vol. 102, p. 518). For $f = 0$ the assertion is clear. In the case $f \neq 0$ we adjoin a new variable z. The polynomials

$$f_1, \ldots, f_r, 1 - zf$$

then have no common zero in $A_{n+1}(\Omega)$. By the theorem just proved, therefore,

$$1 = g_1 f_1 + \cdots + g_r f_r + g \cdot (1 - zf). \tag{16.10}$$

In this identity we make the substitution $z = 1/f$ and remove the resulting fractions by multiplying by a power f^q. We obtain

$$f^q = h_1 f_1 + \cdots + h_r f_r, \qquad \text{Q.E.D.}$$

Extension of the Nullstellensatz: *If the polynomials $p_1, \ldots, p_s$ vanish at all the common zeros of $f_1, \ldots, f_r$, then there exists a natural number q such that all products of powers of the p_i of degree q belong to the ideal $(f_1, \ldots, f_r)$ (and conversely).*

Proof: We have

$$p_i^{q_i} \equiv 0(f_1, \ldots, f_r).$$

Let us put

$$q = (q_1 - 1) + (q_2 - 1) + \cdots + (q_s - 1) + 1.$$

Every product of powers $p_1^{h_1} \ldots p_s^{h_s}$ with $h_1 + \cdots + h_s = q$ then contains at least one factor $p_i^{q_i}$; for otherwise $h_1 + \cdots + h_s$ would be at most equal to

$$(q_1 - 1) + \cdots + (q_s - 1) = q - 1.$$

This proves the assertion. The converse is clear.

As an application of this last theorem we shall derive the conditions which a system of forms (homogeneous polynomials) $F_1, \ldots, F_r$ must satisfy in order that a nontrivial zero [one distinct from $(0, \ldots, 0)$] should exist in the field Ω.

If $(0, \ldots, 0)$ is the only zero, then the monomials $x_1, \ldots, x_n$ vanish at all zeros of the ideal $(F_1, \ldots, F_r)$. Thus any product of powers X_j of the $x_1, \ldots, x_n$ of degree q is contained in the ideal

$$X_j = G_{j1} F_1 + \cdots + G_{jr} F_r. \tag{16.11}$$

Let the degrees of the forms $F_1, \ldots, F_r$ be $g_1, \ldots, g_r$. The terms of degree q on the right-hand side of (16.11) are retained if only the terms of degree $q-g_i$ in the G_{ji} are kept and the others are omitted. We thus obtain a form H_{ji} of degree $q-g_i$ in place of G_{ji}. Equating the members of degree q on the left and right in (16.11) gives

$$X_j = H_{j1}F_1 + \cdots + H_{jr}F_r. \tag{16.12}$$

Conversely, if equations of the form (16.12) hold for all products of powers X_j of degree q, then $(0, \ldots, 0)$ is the only common zero of $F_1, \ldots, F_r$.

Let X_{ki} denote the products of powers of the x_j of degree $q-g_i$. The forms H_{ji} in (16.12) are linear combinations of these products (with coefficients in K). Thus (16.12) states that all products X_j of degree q can be linearly expressed in terms of the products $X_{ki}F_i$. We thus obtain the following result.

A necessary and sufficient condition that $F_1, \ldots, F_r$ have only the trivial zero $(0, \ldots, 0)$ is that all products X_j of a sufficiently high degree q can be expressed linearly in terms of the products $X_{ki}F_i$ with coefficients in K.

If N_q is the number of products X_j of degree q, then this result may also be formulated as follows.

A necessary and sufficient condition that $F_1, \ldots, F_r$ have a nontrivial common zero is that for every $q = 1, 2, \ldots$ the number of linearly independent products $X_{ki}F_i$ is less than N_q.

If we express the products $X_{ki}F_i$ as linear combinations of the X_j,

$$X_{ki}F_i = \sum_j a_{kij}X_j,$$

then for each k and i we can form a row vector

$$(a_{ki1}, \ldots, a_{kiN}) \qquad (N = N_q).$$

Our condition now states that among these row vectors fewer than N are linearly independent. This means that all determinants of N such vectors must vanish. Denoting these determinants by D_{qh}, we obtain the following.

The conditions

$$D_{qh} = 0 \qquad (q = 1, 2, \ldots) \tag{16.13}$$

are necessary and sufficient in order that $F_1, \ldots, F_r$ have a nontrivial common zero.

The a_{kij} are coefficients of the forms F_i. The D_{qh} are therefore integral forms in the coefficients of the forms $F_1, \ldots, F_r$.

If we assume that $F_1, \ldots, F_r$ are general forms of degrees $g_1, \ldots, g_r$ with undetermined coefficients a_j, then there are infinitely many polynomials $D_{qh}(a_j)$ in these coefficients. By Hilbert's basis theorem, however, there exist finitely many of these polynomials in terms of which all other may be linearly expressed (with integral polynomials as coefficients). If (for special forms $F_1, \ldots, F_r$) these finitely many D_{qh} are zero, then they are all zero and the system of equations (16.13) is satisfied. Hence *there exist finitely many integral forms in the a_j*:

$$R_1(a_j), \ldots, R_m(a_j),$$

which are all zero if and only if the forms $F_1, \ldots, F_r$ have a nontrivial common zero.[3]

A system of forms $R_1, \ldots, R_m$ with the property above is called a *resultant system* for the forms $F_1, \ldots, F_r$. If the F_i are linear forms, then the n-rowed determinants which can be formed from n of the r forms form a resultant system. For two forms F_1, F_2 in two variables x_1, x_2 the usual resultant forms a resultant system. Similarly, for n forms in n variables a single resultant R is sufficient. In this regard, see A. Hurwitz, "Über Trägheitsformen," *Ann. di. Mat.* 3ª Serie, Vol. 20 (1913).

16.6 PRIMARY IDEALS

The main problem of ideal theory in polynomial rings is *to determine whether a polynomial f belongs to a given ideal*

$$\mathfrak{m} = (f_1, \ldots, f_r).$$

Here we do not mean a computational decision method with a finite number of actually performable operations, although such a procedure is always possible,[4] but rather a method which affords an insight into the structure of the ideal and expresses the relation between the zeros of the ideal and its element f as clearly as possible. Such a method was first given by E. Lasker;[5] it depends on the decomposition of ideals into primary components.

The basic idea of the Lasker method is the following. According to the decomposition theorem of Section 15.4, every ideal $\mathfrak{m}$ can be represented as the intersection of primary ideals:

$$\mathfrak{m} = [q_1, \ldots, q_s].$$

In order that a polynomial f belong to the ideal $\mathfrak{m}$, it is necessary and sufficient that f belong to all the primary ideals q_ν. In principle then, in order to solve the above problem, we have only to establish the conditions which a polynomial must satisfy in order that it belong to a primary ideal.

By Section 15.3, to every primary ideal $\mathfrak{q}$ there belongs a prime ideal $\mathfrak{p}$ and an exponent ϱ with the following properties.

1. $\mathfrak{p}^\varrho \equiv 0(\mathfrak{q}) \equiv 0(\mathfrak{p})$.
2. $fg \equiv 0(\mathfrak{q})$ and $f \not\equiv 0(\mathfrak{p})$ imply $g \equiv 0(\mathfrak{q})$.

In the case $\mathfrak{q} \neq \mathfrak{o}$ the prime ideal in turn belongs to an irreducible variety M. All the zeros of $\mathfrak{q}$ are at the same time zeros of $\mathfrak{p}$ by property 1, and conversely.

[3]This theorem, which plays an important role in algebraic geometry, is due to F. Mertens (*Sitzungsber. Wiener Akad.*, Vol. 108, p. 1174). Another proof has been given by H. Kapferer (*Sitzungsber. Bayer. Akad. München*, 1929, p. 179).

[4]Cf. J. König, *Einleitung in die Allgemeine Theorie der Algebraischen Grössen* (Leipzig, B. G. Teubner, 1903) and G. Hermann, "Die Frage der Endlich Vielen Schritte in der Theorie der Polynomideale," *Math. Ann.*, Vol. 95, pp. 736–788.

[5]E. Lasker, "Zur Theorie der Moduln und Ideale," *Math. Ann.*, **60**, 20–116 (1905).

The variety of a primary ideal $\mathfrak{q} \neq \mathfrak{o}$ *is therefore irreducible and equal to the variety of its associated prime ideal.*

Let $\mathfrak{q}$ be a primary ideal with associated prime ideal $\mathfrak{p}$ and exponent ϱ, and let M be its variety. If now f is a polynomial which contains M, then $f \equiv 0(\mathfrak{p})$ and hence $f^{\varrho} \equiv 0(\mathfrak{q})$. If, however, f does not contain M, then in every congruence modulo $\mathfrak{q}$ the factor f may be canceled by property 2 above. These are two very important tools by means of which it can frequently be determined whether $f^{\varrho} \equiv 0(\mathfrak{q})$ or $g \equiv 0(\mathfrak{q})$. They can be extended immediately to arbitrary ideals $\mathfrak{m} = [\mathfrak{q}_1, \ldots, \mathfrak{q}_s]$ by means of the decomposition theorem. If f is a polynomial which contains the variety M of $\mathfrak{m}$ and if ϱ is the greatest of the exponents of the primary ideals $\mathfrak{q}_1, \ldots, \mathfrak{q}_s$, then it follows immediately that

$$f^{\varrho} \equiv 0(\mathfrak{q}_i) \qquad (\text{for} \quad i = 1, \ldots, s),$$

and hence

$$f^{\varrho} \equiv 0(\mathfrak{m}).$$

We have thus obtained another proof of *Hilbert's Nullstellensatz* (Section 16.5), which is even sharper in that it shows that the exponent ϱ depends only on the ideal $\mathfrak{m}$.

If f is a polynomial which contains none of the manifolds of the primary ideals $\mathfrak{q}_1, \ldots, \mathfrak{q}_s$, then in any congruence

$$fg \equiv 0(\mathfrak{m})$$

f may be canceled, and we conclude that

$$g \equiv 0(\mathfrak{m})$$

since this is true for all primary ideals $\mathfrak{q}_\nu$. The cancellation possibility can be briefly and precisely expressed by the equation

$$\mathfrak{m} : (f) = \mathfrak{m},$$

which by Section 15.5 holds if and only if f is not divisible by any of the associated prime ideals $\mathfrak{p}_1, \ldots, \mathfrak{p}_s$ of $\mathfrak{m}$ (f thus contains none of their irreducible varieties).

Somewhat more generally, for any ideal $\mathfrak{a}$, by Section 15.5,

$$\mathfrak{m} : \mathfrak{a} = \mathfrak{m} \tag{16.14}$$

if and only if $\mathfrak{a}$ is not divisible by any of the $\mathfrak{p}_1, \ldots, \mathfrak{p}_s$ or, what is the same thing, *if the manifold of* $\mathfrak{a}$ *contains none of the manifolds of the prime ideals* $\mathfrak{p}_1, \ldots, \mathfrak{p}_s$. This theorem is often useful in finding the associated prime ideals $\mathfrak{p}_1, \ldots, \mathfrak{p}_s$ of an ideal $\mathfrak{m}$. If we suspect that some prime ideal $\mathfrak{p}$ is among the $\mathfrak{p}_\nu$, then we may take an ideal $\mathfrak{a}$ divisible by $\mathfrak{p}$, for example, $\mathfrak{a} = \mathfrak{p}$, and check to see if the relation (16.14) or its negation can be proved, that is, whether or not $g\mathfrak{a} \equiv 0(\mathfrak{m})$ implies $g \equiv 0(\mathfrak{m})$. If (16.14) holds, then $\mathfrak{p}$ is not one of the $\mathfrak{p}_\nu$.

The *dimension* of a primary ideal is defined as the dimension of the associated prime ideal (or the dimension of the associated variety). The dimension or *highest dimension* of any ideal $\mathfrak{a} \neq \mathfrak{o}$ is the highest of the dimensions of its primary components (or of its associated prime ideals). If the dimensions of the associated primary ideals of $\mathfrak{a}$ are all the same, say equal to d, then $\mathfrak{a}$ is called an *unmixed d-dimensional ideal.*

Exercises

16.5. The ideal (x_1^2, x_2x_3+1) is primary with exponent 2 and associated prime ideal (x_1, x_2x_3+1).

16.6. Every power p^ϱ of a nonconstant prime polynomial p generates an $(n-1)$-dimensional primary ideal. Every nonconstant polynomial f generates an unmixed $(n-1)$-dimensional ideal.

16.7. If $\mathfrak{p}$ is the prime ideal of Exercise 16.1, then $\mathfrak{p}^2$ is not primary. (The polynomial $(x_2x_3-x_1^3)^2-(x_2^2-x_1x_3)(x_3^2-x_1^2x_2)$ has a factor x_1, and the other factor does not belong to $\mathfrak{p}^2$.)

16.7 NOETHER'S THEOREM

Using the primary ideal decomposition, we shall first completely solve for zero-dimensional ideals the problem of finding the conditions a polynomial must satisfy in order to belong to an ideal $\mathfrak{m}$. We begin with a lemma which is also useful in other contexts.

Lemma: *If Σ is an extension field of* K *and if $f, f_1, \ldots, f_r$ are polynomials of* $\mathsf{K}[x] = \mathsf{K}[x_1, \ldots, x_n]$, *then from*

$$f \equiv 0(f_1, \ldots, f_r) \quad \text{in} \quad \Sigma[x]$$

it follows that

$$f \equiv 0(f_1, \ldots, f_r) \quad \text{in} \quad \mathsf{K}[x].$$

Proof: Let

$$f = \sum g_i f_i, \tag{16.15}$$

where the g_i are polynomials with coefficients in Σ. We express these coefficients linearly in terms of finitely many linearly independent elements $1, \omega_1, \omega_2, \ldots$ of Σ with coefficients in K. Each term $g_i f_i$ in (16.15) acquires the form

$$(g_{i0}+g_{i1}\omega_1+g_{i2}\omega_2+\cdots)f_i,$$

where the g_{ik} are polynomials with coefficients in K. From (16.15) it follows then that

$$f = \sum g_{i0}f_i+\omega_1\sum g_{i1}f_i+\omega_2\sum g_{i2}f_i+\cdots$$

or—since the field elements $1, \omega_1, \omega_2, \ldots$ are linearly independent and the terms containing $1, \omega_1, \omega_2, \ldots$ on the right and left must coincide—that

$$f = \sum g_{i0}f_i, \qquad \text{Q.E.D.}$$

On the basis of this theorem, in answering the question of whether $f \equiv 0(f_1, \ldots, f_r)$ we may always extend the base field K, for example, by adjoining the zeros of the ideal $(f_1, \ldots, f_r)$. If the congruence holds in the extended ring $\Sigma[x]$, then it also holds before the extension.

Under appropriate extension of the base field a zero-dimensional variety always decomposes into single points; we may therefore always assume, whenever it is advantageous to do so, that all zero-dimensional ideals which occur have only a single point as a zero (rather than a system of conjugate points).

A zero-dimensional prime ideal is maximal, since the residue class ring $\mathfrak{o}/\mathfrak{p}$ is a field by Section 16.4. This implies that every zero-dimensional primary ideal is single-primed, for a primary ideal whose associated prime ideal is maximal is always single-primed by Section 15.8. It further follows from the theorems of Section 15.8 that every zero-dimensional isolated primary component $\mathfrak{q}$ of an ideal $\mathfrak{m}$ can be represented by

$$\mathfrak{q} = (\mathfrak{m}, \mathfrak{p}^{\varrho}). \tag{16.16}$$

The exponent ϱ is here the smallest integer σ with the property

$$\mathfrak{p}^{\varrho} \equiv 0(\mathfrak{m}, {}^{\sigma+1}). \tag{16.17}$$

Let us now clarify the meaning of relation (16.16) in the case in which the base field has been extended so that the single-primed ideals $\mathfrak{q}$ under consideration each have only one zero $a = \{a_1, \ldots, a_n\}$. Relation (16.16) states that for $f \equiv 0(\mathfrak{q})$ it is necessary and sufficient that

$$f \equiv 0(\mathfrak{m}, \mathfrak{p}^{\varrho}). \tag{16.18}$$

If now $\mathfrak{m}$ is given by a basis $(f_1, \ldots, f_r)$ and if we put $y_\nu = x_\nu - a_\nu$, then $\mathfrak{p} = (y_1, \ldots, y_n)$. If we suppose that all polynomials which occur are ordered according to increasing powers of y_ν, then $\mathfrak{p}^{\varrho}$ consists of all those polynomials which contain only products of powers of the y_ν of degree $\geqq \varrho$. Relation (16.18) implies then that f coincides with a linear combination $\sum g_\nu f_\nu$ up to terms of degree ϱ and higher. If we then suppose $f_1, \ldots, f_r$ to be multiplied by 1 and by all products of the y_ν of degree less than ϱ, and if we denote by $h_1, \ldots, h_k$ the polynomials formed by omitting all terms of degree $\geqq \varrho$, then (16.18) states that up to terms of degree $\geqq \varrho$ f is a linear combination of $h_1, \ldots, h_k$ with constant coefficients. This is a state of affairs whose existence or nonexistence can actually be established in any given case (for given $\varrho, f_1, \ldots, f_r$, and f). In particular, it exists if there are formal power series $P_1(y), \ldots, P_r(y)$[6] such that

$$f = P_1 f_1 + \cdots + P_r f_r.^{7} \tag{16.19}$$

Indeed, for any value of σ we can break these power series off at terms of degree σ and check to see if both sides coincide mod $\mathfrak{p}^{\sigma}$. The power series criterion (16.19)

[6]Of course, nothing is assumed with regard to convergence.

[7]This means that in a formal expansion in power products of the y_ν the two sides of (16.19) coincide.

actually still requires too much: the two sides of (16.19) need not coincide exactly but rather just up to terms of degree $\geqq \varrho$.

Similarly, the validity or nonvalidity of relation (16.17) can be determined for each σ: it means that all power products of degree σ can be represented by the polynomials $\sum g_\nu f_\nu$ by omitting the power products of degree $>\sigma$. For given $f_1, \ldots, f_r$ we may then test for each zero a the values $\sigma = 1, 2, 3, \ldots$ in succession until a σ is found for which (16.17) holds: this σ is then the exponent of $\mathfrak{q}$.

In the case of a zero-dimensional ideal $\mathfrak{m}$ all primary components are zero-dimensional and isolated; the above criterion for $f = 0(\mathfrak{q})$ may therefore be applied to all the components. If it is satisfied for all zeros, then it follows that $f \equiv 0(\mathfrak{m})$. This implies the following theorem.

Theorem: *If for each zero $a = \{a_1, \ldots, a_n\}$ of a zero-dimensional ideal $\mathfrak{m}$ the exponent ϱ is determined as the least natural number σ for which* (16.17) *holds with $\mathfrak{p} = (x_1 - a_1, \ldots, x_n - a_n)$ and if a polynomial f satisfies condition* (16.18) *for all these $\mathfrak{p}$, then $f \equiv 0(\mathfrak{m})$.*

This theorem was first formulated by Max Noether for the case $\mathfrak{m} = (f_1, f_2)$, where f_1 and f_2 are polynomials in two variables.[8] This was the celebrated *Noethersche Fundamentalsatz*, which formed the basis for the "geometrical trend" in the theory of algebraic functions. Noether actually assumed that the power series condition (16.19) was satisfied at all zeros rather than the weaker relation (16.18). The formulation presented here, in which only the terms up to degree $\varrho - 1$ in $y_1, \ldots, y_n$ are required to agree, is due to Bertini,[9] who also gave a bound for the exponent ϱ.[10] The n-dimensional generalization is due to Lasker and Macaulay. Following Macaulay, we call the sufficient condition $f \equiv 0(\mathfrak{m}, \mathfrak{p}^\varrho)$ for $f \equiv 0(\mathfrak{q})$ the *Noether condition at the point a.*

To illustrate the application of Noether's theorem, we now treat a special case in which the Noether conditions turn out to be especially simple.

Each of the polynomials $f_1, \ldots, f_r$ determines a manifold (hypersurface) $f_\nu = 0$ in n-dimensional space. The polynomial f likewise determines a hypersurface $f = 0$. If f splits into irreducible factors: $f = p_1^{\varrho_1} p_2^{\varrho_2} \ldots$, then the manifold $f = 0$ also decomposes into irreducible components $p_1 = 0, p_2 = 0, \ldots$, each of which we count as often as the exponent in the decomposition of f indicates.

If f is expanded in powers of $y_\nu = x_\nu - a_\nu$ at a point a and if the expansion begins with terms of order s $(s \geqq 0)$

$$f = c_0 y_1^s + c_1 y_1^{s-1} y_2 + \cdots + c_\omega y_n^s + \cdots,$$

then the hypersurface $f = 0$ is said to *have an s-fold point at a*. The terms $c_0 y_1^s + \cdots + c_\omega y_n^s$ of order s also define, when set equal to zero, a hypersurface which

[8] M. Noether, "Über einen Satz aus der Theorie der Algebraischen Funktionen," *Math. Ann.*, **6**, 351–359 (1873).

[9] E. Bertini, "Zum Fundamentalsatz aus der Theorie der Algebraischen Funktionen," *Math. Ann.*, **34**, 447–449 (1889).

[10] Sharper bounds were found by P. Dubreil, Doctoral Thesis, Paris, 1930.

consists solely of "straight lines" through a: the *tangent cone* of the hypersurface $f = 0$ at the point a.

The simplest case of Noether's theorem is that in which, among the hypersurfaces, $f_1 = 0, \ldots, f_r = 0$, which determine the zero-dimensional ideal $\mathfrak{m}$, such $f_1 = 0, \ldots, f_n = 0$ occur which at a all have a simple point and whose tangent hyperplanes have only the point a in common:

$$f_1 = c_{11}y_1 + \cdots + c_{1n}y_n + \cdots$$
$$f_2 = c_{21}y_1 + \cdots + c_{2n}y_n + \cdots$$
$$\cdots$$
$$f_n = c_{n1}y_1 + \cdots + c_{nn}y_n + \cdots$$

The linear forms $\sum_{\mu=1}^{n} c_{\lambda\mu}y_\mu$ are linearly independent.

If the prime ideal $(x_1 - a_1, \ldots, x_n - a_n)$ is denoted by $\mathfrak{p}$, then in this case among the linear combinations of $f_1, \ldots, f_n$ modulo $\mathfrak{p}^2$ (that is, neglecting terms of second and higher order) $y_1, \ldots, y_n$ themselves occur; that is,

$$(y_1, \ldots, y_n) \equiv 0((f_1, \ldots, f_n), \mathfrak{p}^2)$$

and hence

$$\mathfrak{p} \equiv 0(\mathfrak{m}, \mathfrak{p}^2).$$

Thus, at the point a the ideal $\mathfrak{m}$ has an isolated primary component $\mathfrak{q}$ with exponent 1, that is, $\mathfrak{q} = \mathfrak{p}$. Any polynomial with the zero a is therefore divisible by $\mathfrak{q}$.

For further special cases and applications of Noether's theorem see my *Einführung in die Algebraische Geometrie* (Grundlehren, Springer, 1939).

16.8 REDUCTION OF MULTIDIMENSIONAL IDEALS TO ZERO-DIMENSIONAL IDEALS

In this section we shall seek to extend to multidimensional ideals the theorems proved in Section 16.7 for zero-dimensional ideals.

The method is as follows. If $\mathfrak{q}$ is a primary ideal in $K[x]$ of dimension d, $\mathfrak{p}$ the associated prime ideal, and $\{\xi_1, \ldots, \xi_n\}$ its generic zero, and if (say) $\xi_1, \ldots, \xi_d$ are algebraically independent, then we make $\mathfrak{q}$ and $\mathfrak{p}$ zero-dimensional ideals by the substitution $x_1 = \xi_1, \ldots, x_d = \xi_d$. We make this substitution in all polynomials g of the ideal $\mathfrak{q}$; the polynomials q hereby go over into polynomials q' of $K(\xi_1, \ldots, \xi_d)[x_{d+1}, \ldots, x_n]$ which generate an ideal $\mathfrak{q}'$. It is clear that it suffices to make the substitution $x_1 = \xi_1, \ldots, x_d = \xi_d$ in the basis polynomials $q_1, \ldots, q_r$; the resulting polynomials $q_1', \ldots, q_r'$ then generate the ideal $\mathfrak{q}'$:

$$\mathfrak{q}' = (q_1', \ldots, q_r').$$

The ideal $\mathfrak{q}'$ clearly consists of the polynomials q' divided by arbitrary nonzero polynomials φ in $\xi_1, \ldots, \xi_d$; for the polynomials q' form an ideal in $K[\xi_1, \ldots, \xi_d, x_{d+1}, \ldots, x_n]$ and in order to obtain the ideal generated in $K(\xi_1, \ldots, \xi_d)[x_{d+1}, \ldots, x_n]$ we have only to admit the denominators φ.

An ideal $\mathfrak{p}'$ arises from $\mathfrak{p}$ in the same way that $\mathfrak{q}'$ arises from $\mathfrak{q}$; in general, every ideal $\mathfrak{m} = (f_1, \ldots, f_r)$ gives rise to an ideal $\mathfrak{m}' = (f_1', \ldots, f_r')$.

Geometrically the substitution $x_1 = \xi_1, \ldots, x_d = \xi_d$ means that all manifolds which occur are cut by the linear space $x_1 = \xi_1, \ldots, x_d = \xi_d$ which passes through the generic point of the manifold of $\mathfrak{q}$.

If $f(x_1, \ldots, x_n)$ is a polynomial and if $f(\xi_1, \ldots, \xi_d, x_{d+1}, \ldots, x_n)$ belongs to $\mathfrak{q}'$, then by the preceding remarks

$$f(\xi, x) = \frac{q'}{\varphi(\xi_1, \ldots, \xi_d)} = \frac{q(\xi, x)}{\varphi(\xi)} \qquad \text{with} \quad q(x) \equiv 0(\mathfrak{q}),$$

and hence

$$q(\xi, x) = \varphi(\xi) f(\xi, x).$$

Because of the algebraic independence of $\xi_1, \ldots, \xi_d$, this implies

$$q(x) = \varphi(x) f(x) \equiv 0(\mathfrak{q}).$$

However, $\varphi(\xi) \neq 0$ implies $\varphi(x) \not\equiv 0(\mathfrak{p})$; therefore

$$f(x) \equiv 0(\mathfrak{q}).$$

Hence, in order to decide if a polynomial $f(x)$ belong to $\mathfrak{q}$, we need only investigate whether the corresponding $f' = f(\xi_1, \ldots, \xi_d, x_{d+1}, \ldots, x_n)$ belongs to $\mathfrak{q}'$. Thus $\mathfrak{q}'$ determines $\mathfrak{q}$ uniquely.

We now state the following theorem.

Theorem: *The ideal* $\mathfrak{q}'$ *in* $K(\xi_1, \ldots, \xi_d)[x_{d+1}, \ldots, x_n]$ *is primary; the associated prime ideal is* $\mathfrak{p}'$; *the exponent of* $\mathfrak{q}'$ *is equal to that of* $\mathfrak{q}$; *the generic zero of* $\mathfrak{p}'$ *is* $\{\xi_{d+1}, \ldots, \xi_n\}$, *and the dimension of* $\mathfrak{p}'$ *is zero.*

Proof: In order to show that $\mathfrak{q}'$ is primary and $\mathfrak{p}'$ is the associated prime ideal, it suffices to prove the following three properties.

1. $f(\xi, x) g(\xi, x) \equiv 0(\mathfrak{q}')$ and $f(\xi, x) \not\equiv 0(\mathfrak{p}')$ imply $g(\xi, x) \equiv 0(\mathfrak{q}')$.
2. $f(\xi, x) \equiv 0(\mathfrak{q}')$ implies $f(\xi. x) \equiv 0(\mathfrak{p}')$.
3. $f(\xi, x) \equiv 0(\mathfrak{p}')$ implies $f(\xi, x)^\varrho \equiv 0(\mathfrak{q}')$.

In establishing all three properties we may assume that f and q are integral in $\xi_1, \ldots, \xi_d$, for otherwise we have only to multiply by a suitable $\varphi(\xi)$. By the remarks above, we may everywhere replace ξ by x, $\mathfrak{q}'$ by $\mathfrak{q}$ and $\mathfrak{p}'$ by $\mathfrak{p}$; then, for example, $f(\xi, x) \equiv 0(\mathfrak{q}')$ is equivalent to $f(x) \equiv 0(\mathfrak{q})$, and so on. After these substitutions, properties 1, 2, and 3 state nothing more than that $\mathfrak{q}$ is primary and $\mathfrak{p}$ is its associated prime ideal, and this we already know. This shows at the same time that the exponents of $\mathfrak{q}$ and $\mathfrak{q}'$ are the same.

In order to show that $\{\xi_{d+1}(\ldots, \xi_n\}$ is the generic zero of $\mathfrak{p}'$, we must prove that

$$f(\xi_1, \ldots, \xi_d, \xi_{d+1}, \ldots, \xi_n) = 0,$$

where f is rational in $\xi_1, \ldots, \xi_d$ and integral in $\xi_{d+1}, \ldots, \xi_n$, implies that

$$f(\xi, x) \equiv 0(\mathfrak{p}')$$

and conversely. We may again assume that f is integral in $\xi_1, \ldots, \xi_d$. But then $f(\xi, x) \equiv 0(\mathfrak{p}')$ is equivalent to $f(x) \equiv 0(\mathfrak{p})$; this part of the proof is therefore completed simply by the remark that $\{\xi_1, \ldots, \xi_n\}$ is the generic zero of $\mathfrak{p}$.

The zero-dimensionality of $\mathfrak{p}'$ follows finally from the fact that $\xi_{d+1}, \ldots, \xi_n$ are algebraic over $\mathsf{K}(\xi_1, \ldots, \xi_d)$. This completes the proof of the theorem.

In the same way we can show that *if* $\mathfrak{q}$ *is a primary component of the ideal* $\mathfrak{m} = (f_1, \ldots, f_r)$, *then* $\mathfrak{q}'$ *is a primary component of the corresponding ideal* $\mathfrak{m}' = (f_1' \ldots, f_r')$. *If* $\mathfrak{q}$ *is an isolated component of* $\mathfrak{m}$, *then* $\mathfrak{q}'$ *is also an isolated component of* $\mathfrak{m}'$.

The method developed for reducing all primary ideals to zero-dimensional ideals affords us the means of determining whether a polynomial f belongs to a given ideal $\mathfrak{m} = (f_1, \ldots, f_r)$, assuming that the decomposition of $\mathfrak{m}$ into primary components

$$\mathfrak{m} = [\mathfrak{q}_1, \ldots, \mathfrak{q}_s]$$

is given. We find for each primary component $\mathfrak{q}$ the corresponding zero-dimensional ideal $\mathfrak{q}'$and extend the field $\mathsf{K}(\xi_1, \ldots, \xi_d)$ so that $\mathfrak{q}'$ decomposes into primary ideals $\mathfrak{q}_\nu'$, each having only one zero $a^{(\nu)}$; using the methods of Section 16.7, we then determine by means of the "Noether conditions"

$$f' \equiv (0\mathfrak{q}', \mathfrak{p}_\nu'^{\varrho}), \qquad \mathfrak{p}_\nu' = (x_{d+1} - a_{d+1}^{(\nu)}, \ldots, x_n - a_n^{(\nu)}), \tag{16.20}$$

if the polynomial f' belongs to the ideals $\mathfrak{q}_\nu' = (\mathfrak{q}', \mathfrak{p}_\nu'^{\varrho})$ and hence also to the ideal $\mathfrak{q}'$. Since the zeros of the $\mathfrak{p}_\nu'$ are conjugate with respect to $\mathsf{K}(\xi_1, \ldots, \xi_d)$, the $\mathfrak{p}_\nu'$ and hence also the $\mathfrak{q}_\nu'$ are conjugate with respect to $\mathsf{K}(\xi_1, \ldots, \xi_d)$; it therefore suffices to investigate a *single* $\mathfrak{q}_\nu'$ for each $\mathfrak{q}'$. We thus need adjoin only one zero of each $\mathfrak{q}'$. Now $\{\xi_{d+1}, \ldots, \xi_n\}$ is such a zero. Hence, $\mathfrak{p}_\nu'$ is replaced by the prime ideal

$$\mathfrak{p}_\xi = (x_{d+1} - \xi_{d+1}, \ldots, x_n - \xi_n),$$

and we may use the the more convenient condition

$$f' \equiv 0(\mathfrak{m}', \mathfrak{p}_\xi^{\varrho}) \tag{16.21}$$

in place of condition (16.20), since (16.21) is also necessary for $f \equiv 0(\mathfrak{m})$, and (16.20) follows immediately from (16.21). Condition (16.21), which must be satisfied by every primary component $\mathfrak{q}$ of $\mathfrak{m}$, goes by the name of the *Hentzelt criterion* or *Hentzelt's Nullstellensatz*.

In particular, if $\mathfrak{q}$ is an isolated component of $\mathfrak{m}$, and hence $\mathfrak{q}'$ is also an isolated component of $\mathfrak{m}'$, then the exponent ϱ can be determined from the condition

$$\mathfrak{p}_\xi^\varrho \equiv 0(\mathfrak{m}', \mathfrak{p}_\xi^{\varrho+1})$$

as in Section 15.8.

The real geometric meaning of primary ideals is most clearly seen from conditions (16.20) for $f \equiv 0(\mathfrak{q})$. Membership in a primary ideal imposes certain conditions on the initial terms of the expansion of a polynomial f in powers of $x_1 - \xi_1, \ldots, x_n - \xi_n$ for a generic point ξ of an irreducible manifold M. For example, the condition may be that f should vanish at this generic point or that the hypersurface $f = 0$ at this generic point should intersect another hypersurface containing M, and so on.

Exercises

16.8. Using the method of reduction to zero-dimensional ideals, prove that every $(n-1)$-dimensional primary ideal in $K[x_1, \ldots, x_n]$ is a principal ideal.

16.9. Every unmixed $(n-1)$-dimensional ideal in $K[x_1, \ldots, x_n]$ is a principal ideal, and conversely.

Chapter 17

INTEGRAL ALGEBRAIC ELEMENTS

Historically the development of ideal theory has two starting points: the theory of algebraic integers and the theory of polynomial ideals. These two theories have, however, been developed to deal with entirely different problems. Whereas in the case of polynomial ideals the central problems have been to determine the zeros of an ideal and establish the necessary and sufficient conditions for a polynomial to belong to an ideal, the theory of algebraic integers arises from the question of factorization. We arrive at this question through the following considerations, for example.

In the ring of elements $a+b\sqrt{-5}$, where a and b are integers, the unique factorization theorem does not hold. The number 9, for example, admits two essentially different factorizations into indecomposable factors:[1]

$$9 = 3\cdot 3 = (2+\sqrt{-5})\,(2-\sqrt{-5}).$$

This fact led Dedekind (following Kummer, who achieved unique factorization in cyclotomic fields by introducing certain "ideal numbers") to extend the domain of elements to that of ideals (he was the first to use this term). He was able to show that in this domain each ideal is equal to a uniquely determined product of prime ideals. Indeed, if the prime ideals

$$\mathfrak{p}_1 = (3, 2+\sqrt{-5}), \qquad \mathfrak{p}_2 = (3, 2-\sqrt{-5})$$

are introduced in the case above, then

$$(3) = \mathfrak{p}_1\mathfrak{p}_2; \qquad (2+\sqrt{-5}) = \mathfrak{p}_1{}^2; \qquad (2-\sqrt{-5}) = \mathfrak{p}_2{}^2,$$

[1]That the numbers 3 and $2\pm\sqrt{-5}$ are indecomposable follows easily from the fact that their norm is 9 (cf. Section 6.11). If they were decomposable, then either both factors would have norm ± 3, or one factor would have norm ± 1. There is no number $a+b\sqrt{-5}$ with norm ± 3, for then

$$a^2+5b^2 = \pm 3,$$

which is impossible for integers a and b. A number with norm ± 1 is necessarily one of the units ± 1, since

$$a^2+5b^2 = \pm 1$$

is satisfied only by $a = \pm 1$, $b = 0$.

as is easily checked; the principal ideal (9) then has the (unique) factorization

$$(9) = \mathfrak{p}_1{}^2\mathfrak{p}_2{}^2.$$

In this chapter the "classical" (Dedekind) ideal theory of the integral elements of a field will be presented in a modern axiomatic form due to Emmy Noether.[2]

17.1 FINITE $\mathfrak{R}$-MODULES

We consider modules over a (not necessarily commutative) ring $\mathfrak{R}$, that is, modules with $\mathfrak{R}$ as (left) multiplier domain. The modules considered are usually contained either in $\mathfrak{R}$ itself (and thus are left ideals in $\mathfrak{R}$) or in an extension ring $\mathfrak{S}$.

A *finite $\mathfrak{R}$-module* is a module $\mathfrak{M}$ which is generated by a finite *module basis* $(a_1, \ldots, a_h)$, that is, one whose elements can be linearly expressed in terms of $a_1, \ldots, a_h$ with coefficients in $\mathfrak{R}$ and with integer coefficients

$$m = r_1a_1 + \cdots + r_ha_h + n_1a_1 + \cdots + n_ha_h \qquad (r_\nu \in \mathfrak{R},\ n_\nu \text{ integers}). \quad (17.1)$$

In this case we write $\mathfrak{M} = (a_1, \ldots, a_h)$.

The *ascending chain condition* is said to hold in a module $\mathfrak{M}$ if every chain of submodules $\mathfrak{M}_1, \mathfrak{M}_2, \ldots$ of $\mathfrak{M}$ in which each successive one properly contains (divides) the preceding

$$\mathfrak{M}_1 \subset \mathfrak{M}_2 \subset \cdots,$$

breaks off after a finite number of terms.

Theorem: *If the ascending chain condition holds in $\mathfrak{M}$, then every submodule of $\mathfrak{M}$ has a finite basis, and conversely.*

This theorem is a generalization of the theorem of Section 15.1 concerning an ideal basis and the ascending chain condition. The proof is similar. To find a basis for a submodule $\mathfrak{N}$, we first choose an element a_1 in $\mathfrak{N}$. If $(a_1) = \mathfrak{N}$, we are through; otherwise we choose an element a_2 in $\mathfrak{N}$ which is not contained in (a_1). If $(a_1, a_2) = \mathfrak{N}$, we are through; otherwise we choose an a_3, and so forth. If it is known that the chain of modules

$$(a_1) \subset (a_1, a_2) \subset (a_1, a_2, a_3) \subset \cdots$$

must break off after a finite number of terms, then $\mathfrak{N}$ has a finite basis.

Conversely, if every submodule of $\mathfrak{M}$ has a finite basis and if

$$\mathfrak{M}_1 \subset \mathfrak{M}_2 \subset \cdots$$

is a chain of submodules of $\mathfrak{M}$, then the union T of all the $\mathfrak{M}_\nu$ is again a submodule and hence has a finite basis:

$$\mathfrak{B} = (a_1, \ldots, a_r).$$

[2]E. Noether, "Abstrakter Aufbau der Idealtheorie in Algebraischen Zahl- und Funktionenkörpern," *Math. Ann.*, **96**, 26–61 (1926).

All the a_ν are already contained in some $\mathfrak{M}_\omega$ of the chain; hence $\mathfrak{B} \subseteq \mathfrak{M}_\omega$, so that $\mathfrak{B} = \mathfrak{M}_\omega$. The chain therefore breaks off at $\mathfrak{M}_\omega$.

A sufficient condition for the ascending chain condition to hold in $\mathfrak{M}$ is given by the following.

Theorem: *If the ascending chain condition for left ideals holds in $\mathfrak{R}$ and if $\mathfrak{M}$ is a finite $\mathfrak{R}$-module, then the ascending chain condition for $\mathfrak{R}$-modules holds in $\mathfrak{M}$.*

An equivalent formulation (on the basis of the preceding theorem) is the following.

Theorem: *If in $\mathfrak{R}$ every left ideal has a finite ideal basis and if $\mathfrak{M}$ has a finite $\mathfrak{R}$-module basis, then every submodule of $\mathfrak{M}$ has a finite $\mathfrak{R}$-module basis.*

The proof is entirely analogous to the proof of the Hilbert basis theorem (Section 15.1). Let $\mathfrak{M} = (a_1, \ldots, a_h)$, and let $\mathfrak{N}$ be a submodule of $\mathfrak{M}$. Every element of $\mathfrak{N}$ can be written in the form (17.1). If in (17.1) the last $2h-l$ coefficients (from the $(l+1)$th to the $2h$th) of the $2h$ coefficients $r_1, \ldots, n_h$ are all zero, we say that the *expression has length* $\leqq l$. We now consider all expressions in $\mathfrak{N}$ of length $\leqq l$. Their lth coefficients (r_l or n_{l-h}) form a left ideal in $\mathfrak{R}$ or in the ring $\mathbb{Z}$ of integers. This ideal has a finite basis:

$$(b_{l1}, \ldots, b_{ls_l}).$$

Each $b_{l\nu}$ is the last (lth) coefficient (r_l or n_{l-h}) of a certain expression of the form (17.1) which we denote by $B_{l\nu}$:

$$B_{l\nu} = r_1a_1 + \cdots + b_{l\nu}a_l \qquad \text{or} \quad = r_1a_1 + \cdots + b_{l\nu}a_{l-h}.$$

We assert that all these $B_{l\nu}$ ($l = 1, \ldots, 2_h$; $\nu = 1, \ldots, s_l$) together form a basis for $\mathfrak{N}$. Indeed, from any element (17.1) of $\mathfrak{N}$ of length l the last (lth) coefficient can be eliminated by subtracting a linear combination of $B_{l_1}, \ldots, B_{ls_l}$ (with coefficients in $\mathfrak{R}$ or $\mathbb{Z}$ according to the value of l); this means that any such expression can be reduced to an expression of smaller length; this resulting expression can again be reduced in length in the same way until finally, by repeated subtraction of linear combinations of the $B_{l\nu}$, only zero remains. Any element of $\mathfrak{N}$ can thus be written as a linear combination of the $B_{l\nu}$, Q.E.D. If one of the ideals $(b_{l_1}, \ldots, b_{ls_l})$ is the null ideal, then the corresponding $B_{l\nu}$ are redundant in the basis.

17.2 INTEGRAL ELEMENTS OVER A RING

Let $\mathfrak{R}$ be a subring of a ring $\mathfrak{T}$.

An element t of $\mathfrak{T}$ is said to be *integral over* $\mathfrak{R}$ if all powers[3] of t belong to a finite $\mathfrak{R}$-module $(a_1, \ldots, a_m)$, that is, all powers of t can be linearly represented in terms of finitely many elements $a_1, \ldots, a_m$ of $\mathfrak{T}$ in the form

$$t^\varrho = r_1a_1 + \cdots + r_ma_m + n_1a_1 + \cdots n_ma_m \qquad (r_\nu \in \mathfrak{R},\ n_\nu \text{ integers}). \quad (17.2)$$

[3]In this section we shall always mean powers with positive exponents.

In particular, every element r or $\mathfrak{R}$ is integral over $\mathfrak{R}$, since $r, r^2, r^3, \ldots$ belong to the $\mathfrak{R}$-module (r). The identity element of $\mathfrak{T}$, if it exists, is also integral over $\mathfrak{R}$.

If $\mathfrak{T}$ is a field, which thus contains the quotient field P of $\mathfrak{R}$, then all powers of an integral element t depend linearly on finitely many elements $a_1, \ldots, a_m$ with coefficients in P, for P contains not only the ring $\mathfrak{R}$, but also the identity element. Therefore, only finitely many of the powers of t are linearly independent over P; t is thus algebraic over P. For this reason we also speak of *integral algebraic elements* instead of "integral elements."

If $\mathfrak{R}$ is a ring in which the ascending chain condition for left ideals holds, then by Section 17.1 the ascending chain condition also holds for submodules of the finite $\mathfrak{R}$-module $(a_1, \ldots, a_m)$. In particular, a chain of modules

$$(t) \subseteq (t, t^2) \subseteq \cdots$$

cannot consist entirely of distinct modules; that is, some power of t can be expressed linearly in terms of lower powers of t:

$$t^h = r_1 t + \cdots + r_{h-1} t^{h-1} + n_1 t + \cdots + n_{h-1} t^{h-1}. \tag{17.3}$$

Conversely, if t is an element of $\mathfrak{T}$ which admits a representation of the form (17.3) with coefficients in $\mathfrak{R}$ or $\mathbb{Z}$ for some h, then all higher powers of t can successively be linearly expressed in terms of the finite number of elements t, $t^2, \ldots, t^{h-1}$ by means of (17.3); t is therefore integral according to our definition. This proves the following theorem.

Theorem: *If the ascending chain condition for left ideals holds in the ring* $\mathfrak{R}$, *then the existence of an equation of the form* (17.3) *is necessary and sufficient in order that t be integral over* $\mathfrak{R}$.

If $\mathfrak{T}$ is a field equation, (17.3) again expresses the fact that t is algebraic. If $\mathfrak{R}$ has an identity element, then $t^0 = 1$ may be included among the powers of t and, moreover, in (17.3) the tail $n_1 t + \cdots + n_{h-1} t^{h-1}$ may be omitted. In place of (17.3) we then have the simpler equation

$$t^h - r_{h-1} t^{h-1} \cdots - r_0 = 0,$$

whose characteristic feature is that the coefficient of the highest power of t is 1.

Examples: *Algebraic integers* are those algebraic numbers which are integral over the ring $\mathbb{Z}$ of ordinary integers; that is, they satisfy an equation with integer coefficients and leading coefficient 1. *Integral algebraic functions* of $x_1, \ldots, x_n$ are those functions of an algebraic extension field of $\mathsf{K}(x_1, \ldots, x_n)$ which are integral over the polynomial ring $\mathsf{K}[x_1, \ldots, x_n]$; K is here a fixed base field. *Absolutely integral algebraic functions* of $x_1, \ldots, x_n$ are such functions which are integral over the ring $\mathbb{Z}[x_1, \ldots, x_n]$.

In a commutative ring $\mathfrak{T}$ *the sum, difference, and product of two integral elements over* $\mathfrak{R}$ *are again integral.* Or: *The elements in* $\mathfrak{T}$ *which are integral over* $\mathfrak{R}$ *form a ring* $\mathfrak{S}$.

Proof: If all powers of s can be linearly expressed in terms of $a_1, \ldots, a_m$ and all powers of t in terms of $b_1, \ldots, b_n$, then all powers of $s+t$, $s-t$, and $s \cdot t$ can be linearly expressed in terms of $a_1, \ldots, a_m, b_1, \ldots, b_n, \ldots, a_1b_1, a_1b_2, \ldots, a_mb_n$.

If we assume the ascending chain condition for the ideals of the ring $\mathfrak{S}$, then we can prove the *theorem on the transitivity of integral dependence.*

If $\mathfrak{S}$ *is the ring of integral elements in a commutative ring* $\mathfrak{T}$ *(over a subring* $\mathfrak{R}$*) and if the element* t *of* $\mathfrak{T}$ *is integral over* $\mathfrak{S}$*, then* t *is also integral over* $\mathfrak{R}$ *(that is, contained in* $\mathfrak{S}$*).* Or, expressed in another manner: *If* t *satisfies an equation of the form* (17.3) *whose coefficients* r_ν *are integral over* $\mathfrak{R}$*, then* t *itself is integral over* $\mathfrak{R}$.

Proof: By repeated application of equation (17.3) all powers $t^{h+\lambda}$ can be linearly expressed in terms of $t, t^2, \ldots, t^{h-1}$ with coefficients which either are integers or are integral rational functions in the products of powers of the r_ν. For each r_ν there are finitely many elements of $\mathfrak{T}$ in terms of which all powers of r_ν can be linearly expressed with coefficients in $\mathfrak{R}$ or with integer coefficients; all products of powers of the r_ν can therefore be linearly expressed in terms of finitely many products of these finitely many elements. If we now multiply this finite number of products by $t, t^2, \ldots, t^{h-1}$ and include finally, $t, t^2, \ldots, t^{h-1}$ as well, we again obtain finitely many elements in terms of which all powers of t can now be linearly expressed with integer coefficients or coefficients in $\mathfrak{R}$.

A ring $\mathfrak{S}$ is said to be *integrally closed in an overring* $\mathfrak{T}$ if every element of $\mathfrak{T}$ which is integral over $\mathfrak{S}$ is contained in $\mathfrak{S}$. In particular, an integral domain $\mathfrak{S}$ which is integrally closed in its quotient field Σ is simply called *integrally closed.* As is easily seen, this means that each element t of Σ, all of whose powers t^ϱ admit a representation as fractions with a fixed denominator from $\mathfrak{S}$, themselves belong to $\mathfrak{S}$. The finite number of elements, in terms of which all powers of an integral t can be represented, can always be brought to a common denominator; conversely, if all powers of t can be expressed as fractions with denominator s, then they can all be expressed linearly in terms of the element s^{-1}.

It now follows from the theorem above that if $\mathfrak{T}$ is commutative *the ring* $\mathfrak{S}$ *of all elements of* $\mathfrak{T}$ *which are integral over* $\mathfrak{R}$ *is always integrally closed in* $\mathfrak{T}$ *if the ascending chain condition holds for the ideals of* $\mathfrak{S}$.

This theorem can also be proved without assuming the ascending chain condition if, instead, it is assumed that $\mathfrak{R}$ is integrally closed in its quotient field P and $\mathfrak{T}$ is a finite extension field of P. For the proof, $\mathfrak{T}$ is extended to a Galois field $\mathfrak{T}'$ over P, and $\mathfrak{S}$ is extended to the ring $\mathfrak{S}'$ of integral elements of $\mathfrak{T}'$. If an element t is integral over $\mathfrak{S}$, and hence also over $\mathfrak{S}'$, then so are the conjugate elements of t (over P), and hence also the elementary symmetric functions of these conjugate elements, that is, the coefficients of the defining equation of t. Since $\mathfrak{R}$ is integrally closed, these coefficients then belong to $\mathfrak{R}$; t is therefore integral over $\mathfrak{R}$ and hence $t \in \mathfrak{S}$.

A sufficient (but by no means necessary) criterion for the integral closure of an integral domain is given by the following theorem.

Theorem: *An integral domain with identity in which the unique factorization theorem holds is integrally closed in its quotient field.*

Proof: Each element of the quotient field can be represented as a fraction a/b

in which a and b have no common prime factor. If the denominator of all powers of a/b can be eliminated by multiplication by a single element c, then ca^n, and hence c also, must be divisible by b^n for every n. This is possible only if b is a unit and hence $a/b = ab^{-1}$ belongs to the integral domain.

From this theorem it follows that every principal ideal ring (in particular, the ring $\mathbb{Z}$ of integers), every polynomial ring over the ring of integers, and every polynomial ring over a commutative field K is integrally closed.

Exercises

17.1. The roots of unity of a field are always integral over every subring.

17.2. Which numbers of the Gaussian number field $\mathbb{Q}(i)$ are integral over $\mathbb{Z}$? Which numbers of the field $\mathbb{Q}(\varrho)$ of third roots of unity ($\varrho = -\frac{1}{2}+\frac{1}{2}\sqrt{-3}$)?

17.3. If the integral domain $\mathfrak{R}$ is integrally closed, then the polynomial ring $\mathfrak{R}[x]$ is also integrally closed.

17.3 THE INTEGRAL ELEMENTS OF A FIELD

Let $\mathfrak{R}$ be an integral domain, P its quotient field, Σ a finite commutative extension field of P, and $\mathfrak{S}$ the ring of elements of Σ which are integral over $\mathfrak{R}$. Obviously $\mathfrak{S}$ is an extension ring of $\mathfrak{R}$. The relations between the rings $\mathfrak{R}$, $\mathfrak{S}$ and the fields P, Σ may be schematically represented as follows:

$$\begin{array}{ccc} \mathsf{W} & \subseteq & \mathfrak{S} \\ \cap & & \cap \\ \mathsf{P} & \subseteq & \Sigma. \end{array}$$

These designations will be the same throughout this section. By "integral" we shall always mean integral over $\mathfrak{R}$.

Examples: If $\mathfrak{R}$ is the ring of ordinary integers, then P is the field of rational numbers, Σ is a number field (finite over P), and $\mathfrak{S}$ is the ring of algebraic integers of the field Σ.

If $\mathfrak{R}$ is a polynomial ring, $\mathfrak{R} = \mathsf{K}[x_1, \ldots, x_n]$, then P is the field of rational functions; Σ arises by adjunction of finitely many algebraic functions, and $\mathfrak{S}$ consists of the integral algebraic functions of the field Σ; and so on.

Our aim is to study the ideal theory of $\mathfrak{S}$. To this end, as we know, we must first investigate the question of the ascending chain condition for the ideals of $\mathfrak{S}$. More precisely, we ask if the ascending chain condition is inherited by $\mathfrak{S}$ if it holds for $\mathfrak{R}$. By the theorems of Section 17.1, this is so if an $\mathfrak{R}$-module basis can be found for $\mathfrak{S}$. This will be our first objective.

We begin with a preparatory theorem.

Theorem: *If σ is an element of Σ, then $\sigma = s/r$, where $s \in \mathfrak{S}$ and $r \in \mathfrak{R}$.*

Proof: The element σ satisfies an equation with coefficients in P. These coefficients are fractions over $\mathfrak{R}$. They become elements of $\mathfrak{R}$ on multiplication with the product of their denominators:

$$r_0\sigma^m + r_1\sigma^{m-1} + \cdots + r_m = 0.$$

Putting $r_0 = r$ and multiplying by r^{m-1}, we obtain

$$(r\sigma)^m + r_1(r\sigma)^{m-1} + r_2 r(r\sigma)^{m-2} + \cdots + r_m r^{m-1} = 0.$$

Thus, $r\sigma$ is integral over $\mathfrak{R}$. The assertion follows on putting $r\sigma = s$.

It follows from this theorem that Σ is the quotient field of $\mathfrak{S}$.

Theorem: *If an element ξ is integral, then all conjugates of ξ (in a Galois extension field of Σ over* P*) are also integral.*

Proof: The finitely many elements of Σ in terms of which all powers of ξ can be expressed go over, under an isomorphism of Σ, into finitely many elements in terms of which all powers of some conjugate of ξ can be linearly expressed.

Since sums and products of integral elements are again integral, it follows that the elementary symmetric functions of ξ and its conjugates are integral. This implies the following theorem.

Theorem: *In the irreducible equation over* P *which an integral element ξ satisfies, if the leading coefficient is chosen equal to* 1, *then all other coefficients are integral over $\mathfrak{R}$. In particular, if $\mathfrak{R}$ is integrally closed in* P, *then all these coefficients are contained in $\mathfrak{R}$.*

In the case where $\mathfrak{R}$ is integrally closed, this theorem affords the most convenient means of investigating whether an element ξ is integral. We no longer need form all equations which ξ satisfies and check to see if among them there is an equation with integral coefficients; it suffices rather to consider the single irreducible equation with leading coefficient 1. If all the coefficients of this equation are integral, then ξ is also integral; otherwise ξ is not integral.

We now make the following restrictive assumptions:

I. $\mathfrak{R}$ *is integrally closed in its quotient field* P.
II. *The ascending chain condition for ideals holds in* $\mathfrak{R}$.
III. Σ *is a separable extension of* P.

It follows from III by Section 6.10 that Σ is generated by a "primitive element" $\sigma : \Sigma = \mathrm{P}(\sigma)$. By the theorem above, $\sigma = s/r$ ($s \in \mathfrak{S}$, $r \in \mathfrak{R}$); the integral element s therefore generates the field. Here s satisfies an equation of degree n, where n is the field degree (Σ/P). Each element ξ of Σ can be represented in the form

$$\xi = \sum_0^{n-1} \varrho_k s^k \qquad (\varrho_k \in \mathrm{P}). \tag{17.4}$$

If in (17.4) s is replaced by its conjugates s_ν (in a Galois extension field of P containing Σ), of which there are precisely n by Section 6.8, then the equations

$$\xi_\nu = \sum_0^{n-1} \varrho_k s_\nu^{\,k} \qquad (\nu = 1, 2, \ldots, n) \tag{17.5}$$

are obtained for the conjugates ξ_ν of ξ. The determinant of this system of equations is

$$D = |s_\nu^{\,k}| = \prod_{\lambda<\mu} (s_\lambda - s_\mu)$$

according to the Vandermonde determinant theorem. Its square is a symmetric function of the s_ν and therefore contained in P. Furthermore, since the conjugates s_ν are all distinct, $D \neq 0$. The system of equations (17.5) may therefore be solved:

$$\varrho_k = \frac{\sum S_{k\nu}\xi_\nu}{D},$$

where the $S_{k\nu}$ and D are polynomials in the s_ν and are therefore integral over $\mathfrak{R}$. Multiplication of this equation by D^2 gives

$$D^2\varrho_k = \sum_\nu DS_{k\nu}\xi_\nu. \tag{17.6}$$

If we assume now that ξ is an element of $\mathfrak{S}$ and is thus integral, then the ξ_ν are also integral; the right side of (17.6) is therefore integral. The left side, however is an element of P. Since $\mathfrak{R}$ is integrally closed in P, it follows that $D_{\varrho k}{}^2$ lies in $\mathfrak{R}$. If we put $D^2\varrho_k = r_k$, then $\varrho_k = r_k D^{-2}$ and so, by (17.4),

$$\xi = \sum_0^{n-1} r_k D^{-2} s^k.$$

Each element ξ of $\mathfrak{S}$ can therefore be linearly expressed in terms of $D^{-2}s^0$, $D^{-2}s^1, \ldots, D^{-2}s^{n-1}$ with coefficients in $\mathfrak{R}$. In other words: $\mathfrak{S}$ is contained in the finite $\mathfrak{R}$-module

$$\mathfrak{M} = (D^{-2}s^0, D^{-2}s^1, \ldots, D^{-2}s^{n-1}).$$

The theorems of Section 17.1 now imply that $\mathfrak{S}$ *as well as every submodule of* $\mathfrak{S}$ *and, in particular, every ideal in* $\mathfrak{S}$ *has a finite module basis over* $\mathfrak{R}$, or equivalently, *the ascending chain condition holds for the* $\mathfrak{R}$*-modules and, in particular, for the ideals in* $\mathfrak{S}$. If $\mathfrak{R}$ is a principal ideal ring, then $\mathfrak{S}$ and every submodule of $\mathfrak{S}$ have a linearly independent $\mathfrak{R}$-module basis.

An $\mathfrak{R}$-*order* in Σ is a subring of Σ which contains $\mathfrak{R}$ and is a finite $\mathfrak{R}$-module. By the above theorem $\mathfrak{S}$ is an $\mathfrak{R}$-order, and every ring between $\mathfrak{R}$ and $\mathfrak{S}$ is also. Conversely, it follows from the definition of integral elements that every $\mathfrak{R}$-order $\mathfrak{I}$ in Σ consists solely of integral elements, that is, is contained in $\mathfrak{S}$. Therefore $\mathfrak{S}$ may be characterized as the largest $\mathfrak{R}$-order in Σ. Also, $\mathfrak{S}$ is called the *principal order* of the field Σ. When terms such as "ideals of the field," "units of the field," and so on are used, the ideals of $\mathfrak{S}$, the units of $\mathfrak{S}$, and so on are always meant. By Section 17.2, $\mathfrak{S}$ is integrally closed in Σ.

The results of this section no longer hold for noncommutative algebras over P; the main difficulty is that the sum of two integral elements need not be integral. The set of integral elements is therefore not an order. Although every order consists only of integral elements, there is no principal order containing all the orders. Under the appropriate conditions on Σ there are various maximal

$\mathfrak{R}$-orders such that every $\mathfrak{R}$-order and also every integral element is contained in at least one maximal $\mathfrak{R}$-order. Concerning the ideal theory of these maximal $\mathfrak{R}$-orders, see M. Deuring, "Algebren," *Ergebn. Math.*, **4**, No. 1 (1935).

The ascending chain condition holds in all $\mathfrak{R}$-orders of Σ by what has just been proved. The decomposition and uniqueness theorems of Sections 15.4 and 15.5 (representation of all ideals as intersections of primary ideals) therefore also hold for these orders.

This ideal theory is greatly simplified, according to the final considerations of Section 15.8, if every nonzero prime ideal of the order $\mathfrak{o}$ is maximal. The following theorem indicates when this is the case.

Theorem: *If in $\mathfrak{R}$ every prime ideal $\neq (0)$ is maximal, then in every $\mathfrak{R}$-order $\mathfrak{o}$ every prime ideal $\neq (0)$ is also maximal.*

Proof: Let $\mathfrak{p}$ be a prime ideal in $\mathfrak{o}$ containing a nonzero element t. Now t satisfies an equation of least degree with coefficients in $\mathfrak{R}$ and leading coefficient 1:

$$t^h + a_1 t^{h-1} + \cdots + a_h = 0$$

in which $a_h \neq 0$ since otherwise t could be canceled. It follows that $a_h \equiv 0(t) \equiv 0(\mathfrak{p})$, and a_h therefore belongs to the intersection $\mathfrak{p} \cap \mathfrak{R}$. This intersection is a prime ideal in $\mathfrak{R}$, for if a product of two elements of $\mathfrak{R}$ belongs to $\mathfrak{R} \cap \mathfrak{p}$, and hence to $\mathfrak{p}$, then one factor must belong to $\mathfrak{p}$, and hence to $\mathfrak{R} \cap \mathfrak{p}$. Since a_h belongs to the prime ideal $\mathfrak{R} \cap \mathfrak{p}$, this prime ideal is distinct from the null ideal and is therefore maximal.

If now $\mathfrak{a}$ is a proper divisor of $\mathfrak{p}$ and u is an element of $\mathfrak{a}$ which does not belong to $\mathfrak{p}$, then u satisfies an equation

$$u^l + b_1 u^{l-1} + \cdots + b_l = 0,$$

and hence a congruence of least degree

$$u^k + c_1 u^{k-1} + \cdots + c_k \equiv 0(\mathfrak{p}),$$

in which $c_k \not\equiv 0(\mathfrak{p})$, since otherwise u could be canceled. It now follows that $c_k \equiv 0(u) \equiv 0(\mathfrak{a})$; c_k therefore belongs to the intersection $\mathfrak{a} \cap \mathfrak{R}$ without belonging to $\mathfrak{p} \cap \mathfrak{R}$. This intersection $\mathfrak{a} \cap \mathfrak{R}$ is therefore a proper divisor of $\mathfrak{p} \cap \mathfrak{R}$ and is hence equal to the unit ideal $\mathfrak{R}$. Therefore $\mathfrak{a}$ contains the identity element; hence $\mathfrak{a} = \mathfrak{o}$, Q.E.D.

The hypotheses of this theorem are satisfied, in particular, if $\mathfrak{R}$ is a principal ideal ring (the ring of integers, a polynomial ring in one variable). In this case every ideal distinct from the null and unit ideals can be uniquely represented as a product of relatively prime, primary ideals distinct from $\mathfrak{o}$.

For the principal order $\mathfrak{S}$, as we shall see, it is possible to say still more: the primary ideals are powers of prime ideals, and hence *every ideal is the product of powers of prime ideals*. Because of its importance for the theory of number and function fields, we shall give a direct proof of this central result of the "classical" Dedekind ideal theory without assuming the concept of a primary ideal and the general ideal theory. This will be done in the next section by a method due to W. Krull.[4]

Exercises

17.4. If $\mathfrak{R}$ is a principal ideal ring, $(\omega_1, \ldots, \omega_n)$ is a linearly independent basis of an order $\mathfrak{o}$ (which always exists in this case), and $(\omega_1^{(i)}, \ldots, \omega_n^{(i)})$ are the conjugated bases in a Galois extension field of P, then the "field discriminant"

$$D = \begin{vmatrix} \omega_1^{(1)} \ldots \omega_n^{(1)} \\ \ldots \\ \omega_1^{(n)} \ldots \omega_n^{(n)} \end{vmatrix}^2$$

is integral, rational, and nonzero.

17.5. Let $\Sigma = \mathsf{P}(\sqrt{d})$, and let $\mathfrak{R}$ be integrally closed in P. Show that an element $\xi = a+b\sqrt{d}$ is integral over $\mathfrak{R}$ if and only if its trace and norm

$$S(\xi) = \xi+\xi' = (a+b\sqrt{d})+(a-b\sqrt{d}) = 2a$$

$$N(\xi) = \xi\cdot\xi' = (a+b\sqrt{d})\,(a-b\sqrt{d}) = a^2-b^2d,$$

both belong to $\mathfrak{R}$.

17.6. If $\mathfrak{R} = \mathsf{K}[x]$ in Exercise 17.5 is a polynomial ring in one indeterminate and d is a polynomial which contains no multiple factors, then $\xi = a+b\sqrt{d}$ is integral only if a and b belong to $\mathfrak{R}$.

17.7. In Exercise 17.5, if $\mathfrak{R} = \mathbb{Z}$ is the ring of integers and d is a square-free integer, then if $d \not\equiv 1(4)$ a basis for the principal order consists of the numbers $1, \sqrt{d}$; in the case where $d \equiv 1(4)$ a basis consists of the numbers $1, (1+\sqrt{d})/2$.

17.4 AXIOMATIC FOUNDATION OF CLASSICAL IDEAL THEORY

Let $\mathfrak{o}$ be an integral domain (a commutative ring without zero divisors) in which the following three axioms are satisfied:

I. The ascending chain condition for ideals.
II. All nonnull prime ideals are maximal.
III. $\mathfrak{o}$ is integrally closed in its quotient field Σ.

Examples of such rings are (1) the principal ideal rings; (2) the principal orders obtained from principal ideal rings by extension of the quotient field in the manner of Section 17.3 (in particular, the principal orders in number fields and function fields of one variable).

[4]W. Krull, "Zur Theorie der Allgemeinen Zahlringe," *Math. Ann.*, **99**, 51–70 (1928).

The elements of Σ which are integral over $\mathfrak{o}$, and hence by III lie in $\mathfrak{o}$, will be called simply *integral*. In particular, the identity element of Σ is always integral, and so $\mathfrak{o}$ is an integral domain with identity.

In addition to the ideals of $\mathfrak{o}$ (or $\mathfrak{o}$-modules of $\mathfrak{o}$), we now consider also $\mathfrak{o}$-modules in Σ, that is, subsets of Σ which with a and b also contain $a-b$, and with a also ra (r integral). If such an $\mathfrak{o}$-module has a finite module basis, then it is also called a *fractional ideal*. If an $\mathfrak{o}$-module $\mathfrak{a}$ consists solely of integral elements ($\mathfrak{a} \subseteq \mathfrak{o}$), then it is an ideal in the usual sense or, as we shall now say, an *integral ideal*.

The *sum* or g.c.d. $(\mathfrak{a}, \mathfrak{b})$ of two $\mathfrak{o}$-modules $\mathfrak{a}$ and $\mathfrak{b}$ is (as in the case of ideals) the module of all sums $a+b$ with $a \in \mathfrak{a}$, $b \in \mathfrak{b}$; likewise the product $\mathfrak{ab}$ is the module generated by all products ab or the set of all sums $\sum a_\nu b_\nu$.

Sums and products of $\mathfrak{o}$-modules with finite bases again have finite bases.

In the following theorems the German script letters refer exclusively to *integral nonzero* ideals in $\mathfrak{o}$, while the letter $\mathfrak{p}$ always denotes a *prime ideal* $\neq (0)$.

Lemma 1: *For every ideal* $\mathfrak{a}$ *there exist prime ideals* $\mathfrak{p}_i$ *which are divisors of* $\mathfrak{a}$ *and whose product is divisible by* $\mathfrak{a}$:

$$\mathfrak{p}_1\mathfrak{p}_2\cdots\mathfrak{p}_r \equiv 0(\mathfrak{a}).$$

Proof: If $\mathfrak{a}$ is prime, then the lemma is true. If $\mathfrak{a}$ is not prime, then there is a product of two principal ideals $\mathfrak{bc}$ such that

$$\mathfrak{bc} \equiv 0(\mathfrak{a}), \qquad \mathfrak{b} \not\equiv 0(\mathfrak{a}), \qquad \mathfrak{c} \not\equiv 0(\mathfrak{a}).$$

The ideals $\mathfrak{b}' = (\mathfrak{b}, \mathfrak{a})$ and $\mathfrak{c}' = (\mathfrak{c}, \mathfrak{a})$ are proper divisors of $\mathfrak{a}$, and

$$\mathfrak{b}'\mathfrak{c}' = (\mathfrak{b}, \mathfrak{a})\cdot(\mathfrak{c}, \mathfrak{a}) = (\mathfrak{bc}, \mathfrak{ba}, \mathfrak{ac}, \mathfrak{a}^2) \equiv 0(\mathfrak{a}).$$

If we now assume that the lemma is true for the ideals $\mathfrak{b}'$ and $\mathfrak{c}'$, then there are products such that $\mathfrak{p}_1 \ldots \mathfrak{p}_s \equiv 0(\mathfrak{b}')$ and $\mathfrak{p}_{s+1} \ldots \mathfrak{p}_r \equiv 0(\mathfrak{c}')$. Then the product $\mathfrak{p}_1 \ldots \mathfrak{p}_s\mathfrak{p}_{s+1} \ldots \mathfrak{p}_r \equiv 0(\mathfrak{b}'\cdot\mathfrak{c}') \equiv 0(\mathfrak{a})$, and the assertion is therefore true for $\mathfrak{a}$ also. If the lemma were false for some ideal $\mathfrak{a}$, then it would also not hold for one of the two proper divisors, $\mathfrak{b}'$ or $\mathfrak{c}'$; this divisor would again have a proper divisor for which the lemma were not true, and so on. In this manner we would obtain an infinite chain of proper divisors, which is impossible by Axiom I. Hence the lemma is true for every ideal $\mathfrak{a}$.

Lemma 2: *If* $\mathfrak{p}$ *is prime, then* $\mathfrak{ab} \equiv 0(\mathfrak{p})$ *implies* $\mathfrak{a} \equiv 0(\mathfrak{p})$ *or* $\mathfrak{b} \equiv 0(\mathfrak{p})$.

Proof: If $\mathfrak{a} \not\equiv 0(\mathfrak{p})$ and $\mathfrak{b} \not\equiv 0(\mathfrak{p})$, then there would be an element a in $\mathfrak{a}$ and an element b in $\mathfrak{b}$, neither of which would belong to $\mathfrak{p}$. The product ab would lie in $\mathfrak{ab}$ and therefore in $\mathfrak{p}$, contrary to the fact that $\mathfrak{p}$ is prime.

We denote by $\mathfrak{p}^{-1}$ the set of all (integral or fractional) elements a such that $a\mathfrak{p}$ is integral. $\mathfrak{p}^{-1}$ is clearly an $\mathfrak{o}$-module.

Lemma 3: *If* $\mathfrak{p} \neq \mathfrak{o}$, *then* $\mathfrak{p}^{-1}$ *contains a nonintegral element.*

Proof: Let c be any nonzero element of $\mathfrak{p}$. By Lemma 1 there exists a product of prime ideals such that

$$\mathfrak{p}_1\mathfrak{p}_2\cdots\mathfrak{p}_r \equiv 0(c).$$

We may assume that this product is irredundant; that is, there is no subproduct such as $\mathfrak{p}_2 \ldots \mathfrak{p}_r$ which is $\equiv 0(c)$. Since the product $\mathfrak{p}_1\mathfrak{p}_2 \ldots \mathfrak{p}_r$ is divisible by $\mathfrak{p}$, one of the factors, say $\mathfrak{p}_1$, must be divisible by $\mathfrak{p}$ and hence equal to $\mathfrak{p}$.

Therefore

$$\mathfrak{p}\mathfrak{p}_2 \cdots \mathfrak{p}_r \equiv 0(c)$$

$$\mathfrak{p}_2 \cdots \mathfrak{p}_r \not\equiv 0(c).$$

There is thus an element b in $\mathfrak{p}_2 \ldots \mathfrak{p}_r$ which does not belong to (c). For this element

$$\mathfrak{p}b \equiv 0(\mathfrak{p}\mathfrak{p}_2 \cdots \mathfrak{p}_r) \equiv 0(c).$$

Therefore, $\mathfrak{p}b/c$ is integral, and so b/c lies in $\mathfrak{p}^{-1}$. Since $b \not\equiv 0(c)$, b/c is not integral, Q.E.D.

Theorem 1: *If* $\mathfrak{p} \neq \mathfrak{o}$, *then*

$$\mathfrak{p} \cdot \mathfrak{p}^{-1} = \mathfrak{o}.$$

Proof: $\mathfrak{o} \subseteq \mathfrak{p}^{-1}$ by definition of $\mathfrak{p}^{-1}$; therefore, $\mathfrak{p} = \mathfrak{o}\mathfrak{p} \subseteq \mathfrak{p}^{-1}\mathfrak{p}$. The integral ideal $\mathfrak{p}\mathfrak{p}^{-1}$ is thus a divisor of $\mathfrak{p}$ and so is either equal to $\mathfrak{p}$ or $\mathfrak{o}$. Suppose that

$$\mathfrak{p} \cdot \mathfrak{p}^{-1} = \mathfrak{p}.$$

This would imply that $\mathfrak{p} \cdot (\mathfrak{p}^{-1})^2 = (\mathfrak{p}\mathfrak{p}^{-1})\mathfrak{p}^{-1} = \mathfrak{p}\mathfrak{p}^{-1} = \mathfrak{p}$ and likewise $\mathfrak{p}(\mathfrak{p}^{-1})^3 = \mathfrak{p}$, and so on. If then $a \neq 0$ is any element of $\mathfrak{p}$ and b any element of $\mathfrak{p}^{-1}$, then $ab^e \in \mathfrak{p}(\mathfrak{p}^{-1})^e$ is integral, and therefore all powers of b can be represented as fractions with a fixed denominator a. It follows that b is integral. This holds for any b in $\mathfrak{p}^{-1}$, contrary to Lemma 3.

We are now in a position to prove the principal theorem on factorization.

Theorem 2: *Every ideal* $\mathfrak{a}$ *is a product of prime ideals.*

Proof: We may assume $\mathfrak{a} \neq \mathfrak{o}$. Using Lemma 1, let

$$\mathfrak{p}_1\mathfrak{p}_2 \cdots \mathfrak{p}_r \equiv 0(\mathfrak{a}) \tag{17.7}$$

and let the number r be chosen as small as possible so that no shorter product $\equiv 0(\mathfrak{a})$. Further, let $\mathfrak{p}$ be any prime ideal divisor of $\mathfrak{a}$ distinct from $\mathfrak{o}$ (there exists at least one by Lemma 1). The product $\mathfrak{p}_1 \ldots \mathfrak{p}_r$ is then divisible by $\mathfrak{p}$, and hence (by Lemma 2) some $\mathfrak{p}_i$ is divisible by $\mathfrak{p}$; since this $\mathfrak{p}_i$ is maximal, it follows that $\mathfrak{p}_i = \mathfrak{p}$. We may assume that $\mathfrak{p}_1 = \mathfrak{p}$. Multiplying (17.7) by $\mathfrak{p}^{-1}$, we obtain

$$\mathfrak{p}_2 \cdots \mathfrak{p}_r \equiv 0(\mathfrak{p}^{-1}\mathfrak{a}) \equiv 0(\mathfrak{o});$$

therefore $\mathfrak{p}^{-1}\mathfrak{a}$ is an integral ideal which divides a product of fewer than r prime ideals. If we now proceed by induction on r, that is, if we assume that the theorem has already been proved for ideals which divide a product of fewer than r prime ideals $\neq (0)$ (for ideals which divide only *one* prime ideal $\neq (0)$ the theorem is clear), then the theorem holds, in particular, for $\mathfrak{p}^{-1}\mathfrak{a}$, that is,

$$\mathfrak{p}^{-1}\mathfrak{a} = \mathfrak{p}_2' \cdots \mathfrak{p}_s'.$$

Multiplication of both sides by $\mathfrak{p}$ gives the desired representation of $\mathfrak{a}$.

The uniqueness of this representation follows from the next theorem.

Theorem 3: *If* $\mathfrak{a} \equiv 0(\mathfrak{b})$ *and* $\mathfrak{a} = \mathfrak{p}_1 \ldots \mathfrak{p}_r$, $\mathfrak{b} = \mathfrak{p}'_1 \ldots \mathfrak{p}'_s$, *then any prime ideal distinct from* $\mathfrak{o}$ *which occurs in the representation of* $\mathfrak{b}$ *also occurs in the representation of* $\mathfrak{a}$ *and at least the same number of times.*

Proof: Let $\mathfrak{p}'_1 \neq \mathfrak{o}$. Since $\mathfrak{p}'_1$ is a divisor of $\mathfrak{a}$, we conclude as before that $\mathfrak{p}'_1$ must occur among the $\mathfrak{p}_\nu$. Let, say, $\mathfrak{p}_1 = \mathfrak{p}'_1$. Then

$$\mathfrak{p}_1^{-1}\mathfrak{a} \equiv 0(\mathfrak{p}_1^{-1}\mathfrak{b})$$

$$\mathfrak{p}_1^{-1}\mathfrak{a} = \mathfrak{p}_2 \cdots \mathfrak{p}_r$$

$$\mathfrak{p}_1^{-1}\mathfrak{b} = \mathfrak{p}'_2 \cdots \mathfrak{p}'_s.$$

If we now assume that the assertion has been proved for all smaller values of s (for $s = 0$, $\mathfrak{b} = \mathfrak{o}$ it is trivial), then it follows that each of the ideals $\mathfrak{p}'_2, \ldots, \mathfrak{p}'_s$ distinct from $\mathfrak{o}$ occurs at least as often among the $\mathfrak{p}_2, \ldots, \mathfrak{p}_r$. This completes the proof.

Corollary 1: *Up to the order of the factors and the factor* $\mathfrak{o}$ *the representation of an ideal* $\mathfrak{a}$ *as a product of prime ideals is unique.*

Corollary 2: *Divisibility implies product representation. If* $\mathfrak{a} \equiv 0(\mathfrak{b})$, *then* $\mathfrak{a} = \mathfrak{b}\mathfrak{c}$ *with integral* $\mathfrak{c}$.

We have only to take for $\mathfrak{c}$ the product of those prime factors of $\mathfrak{a}$ which remain after the factors of $\mathfrak{b}$ are deleted from the representation of $\mathfrak{a}$ (each as often as it occurs in $\mathfrak{b}$).

Exercise

17.8. Decompose the principal ideals (2) and (3) in the principal order of the number field $\mathbb{Z}(\sqrt{-5})$ into their prime ideal factors.

17.5 CONVERSE AND EXTENSION OF RESULTS

We saw that Theorems 2 and 3 follow from Axioms 1 to III (Section 17.4); these two theorems together give the unique prime factorization of ideals. This situation can also be inverted, as follows.

Let $\mathfrak{o}$ *be an integral domain with identity element. Suppose that in* $\mathfrak{o}$ *every integral ideal* $\mathfrak{a}$ *can be represented as a product of prime ideals:* $\mathfrak{a} = \mathfrak{p}_1\mathfrak{p}_2 \ldots \mathfrak{p}_r$; *if* $\mathfrak{a}$ *is divisible by* $\mathfrak{b}$, *suppose further that in any decomposition of* $\mathfrak{a}$ *every factor of a decomposition of* $\mathfrak{b}$ *distinct from* $\mathfrak{o}$ *occurs and at least as often. Then Axioms I to III hold in* $\mathfrak{o}$.

Proof: The ascending chain condition (Axiom I) follows immediately from the fact that every integral ideal $\mathfrak{a} = \mathfrak{p}_1^{\varrho_1} \ldots \mathfrak{p}_r^{\varrho_r}$ has only a finite number of divisors $\mathfrak{b} = \mathfrak{p}_1^{\sigma_1} \ldots \mathfrak{p}_r^{\sigma_r}$ $(\sigma_i \leqq \varrho_i)$. In particular, a prime ideal $\mathfrak{p}$ has only the divisors $\mathfrak{p}$ and $\mathfrak{o}$, so that Axiom II also holds.

To prove Axiom III (the integral closure of $\mathfrak{o}$ in its quotient field Σ), we assume that λ is an element of Σ which is integral over $\mathfrak{o}$ so that, say, λ^m can be linearly expressed in terms of $\lambda^0, \ldots, \lambda^{m-1}$, or equivalently, λ^m is contained in the $\mathfrak{o}$-module $\mathfrak{l} = (\lambda^0, \lambda^1, \ldots, \lambda^{m-1})$. If $\lambda = a/b$, then $\mathfrak{l}$ can be made an integral ideal by multiplying with $\mathfrak{b} = (b^{m-1})$. Further, $\mathfrak{l}$ obviously satisfies the equation $\mathfrak{l}^2 = \mathfrak{l}$. Multiplication by $\mathfrak{b}^2$ gives

$$(\mathfrak{l}\mathfrak{b})^2 = (\mathfrak{l}\mathfrak{b})\mathfrak{b}.$$

Uniqueness now implies that

$$\mathfrak{l}\mathfrak{b} = \mathfrak{b}$$

and hence if both sides are multiplied by (b^{-m+1}),

$$\mathfrak{l} = \mathfrak{o}.$$

Therefore λ is an element of $\mathfrak{o}$, Q.E.D.

We now discuss certain extensions of Theorems 2 and 3 which likewise belong to the classical ideal theory.

The fact that divisibility implies product representation makes it possible to find the greatest common divisor and least common multiple of ideals in the same way as in the case of integers—that is, by prime factorization.

Let $\mathfrak{a}$ and $\mathfrak{b}$ be two integral ideals

$$\mathfrak{a} = \mathfrak{p}_1^{\varrho_1} \cdots \mathfrak{p}_r^{\varrho_r},$$
$$\mathfrak{b} = \mathfrak{p}_1^{\sigma_1} \cdots \mathfrak{p}_r^{\sigma_r}$$

(where in each case all prime factors are written which occur in either $\mathfrak{a}$ or $\mathfrak{b}$, possibly with zero exponents). Every common divisor contains only prime factors $\mathfrak{p}_i$ appearing in these sequences and with exponents $\leqq \tau_i$, where τ_i is the smaller of the numbers ϱ_i, σ_i. The greatest common divisor $(\mathfrak{a}, \mathfrak{b})$ must be divisible by every common divisor and, in particular, by $\mathfrak{p}_i$. Therefore, it can only be

$$\mathfrak{p}_1^{\tau_1} \cdots \mathfrak{p}_r^{\tau_r}.$$

Similarly, the least common multiple (intersection) $\mathfrak{a} \cap \mathfrak{b}$ of $\mathfrak{a}$ and $\mathfrak{b}$ is the ideal

$$\mathfrak{p}_1^{\mu_1} \cdots \mathfrak{p}_r^{\mu_r},$$

where μ_i is the greater of the numbers ϱ_i, σ_i.

Theorem 4: *If* $\mathfrak{a} \equiv 0(\mathfrak{d})$, *then there exists an element* d *in* $\mathfrak{d}$ *such that*

$$(\mathfrak{a}, d) = \mathfrak{d}.$$

Proof: Let

$$\mathfrak{a} = \mathfrak{p}_1^{\varrho_1} \cdots \mathfrak{p}_r^{\varrho_r}$$
$$\mathfrak{d} = \mathfrak{p}_1^{\sigma_1} \cdots \mathfrak{p}_r^{\varrho_r} \qquad (0 \leqq \sigma_i \leqq \varrho_i).$$

We must choose d so that d is divisible by $\mathfrak{d}$ but has no other divisors in common with $\mathfrak{a}$ other than the divisors of $\mathfrak{d}$. We put

$$\mathfrak{c} = \mathfrak{p}_1^{\sigma_1+1} \cdots \mathfrak{p}_r^{\sigma_r+1}$$
$$\mathfrak{c}_i = \mathfrak{c} : \mathfrak{p}_i = \mathfrak{p}_1^{\sigma_1+1} \cdots \mathfrak{p}_i^{\sigma_i} \cdots \mathfrak{p}_r^{\sigma_r+1}.$$

Then $\mathfrak{c}_i \equiv 0(\mathfrak{c})$. There is thus an element d_i which is in $\mathfrak{c}_i$ but not in $\mathfrak{c}$, so that

$$d_i \equiv 0(\mathfrak{p}_j^{\sigma_j+1}) \quad \text{for } j \neq i$$

$$d_i \not\equiv 0(\mathfrak{p}_i^{\sigma_i+1}).$$

The sum

$$d = d_1 + \cdots + d_r$$

is divisible by $\mathfrak{d}$ (since all the d_i are). However,

$$d \equiv d_i \not\equiv 0(\mathfrak{p}_i^{\sigma_i+1});$$

and so d has no other factors in common with $\mathfrak{a}$ other than the factors of $\mathfrak{d}$.

Corollary 1: *In the residue class ring $\mathfrak{o}/\mathfrak{a}$ every ideal $\mathfrak{d}/\mathfrak{a}$ is a principal ideal.* In fact, $\mathfrak{d}/\mathfrak{a}$ is generated by the residue class $\mathfrak{a} + d$.

Corollary 2: *Every ideal $\mathfrak{d}$ has a basis (a, d) consisting of two terms, where $a \neq 0$ can be chosen arbitrarily in $\mathfrak{d}$.*

Indeed, let a be any nonzero element of $\mathfrak{d}$ and let $\mathfrak{a} = (a)$. The theorem implies that $(a, d) = \mathfrak{d}$.

Corollary 3: *Every ideal $\mathfrak{d}$ can be transformed into a principal ideal by multiplication with an ideal $\mathfrak{b}$ which is relatively prime to a given ideal $\mathfrak{c}$.*

Proof: We put $\mathfrak{a} = \mathfrak{c}\mathfrak{d}$. The theorem above implies that

$$(\mathfrak{a}, d) = \mathfrak{d}. \tag{17.8}$$

Since d is divisible by $\mathfrak{d}$, we may put

$$(d) = \mathfrak{b}\mathfrak{d}$$

so, by (17.8),

$$(\mathfrak{c}\mathfrak{b}, \mathfrak{b}\mathfrak{d}) = \mathfrak{d}.$$

Therefore $\mathfrak{c}$ and $\mathfrak{b}$ must be relatively prime.

Exercise

17.9. Let $\mathfrak{O}$ be the ring of all quotients a/b, where a and b are integral and b is not divisible by certain prime ideals $\mathfrak{p}_1, \ldots, \mathfrak{p}_r$. Then to every ideal $\mathfrak{a}$ of $\mathfrak{o}$ there corresponds an ideal $\mathfrak{A}$ of $\mathfrak{O}$ consisting of the fractions a/b with $a \in \mathfrak{a}$. To the ideals $\mathfrak{p}_1, \ldots, \mathfrak{p}_r$ there correspond the prime ideals $\mathfrak{p}_1, \ldots, \mathfrak{p}_r$; all other prime ideals of $\mathfrak{o}$ correspond to the unit ideal $\mathfrak{O}$. Every ideal in $\mathfrak{O}$ can be uniquely represented as a product of powers of the ideals $\mathfrak{p}_1, \ldots, \mathfrak{p}_r$. Moreover, in $\mathfrak{O}$ every ideal is a principal ideal.

17.6 FRACTIONAL IDEALS

In Section 17.4 we called an $\mathfrak{o}$-module in the quotient field Σ a *fractional ideal* if it had a finite basis. The ideals in $\mathfrak{o}$ or "integral ideals" are thus special fractional ideals.

If $(\sigma_1, \ldots, \sigma_r)$ is a basis of a fractional ideal, then the entire basis, and hence the ideal itself, can be made integral by multiplying by a suitable denominator.

Conversely, if an $\mathfrak{o}$-module $\mathfrak{a}$ can be made integral by multiplication by an integral element $b \neq 0$, then $b\mathfrak{a}$ has a finite basis

$$b\mathfrak{a} = (a_1, \ldots, a_r)$$

as an integral ideal, and from this it follows that

$$\mathfrak{a} = \left(\frac{a_1}{b}, \ldots, \frac{a_r}{b}\right).$$

This proves the following theorem.

Theorem: *An $\mathfrak{o}$-module in Σ is a fractional ideal if and only if it can be transformed into an integral ideal by multiplication with an integral element $b \neq 0$.*

We have already seen that if $\mathfrak{a}$ and $\mathfrak{b}$ have finite bases then $\mathfrak{a}\cdot\mathfrak{b}$ and $(\mathfrak{a}, \mathfrak{b})$ do also and are therefore fractional ideals. The same also holds for the *module quotient* $\mathfrak{a} : \mathfrak{b}$, where $\mathfrak{a}$ and $\mathfrak{b}$ are integral ideals and $\mathfrak{b} \neq (0)$.[5] For if $b \neq 0$ is any element of $\mathfrak{b}$, then

$$b\cdot(\mathfrak{a} : \mathfrak{b}) \subseteq \mathfrak{b}\cdot(\mathfrak{a} : \mathfrak{b}) \subseteq \mathfrak{a} \subseteq \mathfrak{o};$$

$\mathfrak{a} : \mathfrak{b}$ is thus transformed into an integral ideal by multiplication with b.

In particular, $\mathfrak{o} : \mathfrak{p} = \mathfrak{p}^{-1}$ is a fractional ideal.

Every integral or fraction ideal $\neq(0)$ has an inverse.

Proof: Let $\mathfrak{c}$ be an integral or fractional ideal $\neq(0)$, and let $b \neq 0$ be chosen so that $b\mathfrak{c}$ is integral:

$$b\mathfrak{c} = \mathfrak{a}. \tag{17.9}$$

If now $\mathfrak{a} = \mathfrak{p}_1\mathfrak{p}_2 \ldots \mathfrak{p}_r$, then multiplying (17.9) by $\mathfrak{p}_1{}^{-1}\mathfrak{p}_2{}^{-1} \ldots \mathfrak{p}_r{}^{-1}$ gives

$$(\mathfrak{p}^{-1}\mathfrak{p}_2{}^{-1}\cdots\mathfrak{p}_r{}^{-1}b)\, \mathfrak{c} = \mathfrak{o}$$

by Theorem 1 (Section 17.4), which proves the existence of the inverse

$$\mathfrak{c}^{-1} = \mathfrak{p}_1{}^{-1}\cdots\mathfrak{p}_r{}^{-1}b.$$

From this theorem it follows that *the integral and fractional ideals $\neq(0)$ form an Abelian group.*

The equation $\mathfrak{a}\mathfrak{c} = \mathfrak{b}$ can therefore be solved uniquely for $\mathfrak{c}$. The solution is denoted by $\mathfrak{a}^{-1}\mathfrak{b}$ or $\mathfrak{b}/\mathfrak{a}$.

The previous theorems now imply the following.

Every fractional ideal can be represented as the quotient of two integral ideals:

$$\frac{\mathfrak{p}_1'\cdots\mathfrak{p}_s'}{\mathfrak{p}_1''\cdots\mathfrak{p}_t''}.$$

Here any ideal which occurs both in the numerator and denominator may be canceled.

[5]The module quotient $\mathfrak{a}:\mathfrak{b}$ (in Σ) is the set of all elements λ in Σ such that $\lambda\mathfrak{b} \subseteq \mathfrak{a}$.

Every fractional principal ideal (λ) *admits a representation as a quotient of two integral principal ideals so that no one of any r given prime ideals appears both in the denominator and the numerator.*

Proof: In reduced representation let

$$(\lambda) = \frac{\mathfrak{p}_1' \cdots \mathfrak{p}_s'}{\mathfrak{p}_1'' \cdots \mathfrak{p}_t''},$$

and let $\mathfrak{p}_1 \ldots \mathfrak{p}_r$ be the r given prime ideals. If the denominator is transformed into a principal ideal (d) by multiplying by an ideal $\mathfrak{b}$ relatively prime to the product $\mathfrak{p}_1 \ldots \mathfrak{p}_r$, then

$$(\lambda) = \frac{\mathfrak{b}\mathfrak{p}_1' \cdots \mathfrak{p}_s'}{\mathfrak{b}\mathfrak{p}_1'' \cdots \mathfrak{p}_t''} = \frac{\mathfrak{b}\mathfrak{p}_1' \cdots \mathfrak{p}_s'}{(d)},$$

and hence

$$\mathfrak{b}\mathfrak{p}_1' \cdots \mathfrak{p}_s' = (\lambda d).$$

The numerator thus also becomes a principal ideal. No one of the ideals $\mathfrak{p}_1, , , , \mathfrak{p}$ occurs in both the numerator and the denominator.

Exercise

17.10. The ideal fraction $\mathfrak{a}^{-1}\mathfrak{b}$ is equal to the module quotient $\mathfrak{b} : \mathfrak{a}$.[6]

17.7 IDEAL THEORY OF ARBITRARY INTEGRALLY CLOSED INTEGRAL DOMAINS

There are important integral domains which satisfy Axioms I and III of Section 17.4 but not Axiom II. Examples are the polynomial ring $K[x_1, \ldots, x_n]$ in more than one variable and the ring of integral polynomials $\mathbb{Z}[x_1, \ldots, x_n]$ and its finite, integrally closed extensions (principal orders). In all these rings it happens that a prime ideal distinct from the null and unit ideals has a prime ideal as a proper divisor. The ideal theory of Section 17.4 therefore no longer holds in these rings. We shall show, however, that the main results nevertheless continue to hold if the equality of ideals is replaced by a relation of "quasi-equality" which will presently be defined.[7]

Let $\mathfrak{o}$ be an integral domain which is integrally closed in its field of quotients Σ. German letters shall henceforth denote nonzero fractional ideals, that is, $\mathfrak{o}$-

[6] For further development of ideal theory in number fields, we refer the reader to the book by E. Hecke, *Vorlesungen über die Theorie der Algebraischen Zahlen*, Leipzig, 1923. For ideal theory in function fields and its applications, we refer to the fundamental work of Dedekind and Weber, *Crelles J.*, **92**, 181 (1882).

[7] This theory, which was developed by the author in *Math. Ann.*, Vol. 101 (1929), was put into a more polished form by E. Artin, and it is published here in this form for the first time.

modules in Σ which become integral on multiplying by a nonzero element of $\mathfrak{o}$. The inverse $\mathfrak{a}^{-1}$ of an ideal $\mathfrak{a}$ is the set of all elements r of Σ such that $r\mathfrak{a}$ is integral.

Definition: $\mathfrak{a}$ is *quasi-equal* to $\mathfrak{b}$ if $\mathfrak{a}^{-1} = \mathfrak{b}^{-1}$. In this case we write $\mathfrak{a} \sim \mathfrak{b}$. The relation $\sim$ is obviously reflexive, symmetric, and transitive.

Similarly, $\mathfrak{a}$ is a *quasi-divisor* of $\mathfrak{b}$ and $\mathfrak{b}$ a *quasi-multiple* of $\mathfrak{a}$ if $\mathfrak{a}^{-1} \subseteq \mathfrak{b}^{-1}$ or, equivalently, if $\mathfrak{a}^{-1}\mathfrak{b}$ is integral. We write $\mathfrak{a} \leqq \mathfrak{b}$ or $\mathfrak{b} \geqq \mathfrak{a}$.

The simplest properties of the relations $\leqq$ and $\sim$ are the following.

1. The statement $\mathfrak{a} \supseteq \mathfrak{b}$ implies $\mathfrak{a} \leqq \mathfrak{b}$ (the proof is clear).

2. If $\mathfrak{a}$ is a principal ideal, $\mathfrak{a} = (a)$, then $\mathfrak{a} \leqq \mathfrak{b}$ implies that $\mathfrak{a} \supseteq \mathfrak{b}$. For in this case $\mathfrak{a}^{-1} = (a^{-1})$; the hypothesis that $\mathfrak{a}^{-1}\mathfrak{b}$ is integral implies that $a^{-1}\mathfrak{b}$ is integral; that is, all elements of $\mathfrak{b}$ are divisible by a.

3. If $\mathfrak{a} \leqq \mathfrak{b}$ and $\mathfrak{a} \geqq \mathfrak{b}$, then $\mathfrak{a} \sim \mathfrak{b}$.

4. All quasi-multiples $\mathfrak{b}$ of $\mathfrak{a}$, and in particular all $\mathfrak{b}$ quasi-equal to $\mathfrak{a}$, have the property $\mathfrak{b} \subseteq (\mathfrak{a}^{-1})^{-1}$. (A direct consequence of the fact that $\mathfrak{b}\mathfrak{a}^{-1}$ is integral).

In particular, $\mathfrak{a} \subseteq (\mathfrak{a}^{-1})^{-1}$. Property 1 then implies that $\mathfrak{a} \geqq (\mathfrak{a}^{-1})^{-1}$. On the other hand, $\mathfrak{a}^{-1}(\mathfrak{a}^{-1})^{-1}$ is integral, so that $\mathfrak{a} \leqq (\mathfrak{a}^{-1})^{-1}$; hence we have the following.

5.

$$\mathfrak{a} \sim (\mathfrak{a}^{-1})^{-1}.$$

From properties 4 and 5 it follows that $(\mathfrak{a}^{-1})^{-1}$ is the *greatest ideal quasi-equal to* $\mathfrak{a}$. We denote it by $\mathfrak{a}^*$.

6. If $\mathfrak{a} \leqq \mathfrak{b}$, then $\mathfrak{a}\mathfrak{c} \leqq \mathfrak{b}\mathfrak{c}$. For $(\mathfrak{c}\mathfrak{a})^{-1}\mathfrak{c}\mathfrak{a}$ is integral, and hence $(\mathfrak{c}\mathfrak{a})^{-1}\,\mathfrak{c} \subseteq \mathfrak{a}^{-1} \subseteq \mathfrak{b}^{-1}$; therefore, $(\mathfrak{c}\mathfrak{a})^{-1}\mathfrak{c}\mathfrak{b}$ is integral so that $\mathfrak{c}\mathfrak{a} \leqq \mathfrak{c}\mathfrak{b}$.

7. If $\mathfrak{a} \sim \mathfrak{b}$, then $\mathfrak{a}\mathfrak{c} \sim \mathfrak{b}\mathfrak{c}$. (Consequence of property 6.)

8. If $\mathfrak{a} \sim \mathfrak{b}$ and $\mathfrak{c} \sim \mathfrak{d}$, then $\mathfrak{a}\mathfrak{c} \sim \mathfrak{b}\mathfrak{d}$. (For, by property 7, $\mathfrak{a}\mathfrak{c} \sim \mathfrak{b}\mathfrak{c} \sim \mathfrak{b}\mathfrak{d}$.)

If all ideals quasi-equal to a given ideal are combined to form a class, then the class of the product $\mathfrak{a}\mathfrak{c}$ depends only on the class of $\mathfrak{a}$ and the class of $\mathfrak{c}$, by property 8. We may therefore define the class of the product $\mathfrak{a}\mathfrak{c}$ to be the product of the two classes.

9. The identity class for class multiplication is the class of those ideals which are quasi-equal to the unit ideal $\mathfrak{o}$, since $\mathfrak{a}\mathfrak{o} = \mathfrak{a}$ for every $\mathfrak{a}$.

10. All quasi-multiples of $\mathfrak{o}$, in particular all ideals of the identity class, are integral. (Special case of property 1: put $a = 1$.) As a corollary it follows that all ideals quasi-equal to an integral ideal are again integral.

We now prove the most important property of the inverse:

11.

$$\mathfrak{a}\mathfrak{a}^{-1} \sim \mathfrak{o}.$$

It is clear that $\mathfrak{a}\mathfrak{a}^{-1} \geqq \mathfrak{o}$, for $\mathfrak{a}\mathfrak{a}^{-1}$ is integral. It remains to show that $\mathfrak{a}\mathfrak{a}^{-1} \leqq \mathfrak{o}$ or $(\mathfrak{a}\mathfrak{a}^{-1})^{-1} \subseteq \mathfrak{o}$. If λ belongs to $(\mathfrak{a}\mathfrak{a}^{-1})^{-1}$, then $\lambda\mathfrak{a}\mathfrak{a}^{-1}$ is integral so that $\lambda\mathfrak{a}^{-1} \subseteq \mathfrak{a}^{-1}$, and hence $\lambda^2\mathfrak{a}^{-1} \subseteq \lambda\mathfrak{a}^{-1} \subseteq \mathfrak{a}^{-1}$, and so on. In general, $\lambda^n\mathfrak{a}^{-1} \subseteq \mathfrak{a}^{-1}$, and hence $\lambda^n\mathfrak{a}^{-1}\mathfrak{a}$ is integral. If μ is any element of $\mathfrak{a}^{-1}\mathfrak{a}$, then all powers of λ become integral on multiplication by μ. From the integral closure of $\mathfrak{o}$ it now follows that λ itself is integral just as in the proof of Theorem 1, Section 17.4.

It follows from property 11 that for the class multiplication defined above the class of $\mathfrak{a}^{-1}$ represents an inverse to the class of $\mathfrak{a}$: the product of the classes of $\mathfrak{a}$ and $\mathfrak{a}^{-1}$ is the identity class. This implies the following.

Theorem 1: *The classes of quasi-equal ideals form a group.*

The following rules make it possible to characterize quasi-divisibility and quasi-equality as divisibility and equality up to factors from the identity class.

12. The statement $\mathfrak{a} \geqq \mathfrak{b}$ implies $\mathfrak{a}\mathfrak{c} = \mathfrak{b}\mathfrak{d}$ with $\mathfrak{c} \sim \mathfrak{o}$ and integral $\mathfrak{d}$. In particular, it follows that $\mathfrak{a} \sim \mathfrak{b}\mathfrak{d}$.

13. The statement $\mathfrak{a} \sim \mathfrak{b}$ implies $\mathfrak{a}\mathfrak{c} = \mathfrak{b}\mathfrak{d}$ with $\mathfrak{c} \sim \mathfrak{o}$ and $\mathfrak{d} \sim \mathfrak{o}$.

Indeed, in both cases we have $\mathfrak{a}(\mathfrak{b}\mathfrak{b}^{-1}) = \mathfrak{b}(\mathfrak{a}\mathfrak{b}^{-1})$.

The greatest common divisor $(\mathfrak{a}, \mathfrak{b})$ is of course also a quasi-divisor both of $\mathfrak{a}$ and of $\mathfrak{b}$. We now show the following property.

14. Every common quasi-divisor of $\mathfrak{a}$ and $\mathfrak{b}$ is a quasi-divisor of $(\mathfrak{a}, \mathfrak{b})$. For if $\mathfrak{c}$ is such a quasi-divisor, then $\mathfrak{c}^*$ is a common divisor of $\mathfrak{a}$ and $\mathfrak{b}$ and therefore a divisor of $(\mathfrak{a}, \mathfrak{b})$.

Two integral ideals $\mathfrak{a}$ and $\mathfrak{b}$ are called *quasi-relatively prime* if $(\mathfrak{a}, \mathfrak{b}) \sim \mathfrak{o}$, or equivalently, if every common integral quasi-divisor of $\mathfrak{a}$ and $\mathfrak{b}$ is quasi equal to $\mathfrak{o}$.

15. If $\mathfrak{a}$ is quasi-relatively prime to $\mathfrak{b}$ and to $\mathfrak{c}$, then it is also quasi-relatively prime to the product $\mathfrak{b}\mathfrak{c}$. For

$$(\mathfrak{a}, \mathfrak{b}) \cdot (\mathfrak{a}, \mathfrak{c}) = (\mathfrak{a}^2, \mathfrak{a}\mathfrak{c}, \mathfrak{b}\mathfrak{a}, \mathfrak{b}\mathfrak{c}) \subseteqq (\mathfrak{a}, \mathfrak{b}\mathfrak{c}).$$

The left side is $\sim \mathfrak{o}$, and the right side must be also.

We now prove a result due to E. Artin.

Theorem 2 (Refinement Theorem): *If two factorizations of an integral ideal* $\mathfrak{a}$ *are given,*

$$\mathfrak{a} \sim \mathfrak{b}_1\mathfrak{b}_2 \cdots \mathfrak{b}_m \sim \mathfrak{c}_1\mathfrak{c}_2 \cdots \mathfrak{c}_n, \tag{17.10}$$

then the two products can be further decomposed in such a manner that the resulting decompositions coincide up to the order of the factors and quasi-equality:

$$\mathfrak{b}_\lambda \sim \prod_\mu \mathfrak{b}_{\lambda\mu}, \qquad \mathfrak{c}_\mu \sim \prod_\lambda \mathfrak{b}_{\lambda\mu}. \tag{17.11}$$

Proof: Put $(\mathfrak{b}_1, \mathfrak{c}_1) = \mathfrak{b}_{11}$. Then $\mathfrak{b}_1 \sim \mathfrak{b}_{11}\mathfrak{b}_1'$ and $\mathfrak{c}_1 \sim \mathfrak{b}_{11}\mathfrak{c}_1'$ by property 12. It follows that $\mathfrak{b}_{11} = (\mathfrak{b}_1, \mathfrak{c}_1) \sim (\mathfrak{b}_{11}\mathfrak{b}_1', \mathfrak{b}_{11}\mathfrak{c}_1') = \mathfrak{b}_{11}, (\mathfrak{b}_1'\ \mathfrak{c}_1')$, and hence $(\mathfrak{b}_1', \mathfrak{c}_1') \sim \mathfrak{o}$. Now put $(\mathfrak{b}_1', \mathfrak{c}_2) = \mathfrak{b}_{12}$. Then $\mathfrak{b}_1' \sim \mathfrak{b}_{12}\mathfrak{b}_1''$ and $\mathfrak{c}_2 = \mathfrak{b}_{12}\mathfrak{c}_2'$ by property 12; it again follows that $(\mathfrak{b}_1'', \mathfrak{c}_2') \sim \mathfrak{o}$. We continue in this manner until finally $\mathfrak{b}_1 = \mathfrak{b}_{11}\mathfrak{b}_{12} \ldots \mathfrak{b}_{1n}\mathfrak{d}$ and $\mathfrak{c}_\mu = \mathfrak{b}_{1\mu}\mathfrak{c}_\mu'$ $(\mu = 1, 2, \ldots, n)$. Substituting this into (17.10), it follows that

$$\mathfrak{b}_{11}\mathfrak{b}_{12} \cdots \mathfrak{b}_{1n}\mathfrak{d}\mathfrak{b}_2 \cdots \mathfrak{b}_m \sim \mathfrak{b}_{11}\mathfrak{c}_1'\mathfrak{b}_{12}\mathfrak{c}_2' \cdots \mathfrak{b}_{1n}\mathfrak{c}_n'.$$

By the group property (Theorem 1) $\mathfrak{b}_{11} \ldots, \mathfrak{b}_{1n}$ may be canceled:

$$\mathfrak{d}\mathfrak{b}_2 \cdots \mathfrak{b}_m \sim \mathfrak{c}_1'\mathfrak{c}_2' \cdots \mathfrak{c}_n'.$$

Here $\mathfrak{d}$ is quasi-relatively prime to all the $\mathfrak{c}_\mu'$ and hence also to the product $\mathfrak{c}_1'\mathfrak{c}_2' \ldots \mathfrak{c}_n'$. But $\mathfrak{d}$ occurs as a factor on the left side and is therefore a quasi-divisor of the product $\mathfrak{c}_1'\mathfrak{c}_2' \dots \mathfrak{c}_n'$. Hence $\mathfrak{d} \sim \mathfrak{o}$, and the factor $\mathfrak{d}$ may be omitted:

$$\mathfrak{b}_2 \cdots \mathfrak{b}_m \sim \mathfrak{c}_1'\mathfrak{c}_2' \cdots \mathfrak{c}_n'.$$

The same argument is now repeated with $\mathfrak{b}_2, \dots, \mathfrak{b}_m$ until the asserted decomposition (17.11) is obtained.

German letters shall henceforth denote *integral* nonzero ideals. We call such an ideal *indecomposable* if it is not quasi-equal to $\mathfrak{o}$ and if in every product representation $\mathfrak{p} \sim \mathfrak{a}\mathfrak{b}$ one of the factors belongs to the identity class; by property 12, this is equivalent to requiring that $\mathfrak{p}$, without being quasi-equal to $\mathfrak{o}$, have no other quasi-divisors than those which are either quasi-equal to $\mathfrak{p}$ or quasi-equal to $\mathfrak{o}$.

If an indecomposable ideal $\mathfrak{p}$ is replaced by the greatest quasi-equal ideal $\mathfrak{p}^*$, then every integral proper divisor of $\mathfrak{p}^*$ is necessarily not quasi-equal to $\mathfrak{p}$ and is therefore quasi-equal to $\mathfrak{o}$. Every ideal which is quasi-divisible by $\mathfrak{p}$ or $\mathfrak{p}^*$ is divisible by $\mathfrak{p}^*$, by property 4. This further implies what follows.

16. The term $\mathfrak{p}^*$ is a prime ideal. Indeed, if a product $\mathfrak{b}\mathfrak{c}$ of two principal ideals $\mathfrak{b}$ and $\mathfrak{c}$ is divisible by $\mathfrak{p}^*$ and $\mathfrak{b}$ is not divisible by $\mathfrak{p}^*$, then $(\mathfrak{b}, \mathfrak{p}^*)$ is a proper divisor of $\mathfrak{p}^*$ and so is quasi-equal to $\mathfrak{o}$; therefore,

$$\mathfrak{c} = \mathfrak{o}\mathfrak{c} \sim (\mathfrak{b}, \mathfrak{p}^*)\mathfrak{c} = (\mathfrak{b}\mathfrak{c}, \mathfrak{p}^*\mathfrak{c}) \geqq (\mathfrak{p}^*, \mathfrak{p}^*) = \mathfrak{p}^*,$$

and hence $\mathfrak{c}$ is quasi-divisible by $\mathfrak{p}^*$ and therefore divisible by $\mathfrak{p}^*$.

If we assume that the ascending chain condition holds in $\mathfrak{o}$, we obtain the following property.

17. Every chain of integral ideals $\mathfrak{a}_1 > \mathfrak{a}_2 > \dots$, where each successive ideal is a proper quasi-divisor of the preceding (that is, a quasi-divisor which is not quasi-equal), breaks off after finitely many terms. For if we replace the ideals $\mathfrak{a}_1, \mathfrak{a}_2, \dots$ by their greatest quasi-equal ideals $\mathfrak{a}_1^*, \mathfrak{a}_2^*, \dots$, then we obtain a chain of integral ideals $\mathfrak{a}_1^* \subset \mathfrak{a}_2^* \subset \dots$ which must break off by the ascending chain condition.

The "quasi-ascending chain condition" (property 17) can also be formulated as the "principle of divisor induction" (cf. Section 15.1, fourth formulation of the ascending chain condition). From this principle it then follows without difficulty that every integral ideal is quasi-equal to a product of indecomposable ideals. The uniqueness of the decomposition is obtained as a special case of the refinement theorem (Theorem 2).

Theorem 3: *Every nonzero integral ideal is quasi-equal to a product of indecomposable ideals* $\mathfrak{p}_1, \dots, \mathfrak{p}_r$ *which is uniquely determined up to quasi-equality and the order of the factors.* (The $\mathfrak{p}_1, \dots, \mathfrak{p}_r$ may of course also be chosen to be the prime ideals $\mathfrak{p}_1^*, \dots, \mathfrak{p}_r^*$.)

Corollary: An ideal $\mathfrak{a} \sim \mathfrak{p}_1 \dots \mathfrak{p}_r$ is quasi-divisible by $\mathfrak{b} \sim \mathfrak{p}_1' \dots \mathfrak{p}_s'$ if and only if every factor $\mathfrak{p}_i'$ occurring in the decomposition of $\mathfrak{b}$ occurs at least as often in the decomposition of $\mathfrak{a}$. In particular, if $\mathfrak{b}$ is a principal ideal, then it follows from property 2 that quasi-divisibility implies divisibility. If we take for $\mathfrak{a}$ and $\mathfrak{b}$ principal ideals (a) and (b), we obtain a criterion for the divisibility of a by b or a criterion that ab^{-1} be integral. By including with the principal ideals classes of ideals which are not principal ideals, we have obtained a domain in which (by Theorem 3) unique factorization holds, and the objective of "classical ideal theory" is herewith attained.

Theorem 3 also holds for fractional ideals ab^{-1}; however, we must then admit negative powers

$$\mathfrak{p}^{-k} = (\mathfrak{p}^{-1})^k$$

as factors. Indeed, if

$$\mathfrak{a} \sim \mathfrak{p}_1^{a_1} \cdots \mathfrak{p}_r^{a_r} \quad \text{and} \quad (b) \sim \mathfrak{p}_1^{b_1} \cdots \mathfrak{p}_r^{b_r},$$

then

$$\mathfrak{a}b^{-1} \sim \mathfrak{p}_1^{a_1 - b_1} \cdots \mathfrak{p}_r^{a_r - b_r}, \qquad (17.12)$$

and the exponents $a_i - b_i$ are uniquely determined.

In order to establish the relation of the present theory to the general ideal theory and to the special ideal theory of Section 17.4, we must investigate which prime ideals are indecomposable and which ideals are quasi-equal to $\mathfrak{o}$.

We have seen that if $\mathfrak{p}$ is indecomposable, then $\mathfrak{p}^*$ is prime. We now show the following.

18. No nonzero proper multiple of $\mathfrak{p}^*$ is prime. If $\mathfrak{a}$ is such a multiple, then $\mathfrak{a} \geqq \mathfrak{p}^*$; by property 12, $\mathfrak{a}\mathfrak{c} = \mathfrak{p}^*\mathfrak{d}$ with $\mathfrak{c} \sim \mathfrak{o}$. Then $\mathfrak{d} \not\equiv 0(\mathfrak{a})$, since in the decomposition of $\mathfrak{d}$ one less prime factor occurs than in the decomposition of $\mathfrak{a}$; similarly, $\mathfrak{p}^* \not\equiv 0(\mathfrak{a})$, and $\mathfrak{p}^*\mathfrak{d} \equiv 0(\mathfrak{a})$. Thus $\mathfrak{a}$ is not prime.

We now consider the decomposition of an arbitrary prime ideal $\mathfrak{p}$. Either $\mathfrak{p} \sim \mathfrak{o}$ or in the decomposition $\mathfrak{p} \sim \mathfrak{p}_1\mathfrak{p}_2 \ldots \mathfrak{p}_r$ an indecomposable factor $\mathfrak{p}_1$ occurs. Then $\mathfrak{p} \geqq \mathfrak{p}_1$ and hence $\mathfrak{p} \subseteqq \mathfrak{p}_1^*$; however, since a proper multiple of $\mathfrak{p}_1^*$ cannot be prime, it follows that $\mathfrak{p} = \mathfrak{p}_1^*$. Thus $\mathfrak{p}^* = (\mathfrak{p}_1^*)^* = \mathfrak{p}_1^* = \mathfrak{p}$. We therefore have the following property.

19. Every prime ideal $\mathfrak{p}$ is either quasi-equal to $\mathfrak{o}$ or indecomposable and equal to the associated $\mathfrak{p}^*$.

In the second case $\mathfrak{p}$ has no proper prime-ideal multiple distinct from the null ideal. On the other hand, in the first case we shall show that there is such a multiple.

20. If $\mathfrak{p} \sim \mathfrak{o}$, then there exists an indecomposable proper prime-ideal multiple $\mathfrak{p}_\nu^*$ of $\mathfrak{p}$. Indeed, if $p \neq 0$ is an element of $\mathfrak{p}$ and $(p) \sim \mathfrak{p}_1\mathfrak{p}_2 \ldots \mathfrak{p}_r \sim \mathfrak{p}_1^*\mathfrak{p}_2^* \ldots \mathfrak{p}_r^*$ is its decomposition, then it follows from property 2 that $\mathfrak{p}_1^*\mathfrak{p}_2^* \ldots \mathfrak{p}_r^* \equiv 0(p) \equiv 0(\mathfrak{p})$, and thus some $\mathfrak{p}_\nu^* \equiv 0(\mathfrak{p})$. Now $\mathfrak{p}_\nu^* \neq \mathfrak{p}$, since otherwise $\mathfrak{p}_\nu^* \sim \mathfrak{o}$.

We say that a prime ideal is a *higher* prime ideal if it has no proper prime-ideal multiple distinct from the null ideal; if such a multiple exists we say that it is a *lower* prime ideal. We can now combine properties 18, 19, and 20 in the following.

Theorem 4: *Every higher prime ideal* $\mathfrak{p}$ *is indecomposable and is equal to its* $\mathfrak{p}^*$; *every lower prime ideal is quasi-equal to* $\mathfrak{o}$.

From the decomposition theorem (Theorem 3) an ideal which does not belong to the identity class is divisible by at least one higher prime ideal $\mathfrak{p} = \mathfrak{p}^*$. On the other hand, an ideal of the identity class is not divisible by any higher prime ideal. This gives a purely ideal theory characterization (that is, in the domain of integral ideals) of the identity class.

In the rings studied in Section 17.4 a nonzero prime ideal is divisible only by itself and by $\mathfrak{o}$ on the basis of Axiom II; thus, in that section there are no lower prime ideals but $\mathfrak{o}$. Since every ideal $\mathfrak{a} \neq \mathfrak{o}$ is divisible by a prime ideal distinct from $\mathfrak{o}$ (proof: from among all the divisors of $\mathfrak{a}$ distinct from $\mathfrak{o}$ choose a maximal one; since this ideal is maximal it is also prime), it follows that $\mathfrak{a}$ cannot be quasi-equal to $\mathfrak{o}$. The identity class therefore consists of the unit ideal $\mathfrak{o}$ alone. From property 12 it follows further that quasi-divisibility and divisibility are equivalent, and from this, or from 13, that quasi-equality and equality are likewise equivalent. Therefore, the ideal theory of Section 17.4 is contained in the present theory as a special case.

The connection with the general ideal theory is also easy to establish. First of all, it is easy to see that every prime ideal whose associated prime ideal is a lower ideal must be quasi-equal to $\mathfrak{o}$. These primary ideals we call *lower primary ideals*; all others we call *higher primary ideals.* An ideal $\mathfrak{a}$ is quasi-equal to $\mathfrak{o}$ if and only if all its primary components are lower. If two ideals $\mathfrak{a}$ and $\mathfrak{b}$ have the same higher primary components (but not necessarily the same lower components), then they are quasi-equal. Among the ideals quasi-equal to $\mathfrak{a}$ there is a greatest ideal $\mathfrak{a}^*$; it is obtained by omitting all lower primary components from the decomposition $[\mathfrak{q}_1, \ldots, \mathfrak{q}_r]$. The decomposition and uniqueness theorems of this section may therefore be interpreted as neglecting all lower primary components and considering only the higher ones. The higher primary ideals are each divisible by one higher prime ideal; the factorization of Theorem 2 must therefore involve a power of a prime ideal; that is, *every higher primary ideal is quasi-equal to a power of a prime ideal.*

Conversely, every power of a higher prime ideal is quasi-equal to a higher primary ideal. Indeed, if $\mathfrak{a} = \mathfrak{p}^r$ is a power of a higher prime ideal, then $\mathfrak{a}$ is divisible by no higher prime ideal other than $\mathfrak{p}$; therefore, in the factorization

$$\mathfrak{a} = \mathfrak{p}^r = [\mathfrak{q}_1, \ldots, \mathfrak{q}_r]$$

only one higher primary ideal occurs. If this is, say, $\mathfrak{q}_1$, then it follows that $\mathfrak{a}^* = \mathfrak{q}_1$; hence $\mathfrak{a} = \mathfrak{p}^r$ is quasi-equal to the primary ideal $\mathfrak{q}_1$.

Moreover, $\mathfrak{q}_1$ is precisely the rth symbolic power of the prime ideal $\mathfrak{p}$ defined in Section 15.6. Thus *all higher primary ideals are symbolic powers of higher prime ideals.*

Prüfer has called the ideals $\mathfrak{a}$ with the property $\mathfrak{a}^* = \mathfrak{a}$ *v-ideals.* The integral v-ideals are just those in whose primary ideal decomposition only higher primary ideals occur. All principal ideals are v-ideals. In each class of quasi-equal ideals there is a single v-ideal $\mathfrak{a}_v = \mathfrak{a}^*$. If we restrict ourselves with Prüfer and Krull to v-ideals, then the concept of quasi-equality becomes unnecessary. The main theorem (Theorem 3) may then be formulated as follows.

Theorem: *Every v-ideal can be represented uniquely as the intersection of symbolic powers* $\mathfrak{p}^{(r)}$ *of higher prime ideals.*

Exercises

17.11. All results of this section also hold for rings with zero divisors if the quotient field is replaced by the quotient ring and attention is restricted to ideals which are not zero divisors.

17.12. Theorem 1 implies conversely that the ring $\mathfrak{o}$ is integrally closed (cf. Section 17.5).

17.13. Prove $\mathfrak{a} : \mathfrak{b} \sim \mathfrak{a}\mathfrak{b}^{-1}$.[8]

SUMMARY OF IDEAL THEORY

The following summary shows the significance of Axioms I (ascending chain condition), II (every prime ideal maximal), and III (integral closure) formulated in Section 16.3 for the ideal theory of integral domains.

Axiom I implies that every ideal is an l.c.m. of primary ideals; the associated prime ideals are unique.

Axioms I and II imply that every ideal is the product of single-prime primary ideals; the representation is unique.

Axioms I and III imply that every ideal is quasi-equal to a product of powers of prime ideals; unique up to quasi-equality.

Axioms I, II, III imply that every ideal is a product of powers of prime ideals; unique.

[8]For further generalization of the results of this section, see H. Prüfer, *J. Reine u. Angew. Math.*, Vol. 168, 1932, and also P. Lorenzen, *Math. Z.*, Vol. 45, 1939.

Chapter 18

FIELDS WITH VALUATIONS

18.1 VALUATIONS

The construction of the extension field Ω for a given ordered field K which was presented in Section 11.2 did not make full use of the ordering of the field K; only the ordering of the absolute values $|a|$ of the field elements a was used. It is thus natural to try to extend this construction to fields (other than just order fields) for which a function $\varphi(a)$ with the properties of the absolute value exists.

A field is said to *have a valuation* if for the elements a of K a function $\varphi(a)$ is defined having the following properties:

1. $\varphi(a)$ is an element of an ordered field P.
2. $\varphi(a) > 0$ for $a \neq 0$, $\varphi(0) = 0$.
3. $\varphi(ab) = \varphi(a)\varphi(b)$.
4. $\varphi(a+b) \leqq \varphi(a) + \varphi(b)$.

From properties 2 and 3 it follows immediately that

$$\varphi(1) = 1, \qquad \varphi(-1) = 1, \qquad \varphi(a) = \varphi(-a).$$

Putting $c = a+b$, it follows from property 4 that

$$\varphi(c) - \varphi(a) \leqq \varphi(c-a);$$

similarly,

$$\varphi(a) - \varphi(c) \leqq \varphi(c-a);$$

hence

$$|\varphi(c) - \varphi(a)| \leqq \varphi(c-a).$$

Inequality 4 continues to hold if b is replaced by $-b$; thus

$$\varphi(a-b) \leqq \varphi(a) + \varphi(b).$$

Inequality 4 can be easily extended by induction to sums of n terms.

Every field has the "trivial" valuation $\varphi(a) = 1$ for $a \neq 0$, $\varphi(0) = 0$. We shall not consider this valuation further.

If K is ordered, then we may put $\varphi(a) = |a|$. There are, however, entirely different types of valuations. Let $\mathbb{Q}$ be the field of rational numbers. If p is a fixed prime number and each rational number $a \neq 0$ is written in the form

$$a = \frac{s}{t} p^n$$

with integers s and t not divisible by p, then

$$\varphi_p(a) = p^{-n}, \qquad \varphi_p(0) = 0$$

defines a valuation of $\mathbb{Q}$. Properties 1 to 3 are easy to check. In place of inequality 4, the stronger inequality

$$\varphi_p(a+b) \leqq \max(\varphi_p(a), \varphi_p(b)) \tag{18.1}$$

holds. For if

$$a = \frac{s}{t} p^n, \; b = \frac{u}{v} p^m, \qquad \text{with } s, t, u, \text{ and } v \text{ prime to } p,$$

and, say, $\varphi_p(b) \geqq \varphi_p(a)$, that is, $n \geqq m$, then

$$a+b = \frac{svp^{n-m}+tu}{tv} p^m,$$

and hence

$$\varphi_p(a+b) = p^{-m'} \qquad \text{with} \quad m' \geqq m;$$

thus

$$\varphi_p(a+b) \leqq \varphi_p(b).$$

This is the *p-adic* valuation of $\mathbb{Q}$.

The p-adic valuation can be easily generalized. Let $\mathfrak{o}$ be an integral domain, K its quotient field, and $\mathfrak{p}$ a prime ideal of $\mathfrak{o}$ with the following properties.

A. *All powers* $\mathfrak{p}, \mathfrak{p}^2, \ldots$ *are distinct and their intersection contains only zero.*

B. *If a is exactly divisible by* $\mathfrak{p}^\alpha$, *that is, by* $\mathfrak{p}^\alpha$ *but not by* $\mathfrak{p}^{\alpha+1}$, *and if b is exactly divisible by* $\mathfrak{p}^\beta$, *then ab is exactly divisible by* $\mathfrak{p}^{\alpha+\beta}$.

Here $\mathfrak{p}^\alpha$ denotes the set of all sums $\sum_\nu p_{\nu 1} p_{\nu 2} \cdots p_{\nu\alpha}$, where all the $p_{\nu\kappa}$ are elements of $\mathfrak{p}$. In particular, $\mathfrak{p}^1 = \mathfrak{p}$ and $\mathfrak{p}^0 = \mathfrak{o}$. In a in $\mathfrak{o}$ is exactly divisible by $\mathfrak{p}^\alpha$, we now define

$$\varphi(a) = e^{-\alpha} \qquad \text{and} \quad \varphi(0) = 0,$$

where e is any real number > 1. The valuation is then defined for the elements of $\mathfrak{o}$ and has properties 1 through 4.

But if a valuation is defined for the elements of an integral domain, it can be immediately extended to the elements of the quotient field by

$$\varphi\left(\frac{a}{b}\right) = \frac{\varphi(a)}{\varphi(b)}.$$

This is well defined, since

$$\frac{a}{b} = \frac{c}{d} \qquad \text{or} \quad ad = bc$$

implies

$$\varphi(a)\varphi(d) = \varphi(b)\varphi(c) \qquad \text{or} \qquad \frac{\varphi(a)}{\varphi(b)} = \frac{\varphi(c)}{\varphi(d)}.$$

Furthermore, the valuation $\varphi(a/b)$ also has properties 1 through 4. The first three are obvious. Property 4 is verified as follows:

$$\varphi\left(\frac{a}{b}+\frac{c}{d}\right) = \frac{\varphi(ad+bc)}{\varphi(bd)} \leqq \frac{\varphi(ad)+\varphi(bc)}{\varphi(bd)} = \varphi\left(\frac{a}{b}\right)+\varphi\left(\frac{c}{d}\right).$$

In this manner a valuation of the quotient field K is immediately obtained from the valuation of the integral domain $\mathfrak{o}$ defined by the prime ideal $\mathfrak{p}$. This is called the $\mathfrak{p}$-adic valuation of K.

Properties A and B are satisfied, in particular, if $\mathfrak{p}$ is any prime ideal of an integral domain $\mathfrak{o}$ distinct from the null and unit ideals which satisfies the three axioms of Section 17.4. With each such prime ideal $\mathfrak{p}$ there is thus associated a $\mathfrak{p}$-adic valuation of the quotient field K. This applies then, in particular, to the prime ideals $\mathfrak{p}$ in the ring of integral quantities of an algebraic number field. This reveals the close relationship between classical ideal theory and valuation theory.

More generally, as in Section 17.7, we can start with an integral domain $\mathfrak{o}$ which satisfies only Axioms I and III. We then restrict our consideration to higher prime ideals $\mathfrak{p}$ in the sense of Section 17.7 and form their symbolic powers

$$\mathfrak{q} = \mathfrak{p}^{(r)}$$

in the sense of Section 15.6. The following properties analogous to A and B then hold.

A′. *The $\mathfrak{p}^{(r)}$ are all distinct and their intersection is the null ideal.*

B′. *If a is exactly divisible by $\mathfrak{p}^{(r)}$ and b extacly divisible by $\mathfrak{p}^{(s)}$, then ab is exactly divisible by $\mathfrak{p}^{(r+s)}$.*

If a is exactly divisible by $\mathfrak{p}^{(r)}$, we may thus again define

$$\varphi(a) = e^{-r} \qquad \text{and} \qquad \varphi(0) = 0.$$

A $\mathfrak{p}$-adic valuation is thus again obtained for each higher prime ideal $\mathfrak{p}$.

In the polynomial ring $\Delta[x_1, \ldots, x_r]$ the ideal

$$\mathfrak{p} = (x_1, \ldots, x_n)$$

has properties A and B. The associated valuation $\varphi(f)$ is e^{-s}, where s is the degree of the term of lowest degree occurring in the polynomial f.

Exercises

18.1. In the definition of a valuation, drop the requirement that $\varphi(a)$ should be nonnegative, and prove: if there exists a c in K such that $\varphi(c) < 0$, then

$a \to \varphi(a)$ is an isomorphism of K onto a subfield of the value field P. [Prove that equality holds in property 4 by considering the inequality corresponding to 4 for $\varphi(ac+bc)$.]

18.2. For p-adic valuations, inequality 4 can be strengthened to (18.1).

The most important investigations of fields with valuations concern the case in which the ordering of the value field P is Archimedean. According to Exercise 2 in Section 11.2, P can then be imbedded in the field of real numbers. We shall now assume that the values $\varphi(a)$ are real numbers. We assume the reader is familiar with (natural) logarithms of real numbers and their simplest properties as well as the powers α^β of a positive number α with arbitrary real exponents.

We shall make use of the following lemma concerning real numbers.

Lemma: *If α, β, γ are positive real numbers and*

$$\gamma^\nu \leqq \alpha\nu+\beta$$

for every natural number ν, then $\gamma \leqq 1$.

Proof: Suppose that $\gamma = 1+\delta$, $\delta>0$. Then for $\nu \geqq 2$ $\gamma^\nu = (1+\delta)^\nu = 1+\nu\delta+\frac{1}{2}\nu(\nu-1)\delta^2+\ldots > \nu\delta+\frac{1}{2}\nu(\nu-1)\delta^2$; but for sufficiently large ν, certainly

$$\nu\delta>\beta \quad \text{and} \quad \tfrac{1}{2}(\nu-1)\delta^2>\alpha,$$

and hence

$$\gamma^\nu>\beta+\alpha\nu,$$

contrary to hypothesis.

A real valuation $\varphi(a)$ of a field K is called *non-Archimedean* if for all natural multiples $n = 1+1+\cdots+1$ of the identity the condition

$$\varphi(n) \leqq 1$$

holds. The p-adic valuation of the field $\mathbb{Q}$ is non-Archimedean. The fact that the value field is Archimedean makes no difference.

The valuation φ of K *is non-Archimedean if and only if in place of property* 4 *the sharper inequality*

4′. $$\varphi(a+b) \leqq \max(\varphi(a), \varphi(b))$$

holds.

Proof: (1) If inequality 4′ holds for the sum of two terms, then a corresponding inequality holds for the sum of n terms. In particular, for $n = 1+1+\cdots+1$,

$$\varphi(n) \leqq \max(\ldots, \varphi(1), \ldots) = 1.$$

(2) If φ is non-Archimedean, then, for $\nu = 1, 2, 3, \ldots,$

$$(\varphi(a+b))^\nu = \varphi((a+b)^\nu) = \varphi(a^\nu+\tbinom{\nu}{1}a^{\nu-1}b+\cdots+b^\nu) \leqq$$
$$\leqq \varphi(a)^\nu+\varphi(a)^{\nu-1}\varphi(b)+\cdots+\varphi(b)^\nu \leqq (\nu+1)M^\nu,$$

where $M = \max(\varphi(a), \varphi(b))$. From this it follows by the preceding lemma that

$$\frac{\varphi(a+b)}{M} \leqq 1, \qquad \text{hence} \quad \varphi(a+b) \leqq M,$$

and this is 4′.

We shall henceforth consider inequality 4′ as a characterization of a non-Archimedean valuation even in the case in which the value field P is not the field of real numbers. As Krull has pointed out, an arbitrary ordered Abelian group can then be taken for the range of the valuation, since the values need only be multiplied and compared as to magnitude, and the addition of values is not involved at all.

The following remark is often useful; it applies to all non-Archimedean valuations in the sense just defined.

If $\varphi(a)$ and $\varphi(b)$ *are distinct, then equality holds in* 4′.

Proof: Suppose that $\varphi(a) > \varphi(b)$. We must show that

$$\varphi(a+b) = \varphi(a).$$

Suppose that

$$\varphi(a+b) < \varphi(a);$$

then $\varphi(a+b)$ and $\varphi(-b) = \varphi(b)$ would each be less than $\varphi(a)$. This is contradicted, however, by the inequality

$$\varphi(a) \leqq \max(\varphi(a+b), \varphi(-b)).$$

It is often expedient (and customary in the literature) to introduce another notation for non-Archimedean valuations. In place of the real value $\varphi(a)$ we consider the *exponent* $w(a) = -\log \varphi(a)$. In terms of exponents the defining relations for the valuation are as follows.

1. $w(a)$ is a real number for $a \neq 0$.
2. $w(0)$ is the symbol ∞.
3. $w(ab) = w(a) + w(b)$.
4. $w(a+b) \geqq \min(w(a), w(b))$.

We then speak of an *exponential valuation*. The transition to exponents is made possible by the fact that, because of the sharper inequality 4′, the values $\varphi(a)$ need not be added. Formation of logarithms reverses the sense of the ordering and replaces multiplication by addition.

Example: Let the elements of the field K be meromorphic functions in a region of the z-plane or, more generally, on a Riemann surface. We choose a particular point P of the Riemann surface and make the following definition. The value $w(a)$ of a function a shall be α if the function has a zero of order α at P, it shall be zero if the function has a finite nonzero value there, and it shall be $-\alpha$ if the function has a pole of order α at P. Properties 1 through 4 are then satisfied. Thus to each place P there corresponds a valuation of the field K. This example leads us to suspect the significance of valuation theory for the theory of algebraic functions of a single complex variable.

Among the exponential valuations two types are distinguished: *discrete* valuations, which are characterized by the fact that there is a smallest possible value $w(a)$ of which all values $w(a)$ are multiples (cf. the example above), and *nondiscrete* valuations, in which the values $w(a)$ come arbitrarily close to zero.

Since multiples of a value $w(a)$ are again values, $nw(a) = w(a^n)$, it follows that in the nondiscrete case the values $w(a)$ are dense in the set of real numbers.

The p-adic valuation of the rational number field is discrete; the p-adic valuations are likewise discrete.

In a field K with an exponential valuation, the elements a such that $w(a) \geqq 0$ form a ring $\mathfrak{J}$. For $w(a) \geqq 0$ and $w(b) \geqq 0$ imply $w(a+b) \geqq \min(w(a, w(b)) \geqq 0$ and $w(ab) = w(a)+w(b) \geqq 0$. The set $\mathfrak{p}$ of all elements a of K such that $w(a)>0$ is a prime ideal of $\mathfrak{J}$. For, first of all, $w(a)>0$ and $w(b)>0$ imply $w(a+b) \geqq \min(w(a), w(b))>0$, and hence $\mathfrak{p}$ is a module. Second, it follows from $a \in \mathfrak{p}$, that is, $w(a)>0$, and $w(c) \geqq 0$ that $w(ca) = w(c)+w(a)>0$, and hence $\mathfrak{p}$ is an ideal. Finally, $ab \equiv 0$ (mod $\mathfrak{p}$—that is, $w(ab) = w(a)+w(b)>0$—implies that at least one of the two numbers $w(a)$ and $w(b)$ is positive and hence that at least one of the two elements a and b is contained in $\mathfrak{p}$; $\mathfrak{p}$ is therefore prime.

Here $\mathfrak{J}$ is called the *valuation ring* corresponding to the valuation w. The elements of $\mathfrak{J}$ are called *integral* (with respect to the valuation). An element a is said to be *divisible* by b (with respect to the valuation w) if a/b is integral or if $w(a) \geqq w(b)$.

The elements a with $w(a) = 0$ are the units of the ring $\mathfrak{J}$. Since all elements of $\mathfrak{J}$ not belonging to $\mathfrak{p}$ are units of $\mathfrak{J}$, it follows that $\mathfrak{p}$ is a maximal ideal of $\mathfrak{J}$. The residue class ring $\mathfrak{J}/\mathfrak{p}$ is therefore a field, the *residue class field* of the valuation. If the field K has characteristic p, then the residue class field clearly also has characteristic p. If K has characteristic zero, however, then the residue class field may have either characteristic zero (equal characteristic case) or a prime characteristic (unequal characteristic case). The p-adic valuations are typical examples of the unequal characteristic case. An example of the equal characteristic case is obtained by considering the field of rational functions in one variable and setting the exponential value of a rational function equal to the degree of the numerator minus the degree of the denominator. The $\mathfrak{p}$-adic valuations defined by the ideals of the polynomial ring $K[x_1, \ldots, x_n]$ also belong to the equal characteristic case.[1]

Exercises

18.3. Show that in $\mathfrak{J}$ every ideal is either the set of all a such that $w(a)>\delta$ or the set of all a such that $w(a) \geqq \delta$, where δ is a nonnegative real number. In the case of a discrete valuation we need only consider the case $\geqq$ and a number δ may be chosen which actually occurs in the set of values. In the case of a nondiscrete valuation, δ is uniquely determined by the ideal.

18.4. In the case of a discrete valuation all ideals of $\mathfrak{J}$ are powers of $\mathfrak{p}$, whereas in the case of a nondiscrete valuation all powers of $\mathfrak{p}$ are equal to $\mathfrak{p}$.

[1]For further study of these ideas through a complete classification of all valuations, see the papers of H. Hasse, F. K. Schmidt, O. Teichmüller, and E. Witt (E. Witt, *J. Reine u. Angew. Math.*, **176**, 126–140 (1936) and literature cited there). For generalizations of the valuation concept see the papers of K. Mahler and W. Krull: K. Mahler, "Über Pseudobewertungen," I, *Acta Math.*, **66**, 79–199 (1936); Ia, *Akad. Wetensch. Amsterdam, Proc.*, **39**, 57–65 (1936); II, *Acta Math.*, **67**, 51–80 (1936). W. Krull, "Allgemeine Bewertungstheorie," *J. Reine u. Angew. Math.*, **167**, 160–196 (1932).

18.2 COMPLETE EXTENSIONS

For every field K with a valuation we can construct an extension field Ω_K with a valuation in which *Cauchy's convergence theorem* holds by the procedure of Section 11.2. We shall here again assume that the values $\varphi(a)$ are real numbers. A *fundamental sequence* $\{a_\nu\}$ in K is defined by the property

$$\varphi(a_p - a_q) < \varepsilon \quad \text{for} \quad p > n(\varepsilon), \quad q > n(\varepsilon),$$

where ε is an arbitrary positive real number. The residue class field Ω_K is obtained from the ring of fundamental sequences precisely as in Section 11.2; all proofs carry over verbatim. The only difference is that Ω_K, just as K, is not ordered; it is simply a field with a valuation. The valuation of Ω_K is defined as follows. If α is defined by the fundamental sequence $\{a_\nu\}$ then from an inequality already proved,

$$|\varphi(a_\nu) - \varphi(a_\mu)| \leqq \varphi(a_\nu - a_\mu),$$

the values $\varphi(a_\nu)$ also form a fundamental sequence which therefore has a limit in the field of real numbers. We put

$$\varphi(\alpha) = \omega.$$

All fundamental sequences with the same limit α define the same value $\varphi(\alpha)$, and this valuation satisfies conditions 1 through 4.

The field Ω_K is *complete* in the valuation φ; that is, Cauchy's convergence theorem holds in Ω_K as follows.

Theorem: *Every fundamental sequence of Ω_K has a limit in Ω_K.*

We have called a sequence $\{a_\nu\}$ fundamental if for every $\varepsilon > 0$ of the value field there exists an n such that

$$\varphi(a_p - a_q) < \varepsilon \quad \text{for} \quad p > n, \quad q > n.$$

In the case of a non-Archimedean valuation it is sufficient instead of this to require that

$$\varphi(a_{\nu+1} - a_\nu) < \varepsilon \quad \text{for} \quad \nu > n(\varepsilon).$$

For $a_p - a_q$ is the sum of $|p-q|$ summands $a_{\nu+1} - a_\nu$, and if these all have a value $< \varepsilon$ then the value of the sum is likewise $< \varepsilon$ by (18.1). Thus we have the following.

In a complete field with a non-Archimedean valuation a sequence $\{a_\nu\}$ has a limit if the differences $a_{\nu+1} - a_\nu$ form a null sequence.

This criterion may also be formulated as follows: *for the convergence of an infinite series $a_1 + a_2 + a_3 + \cdots$ it is necessary and sufficient that* $\lim a_\nu = 0$.

If we consider the field $\mathbb{Q}$ of rational numbers with the valuation given by the usual absolute value, $\varphi(a) = |a|$, then of course we obtain the field of real numbers as the complete extension. However, if we begin with the p-adic valuation of $\mathbb{Q}$, then we obtain as complete extension the *field Ω_p of Hensel's p-adic numbers.*

The fields $\Omega_2, \Omega_3, \Omega_5, \Omega_7, \Omega_{11}, \ldots$ thus emerge as complete fields of the same status as the field of real numbers (and for number theory they are just as important).

The elements of the field Ω_p, the p-adic numbers, admit a somewhat more convenient representation than that by fundamental sequences. We consider for $\lambda = 0, 1, 2, 3, \ldots$ the module $\mathfrak{M}_\lambda$ consisting of those rational numbers whose numerators are divisible by p^λ and whose denominators are not divisible by p, that is, those for which $\varphi(a) \leqq p^{-\lambda}$. We call two rational numbers congruent (mod p^λ) if their difference belongs to $\mathfrak{M}_\lambda$. If now $\{r_\mu\}$ is a p-adic fundamental sequence of rational numbers, then for each λ there is a $n = n(\lambda)$ such that

$$\varphi(r_\mu - r_\nu) \leqq p^{-\lambda} \quad \text{for} \quad \mu > n(\lambda), \quad \nu > n(\lambda),$$

that is,

$$r_\mu \equiv r_\nu (\text{mod}\, p^\lambda).$$

All numbers r_μ with $\mu > n(\lambda)$ therefore belong to a single residue class $\mathfrak{R}_\lambda$ modulo $\mathfrak{M}_\lambda$. The fundamental sequence $\{r_\mu\}$ thus defines a sequence of residue classes

$$\mathfrak{R}_0 \supset \mathfrak{R}_1 \supset \mathfrak{R}_2 \supset \mathfrak{R}_3 \supset \mathfrak{R}_4 \supset \cdots,$$

which are nested in the manner indicated. Conversely, every sequence $\{r_1, r_2, \ldots\}$ which defines a sequence $\{\mathfrak{R}_\lambda\}$ of residue classes $\mathfrak{R}_\lambda$ modulo $\mathfrak{M}_\lambda$ nested in this manner such that

$$r_\mu \text{ is in } \mathfrak{R}_\lambda \qquad \text{for all } \mu > n(\lambda)$$

is always a fundamental sequence.

In particular, if $\{r_\mu\}$ is a null sequence, then $\mathfrak{R}_\lambda = \mathfrak{M}_\lambda$ is the null residue class. If two fundamental sequences are added, $\{r_\mu\} + \{s_\mu\} = \{r_\mu + s_\mu\}$, then the associated residue classes are also added to form $\{\mathfrak{R}_\lambda + \mathfrak{S}_\lambda\}$. If a null sequence is added to a fundamental sequence, then the associated residue-class sequence remains unchanged. Conversely, if two sequences $\{r_\mu\}$ and $\{s_\mu\}$ belong to the same residue-class sequence $\{\mathfrak{R}_\lambda\}$, then their difference is a null sequence. *There is thus a one-to-one correspondence between p-adic numbers* $\alpha = \lim r_\nu$ *and residue-class sequences* $\{\mathfrak{R}_\lambda\}$ *of the type indicated.*

This representation of p-adic numbers by residue-class sequences is the convenient representation to which we referred above. To go back from the representation of a p-adic number α by a sequence of residue classes to a (particular) fundamental sequence, we have only to select an r'_λ from each residue class $\mathfrak{R}_\lambda$: then $\alpha = \lim r'_\lambda$. The number α can also be represented as an infinite sum by putting

$$r'_1 = s_0, \qquad r'_{\lambda+1} - r'_\lambda = s_\lambda p^\lambda;$$

then

$$r'_{\lambda+1} = s_0 + s_1 p + s_2 p^2 + \cdots + s_\lambda p^\lambda,$$

and hence

$$\alpha = \lim_{\lambda \to \infty} \sum_{\nu=0}^{\lambda} s_\nu p^\nu = \sum_{\nu=0}^{\infty} s_\nu p^\nu. \tag{18.2}$$

Here $s_1, s_2, \ldots$ are rational numbers whose denominators are not divisible by p.

A p-adic limit of ordinary integers is called a *p-adic integer*. In terms of the residue classes $\mathfrak{R}_0, \mathfrak{R}_1, \ldots$ this means that an integer must occur in each of them. In particular, in the case of a p-adic integer, $\mathfrak{R}_0$ is the null residue class $\mathfrak{M}_0$, the set of all rational numbers whose denominators are not divisible by p. But this is also a sufficient condition for a p-adic integer: if $\mathfrak{R}_0$ is the null residue class modulo $\mathfrak{M}_0$, then all the residue classes $\mathfrak{R}_1, \mathfrak{R}_2, \ldots$ contain integers. Indeed, $\mathfrak{R}_\lambda$ is contained in $\mathfrak{R}_0$ and therefore consists of numbers r/s with $s \not\equiv 0 (\text{mod } p)$. Solving the congruence

$$sx \equiv r(\text{mod } p^\lambda),$$

we obtain

$$x - \frac{r}{s} = \frac{sx-r}{s} \equiv 0\ (\text{mod } \mathfrak{M}_\lambda),$$

and thus the number x belongs to the residue class $\mathfrak{R}_\lambda$.

If α is a p-adic integer, we may thus choose in the series representation (18.2) all the r'_λ, and hence all the s_ν also, to be ordinary integers. Therefore (18.2) is a power series in p with integral coefficients. Every such power series converges in the sense of the p-adic valuation and represents a p-adic integer.

Every p-adic number α with the representation $\{\mathfrak{R}_0, \mathfrak{R}_1, \ldots\}$ as residue classes can be converted to a p-adic integer by multiplication with a power of p. If r'_0 is an element of the residue class $\mathfrak{R}_0$, then after multiplying r'_0 by some power p^m of p the denominator $p^m r'_0$ will contain no factor p, and thus r'_0 belongs to the null residue class modulo $\mathfrak{M}_0$. If we now expand the p-adic integer $p^m\alpha$ in a power series (18.2) with integral $s_0, s_1, \ldots,$ we obtain for α a representation with a finite number of negative exponents:

$$\alpha = a_{-m}p^{-m} + a_{-m+1}p^{-m+1} + \cdots + a_0 + a_1 p + a_2 p^2 + \cdots. \tag{18.3}$$

The representation (18.2) of the p-adic integer α can be normalized by choosing for r'_λ the smallest nonnegative integer from the residue class $\mathfrak{R}_\lambda$. The numbers s_ν then satisfy the condition $0 \leqq s_\nu < p$. Passing now from (18.2) to (18.3), we find that *for every p-adic number there is a uniquely determined expansion* (18.3) *with* $0 \leqq a_\nu < p$.

A complete $\mathfrak{p}$-*adic field* $\Omega_\mathfrak{p}$, the generalization of Hensel's p-adic field, is likewise obtained from the $\mathfrak{p}$-adic valuation of a field K which is defined by a prime ideal $\mathfrak{p}$ of the integral domain $\mathfrak{o}$ in the manner indicated in Section 18.1. For example, if $\mathfrak{p}$ is the ideal $(x-c)$ in the polynomial ring $\Delta[x]$, then $\Omega_\mathfrak{p}$ is the ring of all power series

$$\alpha = a_{-m}(x-c)^{-m} + \cdots + a_0 + a_1(x-c) + a_2(x-c)^2 + \cdots \tag{18.4}$$

with constant a_ν in Δ. The power series *always* converges in the sense of the $\mathfrak{p}$-adic valuation however the coefficients a_ν are chosen. Expressions such as (18.4) are called *formal power series* in $(x-c)$.

Exercises

18.5. Write -1 and $\frac{1}{2}$ as 3-adic normalized power series.

18.6. An equation $f(\xi) = 0$, where f is an integral polynomial, is solvable in the field Ω_p if and only if the congruence

$$f(\xi) \equiv 0 \pmod{p^n}$$

has a rational solution ξ for every natural number n.

18.7. Are the equations

$$x^2 = -1, \qquad x^2 = 3, \qquad x^2 = 7$$

solvable in the field Ω_3?

It is possible that two distinct valuations φ and ψ of a field K may lead to the same complete extension field Ω. This is clearly the case if and only if every sequence $\{a_\nu\}$ of K which is a null sequence for the valuation φ is also a null sequence for ψ and conversely. In this case, when $\lim_{\nu\to\infty} \varphi(a_\nu) = 0$ and $\lim_{\nu\to\infty} \psi(a_\nu) = 0$ have the same meaning, the two valuations φ and ψ are called *equivalent*.

In the case of the valuation $\varphi(a) = |a|$ of the field of complex numbers by the usual absolute value, an infinite number of equivalent valuations may be obtained by putting $\varphi(a) = |a|^\varrho$ where ϱ is a fixed positive real number less than or equal to 1. Conditions 1 through 3 are trivially satisfied. Condition 4 follows from $|a+b| \leqq |a|+|b|$ by the inequality $\varepsilon^\varrho+\delta^\varrho \geqq (\varepsilon+\delta)^\varrho$ which holds for any two real numbers $\varepsilon \geqq 0$ and $\delta \geqq 0$ with $0<\varrho \leqq 1$.[2]

In the case of the p-adic valuation $\varphi_p(a)$ of the field of rational numbers, every valuation $\psi(a) = \varphi_p(a)^\sigma$, where σ is any fixed positive number, is an equivalent valuation.

Let φ and ψ be valuation of a field K. We shall show that the following three assertions are equivalent.

1. φ and ψ are equivalent.
2. $\varphi(a)<1$ implies $\psi(a)<1$.
3. ψ is a power of φ, that is, $\psi(a) = \varphi(a)^\varepsilon$ for all a and fixed $\varepsilon>0$.

We first show that statement 1 implies 2. Here $\varphi(a)<1$ implies that a^n converges to zero in the sense of the valuation φ. But then a^n must also converge to zero in the sense of the valuation ψ. Hence $\psi(a)<1$.

We now show that statement 2 implies 3. We note first of all that $\varphi(a)<\varphi(b)$ implies $\varphi(a/b)<1$; therefore, $\psi(a/b)<1$ and hence $\psi(a)<\psi(b)$. Now let p be any fixed element of K with $\varphi(p)>1$. Then also $\psi(p)>1$. Let a be any element of K

[2]Cf., for example, Hardy–Littlewood–Polya, *Inequalities*, Cambridge, 1934, Chapter II.

and let $\varphi(a) = \varphi(p)^\delta$ and $\psi(a) = \psi(p)^{\delta'}$. We wish to show that $\delta = \delta'$. Let n and m be integers with $n/m < \delta$ and $m > 0$. Then

$$\varphi(p)^{n/m} < \varphi(p)^\delta = \varphi(a), \qquad \text{hence} \quad \varphi(p^n) < \varphi(a^m).$$

But from this it follows that

$$\psi(p^n) < \psi(a^m), \qquad \psi(p)^{n/m} < \psi(a) = \psi(p)^{\delta'}, \qquad \frac{n}{m} < \delta'.$$

Since the least upper bound of all fractions n/m with $n/m < \delta$ is precisely δ, it follows that $\delta \leqq \delta'$; similarly, $\delta' \leqq \delta$ and hence $\delta = \delta'$. Now $\varepsilon = \log\psi(p)/\log\varphi(p)$ is a fixed positive number independent of a, and, since $\delta = \delta'$ for all a,

$$\log\psi(a) = \delta' \log\psi(p) = \delta \log\psi(p) = \delta\varepsilon \log\varphi(p) = \varepsilon \log\varphi(a),$$

and thus

$$\psi(a) = \varphi(a)^\varepsilon.$$

That statement 3 implies 1 is clear. Hence 1, 2, and 3 are equivalent.

If K is a field with valuation φ and K′ a field isomorphic to K with valuation ψ, an isomorphism between K and K′ is said to be *continuous in both directions* or *topological* if it takes φ-null sequences of K into ψ-null sequences of K′ and conversely. The fields K and K′ in this case are called *continuously isomorphic.* Under a topological isomorphism, convergent sequences correspond to convergent sequences and fundamental sequences to fundamental sequences. We thus have the following theorem.

Theorem: *Continuously isomorphic fields* K *and* K′ *with valuations have continuously complete extensions* Ω_K *and* $\Omega_{K'}$.

Exercise

18.8. Show that among the valuations of the field of rational numbers with which we are familiar, namely valuation by absolute value and p-adic valuations, no two are equivalent.

18.3 VALUATIONS OF THE FIELD OF RATIONAL NUMBERS

The following theorem, due to Ostrowski, shows that the valuations of the field of rational numbers with which we are already familiar, that is, the p-adic valuations and absolute value, are essentially the only possible ones. The field of real numbers is again taken as the value field.

Theorem: *A nontrivial valuation* φ *of the field* $\mathbb{Q}$ *of rational numbers is either* $\varphi(a) = |a|^\varrho$ *with* $0 < \varrho \leqq 1$, *and hence equivalent to the usual absolute-value* *valuation, or it is $\varphi(a) = \varphi_p(a)^\sigma$ with fixed prime number p and fixed positive number σ, and hence equivalent to a p-adic valuation.*

Proof: For every integer

$$\varphi(n) \leqq |n| .$$

For

$$\varphi(n) = \varphi(|n|) = \varphi(1+1+\cdots+1) \leqq \varphi(1)+\varphi(1)+\cdots+\varphi(1) = |n| .$$

Let $a>1$ and $b>1$ be any two natural numbers. We expand b^ν in powers of a:

$$b^\nu = c_0+c_1a+\cdots+c_na^n$$

$$0 \leqq c_\nu<a, \qquad c_n \neq 0.$$

The highest power a^n which occurs is at most equal to b^ν:

$$a^n \leqq b^\nu,$$

that is,

$$n \leqq \nu\frac{\log b}{\log a}.$$

Since now

$$\varphi(b^\nu) \leqq \varphi(c_0)+\varphi(c_1)\varphi(a)+\cdots+\varphi(c_n)\varphi(a)^n < $$
$$< a(1+\varphi(a)+\cdots+\varphi(a)^n) \leqq a(n+1)M^n$$

where $M = \max(1, \varphi(a))$, it follows that

$$\varphi(b)^\nu < a\left(\frac{\log b}{\log a}\nu+1\right)M^{(\log b/\log a)\nu}$$

or

$$\left(\frac{\varphi(b)}{M^{\log b/\log a}}\right)^\nu < a\frac{\log b}{\log a}\nu+a.$$

Hence, by the lemma of Section 18.1,

$$\varphi(b) \leqq M^{\log b/\log a},$$

that is,

$$\varphi(b) \leqq \max(1, \varphi(a)^{\log b/\log a}).$$

Case 1: φ is Archimedean. There then exists an integer b such that $\varphi(b)>1$. For any other integer $a>1$, if $\varphi(a) \leqq 1$, then from the inequality just proved we would obtain the contradiction $\varphi(b) \leqq 1$. Therefore, $\varphi(a)>1$ for all integers $a>1$. In the present case the inequality therefore reads

$$\varphi(b) \leqq \varphi(a)^{\log b/\log a}$$

or

$$\varphi(b)^{1/\log b} \leqq \varphi(a)^{1/\log a}.$$

Since we interchanged the roles of a and b, we also have

$$\varphi(a)^{1/\log a} \leqq \varphi(b)^{1/\log b},$$

and hence

$$\varphi(a)^{1/\log a} = \varphi(b)^{1/\log b}.$$

If $\varphi(b) = b^{\varrho}$, then from this it follows that $\varphi(a) = a^{\varrho}$. Hence

$$\varphi(r) = |r|^{\varrho}$$

for every rational number $r = a/b$. Then $\varrho > 0$ since $\varphi(a) > 1$, and $\varrho \leqq 1$ since

$$2^{\varrho} = \varphi(2) = \varphi(1+1) \leqq \varphi(1)+\varphi(1) = 2.$$

Case 2: *φ is non-Archimedean.* In this case $\varphi(a) \leqq 1$ for all integers a. The set of all integers a with $\varphi(a) < 1$ is clearly an ideal in the ring of integers. This ideal is prime, since $\varphi(ab) = \varphi(a)\varphi(b) < 1$ implies $\varphi(a) < 1$ or $\varphi(b) < 1$. Now in the ring of integers every ideal is principal; in particular, every prime ideal is generated by a prime number. The integers a such that $\varphi(a) < 1$ are therefore precisely the multiples of a prime number p. Every rational number r can be written in the form $r = (z/n)p^{\varrho}$ with integers z and n not divisible by p. Since $\varphi(z) = \varphi(n) = 1$, it follows that $\varphi(r) = \varphi(p)^{\varrho} = p^{-\varrho\sigma} = \varphi_p(r)^{\sigma}$, where $\sigma = -\log \varphi(p)/\log p$ is a fixed positive number (since $\varphi(p) < 1$). The valuation φ is therefore equivalent to the p-adic valuation φ_p.

Having thus completely determined the valuations of the field $\mathbb{Q}$ of rational numbers, we proceed to algebraic and transcendental extension fields. We first consider algebraic extensions.

We shall here restrict ourselves mainly to non-Archimedean valuations: Archimedean valuations are less interesting. Indeed, Ostrowski has proved that *a field* K *with an Archimedean valuation is continuously isomorphic to a field of complex numbers with the ordinary absolute-value valuation.* For the proof we refer the reader to the original paper.[3]

We thus set up the following program. We assume that we are given a (non-Archimedean) valuation φ of a field K. We consider an algebraic extension field Λ of K and ask if and in how many ways the valuation φ of K can be extended to a valuation Φ of Λ.

In Section 18.4 it will be assumed that the base field K is complete in the valuation. In Section 18.5 the case of a field which is not complete will be reduced to the complete case by an imbedding. In Section 18.6 the results obtained will be used to find all Archimedean and non-Archimedean valuations of an arbitrary algebraic number field.

Exercise

18.9. If $\varphi_0(a) = |a|$ and $\varphi_p(a)$ are p-adic valuations, then the product of all these values for each fixed element a is equal to 1.

[3]A. Ostrowski, "Über einige Lösungen der Funktionalgleichung $\varphi(x)\varphi(y) = \varphi(xy)$," *Acta Math.*, **41**, 271–284 (1918). Ostrowski's long paper in *Math. Z.*, **39**, 296–404 (1934), is basic for the following discussion.

18.4 VALUATION OF ALGEBRAIC EXTENSION FIELDS: COMPLETE CASE

Let the field K be complete with respect to the exponential valuation $w(a) = -\log \varphi(a)$; that is, Cauchy's convergence criterion holds. We wish to investigate how the exponential valuation can be continued to algebraic extension fields Λ.

We recall that the elements a with $w(a) \geqq 0$ are called *integral* and form a ring; the elements a with $w(a) > 0$ form a prime ideal $\mathfrak{p}$ in this ring.

A *reducibility criterion* in perfect fields due to Hensel is basic for the investigation.

If a_ν is the coefficient with smallest exponential value of the polynomial

$$a_n x^n + a_{n-1}x^{n-1} + \cdots + a_0$$

in a field with exponential valuation, then

$$\frac{a_n}{a_\nu} x^n + \frac{a_{n-1}}{a_\nu} x^{n-1} + \cdots + \frac{a_0}{a_\nu}$$

is a polynomial with *integral* coefficients, not all of which are divisible by $\mathfrak{p}$. A polynomial with this property is called *primitive*.

Hensel's Lemma: *Let K be complete in the exponential valuation w. Let $f(x)$ be a primitive polynomial with integral coefficients in K. If $g_0(x)$ and $h_0(x)$ are two polynomials with integral coefficients in K such that*

$$f(x) \equiv g_0(x)h_0(x) \pmod{\mathfrak{p}},$$

then there exist two polynomials $g(x)$, $h(x)$ with integral coefficients in K such that

$$f(x) = g(x)h(x)$$
$$g(x) \equiv g_0(x) \pmod{\mathfrak{p}}$$
$$h(x) \equiv h_0(x) \pmod{\mathfrak{p}},$$

provided that $g_0(x)$ and $h_0(x)$ are relatively prime modulo $\mathfrak{p}$. Moreover, it is possible to determine $g(x)$ and $h(x)$ so that the degree of $g(x)$ is equal to the degree of $g_0(x)$ modulo $\mathfrak{p}$.

Proof: Since we may simply omit coefficients in $g_0(x)$ and $h_0(x)$ which are divisible by $\mathfrak{p}$ without altering the hypothesis or assertion, we may assume that $g_0(x)$ is a polynomial of degree r and that the leading coefficients of $g_0(x)$ and $h_0(x)$ are units. Since it also makes no difference if we replace $g_0(x)$ by $(1/a)g_0(x)$ and $h_0(x)$ by $ah_0(x)$, we may assume from the beginning that $g_0(x)$ is a *normalized* polynomial of degree r; that is, its leading coefficient is 1: $g_0(x) = x^r + \cdots$. If b is the leading coefficient and s the degree of $h_0(x)$, then the leading coefficient of the product $g_0(x)h_0(x)$ is equal to b and the degree is $r+s \leqq n$. We shall now construct the factors $g(x)$ and $h(x)$ in such a manner that $g(x)$ is a normalized polynomial of degree r and $h(x)$ is a polynomial of degree $n-r$.

All the coefficients c of the polynomial $f(x)-g_0(x)h_0(x)$ have positive values $w(c)$ by hypothesis; let the smallest value be $\delta_1>0$. If $\delta_1=\infty$, then $f(x)=g_0(x)h_0(x)$, and there is nothing more to prove.

Since $g_0(x)$ and $h_0(x)$ are relatively prime modulo $\mathfrak{p}$, there exist two polynomials $l(x)$ and $m(x)$ with integral coefficients in K such that

$$l(x)g_0(x)+m(x)h_0(x)\equiv 1 \pmod{\mathfrak{p}}.$$

Let the smallest of the values of the coefficients in the polynomial

$$l(x)g_0(x)+m(x)h_0(x)-1$$

be $\delta_2>0$. Let the smaller of the two numbers δ_1, δ_2 be ε, and finally let π be an element such that $w(\pi)=\varepsilon$. Then

$$f(x)\equiv g_0(x)h_0(x) \pmod{\pi} \tag{18.5}$$

$$l(x)g_0(x)+m(x)h_0(x)\equiv 1 \pmod{\pi}. \tag{18.6}$$

We now construct $g(x)$ as the limit of a sequence of polynomials $g_\nu(x)$ of degree r which begins with $g_0(x)$; similarly, we construct $h(x)$ as the limit of a sequence of polynomials $h_\nu(x)$ of degree $\leqq n-r$ which begins with $h_0(x)$. Suppose that $g_\nu(x)$ and $h_\nu(x)$ have already been determined so that

$$f(x)\equiv g_\nu(x)h_\nu(x) \quad \pmod{\pi^{\nu+1}} \tag{18.7}$$

$$g_\nu(x)\equiv g_0(x) \qquad \pmod{\pi} \tag{18.8}$$

$$h_\nu(x)\equiv h_0(x) \qquad \pmod{\pi}, \tag{18.9}$$

and moreover that $g_\nu(x)=x^r+\cdots$ has leading coefficient 1. In order to determine $g_{\nu+1}(x)$ and $h_{\nu+1}(x)$, we put

$$g_{\nu+1}(x)=g_\nu(x)+\pi^{\nu+1}u(x) \tag{18.10}$$

$$h_{\nu+1}(x)=h_\nu(x)+\pi^{\nu+1}v(x). \tag{18.11}$$

Then

$$\begin{aligned} g_{\nu+1}(x)h_{\nu+1}(x)-f(x) &= g_\nu(x)h_\nu(x)-f(x)+\\ &\quad +\pi^{\nu+1}\{g_\nu(x)v(x)+h_\nu(x)u(x)\}+\pi^{2\nu+2}u(x)v(x). \end{aligned}$$

If, in accordance with (18.7), we put

$$f(x)-g_\nu(x)h_\nu(x)=\pi^{\nu+1}p(x),$$

then

$$g_{\nu+1}(x)h_{\nu+1}(x)-f(x)\equiv \pi^{\nu+1}\{g_\nu(x)v(x)+h_\nu(x)u(x)-p(x)\} \pmod{\pi^{\nu+2}}.$$

In order that the left side be divisible by $\pi^{\nu+2}$, it is sufficient that the congruence

$$g_\nu(x)v(x)+h_\nu(x)u(x)\equiv p(x) \pmod{\pi} \tag{18.12}$$

be satisfied.

To achieve this, we multiply the congruence (18.6) by $p(x)$,

$$p(x)l(x)g_0(x)+p(x)m(x)h_0(x)\equiv p(x) \pmod{\pi}, \tag{18.13}$$

divide $p(x)m(x)$ by $g_0(x)$ so that the remainder $u(x)$ has degree $<r$,

$$p(x)m(x) = q(x)g_0(x)+u(x), \tag{18.14}$$

and substitute (18.14) into (18.13):

$$\{p(x)l(x)+q(x)h_0(x)\}g_0(x)+u(x)h_0(x) \equiv p(x) \pmod{\pi}.$$

Then all the coefficients divisible by π of the polynomial in the braces are replaced by 0, and so we obtain

$$v(x)g_0(x)+u(x)h_0(x) \equiv p(x) \pmod{\pi}. \tag{18.15}$$

The desired congruence (18.12) follows from (18.15) because of (18.8) and (18.9). Furthermore, $u(x)$ has degree $<r$, and hence, by (18.10), $g_{\nu+1}(x)$ has the same degree and leading term as $g_\nu(x)$. It remains only to show that $v(x)$ has degree $\leqq n-r$. If this were not the case, a highest term of degree $>n$ would occur in the first term of (18.15) but not in the other terms. The coefficient of this term would have to be divisible by π from (18.15), and hence the leading coefficient of $v(x)$ would be divisible by π. But since we have omitted all the coefficients of $v(x)$ divisible by π, it follows that $v(x)$ has degree $\leqq n-r$.

From the congruence (18.12) it follows, as we have seen above, that

$$f(x) \equiv g_{\nu+1}(x)h_{\nu+1}(x) \pmod{\pi^{\nu+2}}. \tag{18.16}$$

From (18.10) it follows that the coefficients of the polynomial $g_{\nu+1}(x)-g_\nu(x)$ are divisible by $\pi^{\nu+1}$ and therefore have limit zero as $\nu\to\infty$. From this it follows by Cauchy's convergence criterion that $g_\nu(x)$ converges to a polynomial

$$g(x) = x^r+\cdots$$

as $\nu\to\infty$. Similarly, $h_\nu(x)$ converges to a polynomial $h(x)$ as $\nu\to\infty$. Passing to the limit in (18.7), it follows finally that

$$f(x) = g(x)h(x).$$

From (18.8) and (18.9) it follows further that

$$g(x) \equiv g_0(x) \pmod{\mathfrak{p}}$$

$$h(x) \equiv h_0(x) \pmod{\mathfrak{p}}.$$

We also obtain the following simple corollary.

Corollary: *If*

$$f(x) = a_0+a_1x+\cdots+a_nx^n$$

is a polynomial irreducible over K, *then*

$$\min(w(a_0), w(a_1), \ldots, w(a_n)) = \min(w(a_0), w(a_n)).$$

For the proof we may assume that $f(x)$ is primitive. The minimum on the left is then zero. If we assume that $w(a_0)$ and $w(a_n)$ are both greater than zero, then there exists an r, $0<r<n$, such that $w(a_r) = 0$, and $w(a_\nu)>0$ for $\nu = r+1, \ldots, n$. Then

$$f(x) \equiv (a_0+a_1x+\cdots+a_rx^r)\cdot 1 \pmod{\mathfrak{p}}, \qquad 0<r<n,$$

and this implies by Hensel's lemma that $f(x)$ can be decomposed into a factor of degree r and one of degree $n-r$.

Exercises

18.10. If a polynomial $f(x) = x^n+a_{n-1}x^{n-1}+\cdots+a_0$ has integral coefficients in K and is irreducible mod $\mathfrak{p}$, the $f(x)$ is also irreducible in the complete field Ω_K.

18.11. If in $f(x) = x^n+a_{n-1}x^{n-1}+\cdots+a_0$ all coefficients $a_{n-1}, \ldots, a_0$ are divisible by $\mathfrak{p}$ and a_0 is not the product of two elements of $\mathfrak{p}$, then $f(x)$ is irreducible (generalization of the Eisenstein irreducibility criterion).

18.12. Investigate the factorization of the rationally irreducible polynomials

$$x^2+1, \qquad x^2+2, \qquad x^2-3$$

in the field of 3-adic numbers. (Use Exercise 18.10, Hensel's lemma, and Exercise 18.11).

The most important application of the foregoing theorem is in the proof of the extendability of complete exponential valuations to algebraic extensions.

Theorem: *Let* K *be complete with respect to the exponential valuation* w, *and let* Λ *be an algebraic extension of* K. *Then there exists an exponential valuation* W *of* Λ *which coincides with* w *on* K.

Proof: Let ξ be an element of Λ and let

$$\xi^n+a_{n-1}\xi^{n-1}+\cdots+a_0 = 0$$

be the irreducible equation for ξ with coefficients in K. We assert that

$$W(\xi) = \frac{1}{n}\, w(a_0)$$

is a valuation of Λ (which clearly coincides with w on K). In order to prove the relations

$$W(\xi\eta) = W(\xi)+W(\eta)$$

$$W(\xi+\eta) \geqq \min(W(\xi), W(\eta))$$

for any two elements ξ, η of Λ, we consider the subfield $\Lambda_0 = K(\xi, \eta)$, which is of finite degree t over K, and form in this field the norm of ξ. By Section 6.11,

$$N(\xi) = (-1)^t a_0{}^r; \qquad r = \frac{t}{n},$$

and hence

$$w(N(\xi)) = w(a_0{}^r) = rw(a_0)$$

$$W(\xi) = \frac{1}{n}\, w(a_0 = \frac{1}{t}\, w(N(\xi)).$$

Since $N(\xi\eta) = N(\xi)N(\eta)$, it follows immediately that

$$W(\xi\eta) = W(\xi)+W(\eta).$$

Since

$$W(\xi+\eta) = W(\eta)+W\left(1+\frac{\xi}{\eta}\right)$$

and

$$\min(W(\xi),\, W(\eta)) = W(\eta)+\min\left(W\left(\frac{\xi}{\eta}\right), 0\right)$$

we may restrict consideration to $\eta = 1$ in proving $W(\xi+\eta) \geqq \min(W(\xi), W(\eta))$.

Now the irreducible equation for $\xi+1$ is

$$(\xi+1)^n+\cdots+(a_0-a_1+a_2-\cdots+(-1)^{n-1}a_{n-1}+(-1)^n) = 0.$$

By the preceding theorem,

$$\begin{aligned} W(\xi+1) &= \frac{1}{n}\, w(a_0-a_1+\cdots) \\ &\geqq \frac{1}{n} \min(w(a_0), w(a_1), \ldots, w(a_{n-1}), w(1)) \\ &= \frac{1}{n} \min(w(a_0), w(1)) = \min(W(\xi), 0). \end{aligned}$$

If we pass from the exponential valuations $w(a)$ and $W(\xi)$ to the ordinary valuations

$$\varphi(a) = e^{-w(a)}, \qquad \Phi(\xi) = e^{-W(\xi)},$$

then the valuation of the extension field Λ is defined by

$$\Phi(\xi) = \sqrt[n]{\varphi(a_0)}$$

or, in the case where Λ has a finite degree m over K, by

$$\Phi(\xi) = \sqrt[m]{\varphi(N_\Lambda(\xi))}.$$

We note that precisely the same formula is also correct in the case of Archimedean valuations. The only nontrivial case is that in which K is the field of real numbers and Λ is the field of complex numbers. The valuation

$$\varphi(\xi) = |\xi|^\varrho$$

of K can immediately be extended to

$$\Phi(\xi) = |\xi|^\varrho.$$

But now, for $\xi = a+bi$,

$$|\xi| = \sqrt{a^2+b^2} = \sqrt{N(\xi)} = \sqrt{|N(\xi)|}\,,$$

and hence

$$\Phi(\xi) = |\xi|^{\varrho} = \sqrt{\varphi(N(\xi))}.$$

Thus we shall henceforth treat Archimedean and non-Archimedean valuations together.

Let Λ *be of finite degree over* K, *and let* $u_1, \ldots, u_n$ *be a basis of* $\Lambda/$K. *Let* K *be complete in the valuation* φ. *If* Φ *is a valuation of* Λ *which coincides with* φ *on* K, *then a sequence*

$$c_\nu = a_1^{(\nu)}u_1 + \cdots + a_n^{(\nu)}u_n, \qquad \nu = 1, 2, 3, \ldots$$

is a fundamental sequence for Φ *if and only if the n sequences* $\{a_i^{(\nu)}\}$ *are fundamental sequences for* φ.

Since the sequences $\{a_i^{(\nu)}\}$ converge to a limit a_i in K, it follows that Λ *is complete with respect to* Φ.

Proof: We prove the convergence of the sequences $\{a_i^{(\nu)}\}$ by induction. If the c_ν have the form

$$c_\nu = a_1^{(\nu)}u_1,$$

then $\{a_1^{(\nu)}\}$ is naturally a fundamental sequence if $\{c_\nu\}$ is. Suppose that the assertion has been proved for all sequences $\{c_\nu\}$ of the form

$$c_\nu = \sum_{i=1}^{m-1} a^{(\nu)}u_i.$$

Let a sequence

$$c_\nu = \sum_{i=1}^{m} a^{(\nu)}u_i$$

be given. If the sequence $\{a_m^{(\nu)}\}$ converges, then $\{c_\nu - a_m^{(\nu)}u_m\}$ is also a fundamental sequence; the $\{a_i^{(\nu)}\}$, $i<m$, converge by the induction hypothesis. Suppose that $\{a_m^{(\nu)}\}$ were not convergent. It would then be possible to choose a sequence $n_1, n_2, n_3, \ldots$ such that $\varphi(a_m^{(\nu)} - a_m^{(\nu+n_\nu)}) > \varepsilon$ for all ν, where ε is a fixed *positive* number. The sequence

$$d_\nu = \frac{c_\nu - c_{\nu+n_\nu}}{a_m^{(\nu)} - a_m^{(\nu+n_\nu)}} = \sum_{i=1}^{m-1} \frac{a_i^{(\nu)} - a_i^{(\nu+n_\nu)}}{a_m^{(\nu)} - a_m^{(\nu+n_\nu)}}\,u_i + u_m = \sum_{i=1}^{m-1} b_i^{(\nu)}u_i + u_m$$

would then have to converge to zero, for the sequence of numerators converges to zero, since $\{c_\nu\}$ is a fundamental sequence. Now

$$d_\nu - u_m = \sum_{i=1}^{m-1} b_i^{(\nu)}u_i.$$

By the induction hypothesis, the sequences $\{b_i^{(\nu)}\}$ thus converge to certain limits b_i and

$$-u_m = \sum_{i=1}^{m-1} b_i u_i,$$

which contradicts the fact that $u_1, \ldots, u_n$ is a basis of Λ over K.

We prove in precisely the same manner that the sequence $\{c_\nu\}$ is a null sequence if and only if the sequences $\{a_i^{(\nu)}\}$ $(i = 1, \ldots, n)$ are null sequences.

This remark forms the basis for the proof of the following uniqueness theorem.

Theorem: *The continuation Φ of the valuation φ of a complete field* K *to an algebraic extension Λ is uniquely determined, and indeed*

$$\Phi(\xi) = \sqrt[n]{\varphi(N(\xi))},$$

where the norm is formed in the field K(ξ) *and n is the degree of this field over* K.

Proof: It suffices to consider a fixed element ξ and the associated field K(ξ); the norms shall then always be norms in this field. If a sequence $\{c_\nu\}$ in this field converges to zero (in the sense of Φ) and if the c_ν are expressed linearly in terms of the basis elements $u_1, \ldots, u_n$ of K(ξ), then the individual coefficients $a_i^{(\nu)}$ also converge to zero by the remark above; hence the norm, which is a homogeneous polynomial in these coefficients, also converges to zero. Suppose now that $\Phi(\xi)^n < \varphi(N(\xi))$ or $\Phi(\xi)^n > \varphi(N(\xi))$; if we consider the element

$$\eta = \frac{\xi^n}{N(\xi)} \quad \text{or} \quad \eta = \frac{N(\xi)}{\xi^n},$$

then in both cases $N(\eta) = 1$ and $\Phi(\eta) < 1$. It follows that $\lim \eta^\nu = 0$, and hence $\lim N(\eta^\nu) = 0$, which contradicts $N(\eta^\nu) = N(\eta)^\nu = 1$.

Exercises

18.13. An isomorphism between two fields Λ, Λ' with valuations, which are algebraic extensions of the complete field K, that takes the elements of K again into elements of K necessarily takes the valuation of Λ into the valuation of Λ'.

18.14. The field of complex numbers has only one valuation Φ which coincides on the field of real numbers with $\varphi(a) = |a|^\varrho$, namely $\Phi(a) = |a|^\varrho$.

18.5 VALUATION OF ALGEBRAIC EXTENSION FIELDS: GENERAL CASE

Let K be an arbitrary field with a valuation, and let Λ be an algebraic extension of K. We again inquire if and in how many ways the given valuation φ of K can be extended to a valuation of Λ.

We first restrict ourselves to simple extensions $\Lambda = \mathsf{K}(\vartheta)$ for ease of notation. Let the quantity ϑ be a zero of an irreducible polynomial $F(t)$ of $\mathsf{K}[t]$.

We first of all extend K to a complete field Ω with a valuation. We then form the splitting field Σ of $F(t)$ over Ω. The valuation φ of Ω can be uniquely extended to a valuation Φ of Σ by Section 18.4.

By an *imbedding* of Λ in Σ we mean an isomorphism σ which takes $\Lambda = \mathsf{K}(\vartheta)$ into a subfield $\Lambda' = \mathsf{K}'(\vartheta')$ of Σ and leaves the elements of the base field K fixed. Of course, the isomorphism σ takes ϑ into a zero ϑ' of $F(t)$ and is hereby defined. We now assert the following.

Every imbedding of Λ *in* Σ *defines a valuation of* Λ. For Λ', as a subfield of Σ, automatically has a valuation, and this valuation is transferred from Λ' to Λ by the isomorphism σ^{-1}. It is clear that the valuation Φ of Λ so obtained is a continuation of the valuation φ of K.

We now make the following statement.

Every valuation Φ *of* Λ *which is a continuation of the valuation* φ *of* K *can be obtained in the manner described above by imbedding* Λ *in* Σ.

Proof: We form the complete extension of Λ. It contains the complete extension Ω of K and also ϑ; it therefore contains the field $\Omega(\vartheta)$. This field can always be extended to a splitting field of the polynomial F which is isomorphic to the splitting field Σ. The isomorphism takes $\Omega(\vartheta)$ into a subfield $\Omega(\vartheta')$ of Σ, leaves all the elements of Ω fixed, and therefore takes the valuation of $\Omega(\vartheta)$ into the only possible valuation of $\Omega(\vartheta')$.

The restriction to simple extensions is not important for the proof. If finitely many algebraic quantities $\zeta_1, \ldots, \zeta_r$ are adjoined in place of the single element ϑ, then these quantities are zeros of polynomials $g_1, \ldots, g_r$ of $\mathsf{K}[t]$; we then take for Σ the splitting field of the product $g_1(t) \ldots g_r(t)$ and continue as above. If Λ is an infinite algebraic extension field of K, then we take for Σ the algebraically closed extension field of Ω. The proof remains the same.

We now return to the case of a simple extension and decompose the defining polynomial $F(t)$ in $\Omega[t]$ into its irreducible factors:

$$F(t) = F_1(t)F_2(t)\cdots F_s(t). \tag{18.17}$$

Every isomorphism σ of $\mathsf{K}(\vartheta)$ takes ϑ into a zero of a polynomial $F_\nu(t)$. To each $F_\nu(t)$ there corresponds an extension field $\Omega(\vartheta_\nu)$, where ϑ_ν is some zero of $F_\nu(t)$: *which* zero is of no importance, since all the zeros of an irreducible polynomial are conjugate.

If an isomorphism σ takes the element ϑ into ϑ_ν and leaves the elements of K fixed, then it takes every polynomial $g(\vartheta)$ into $g(\vartheta_\nu)$ and is hereby defined. All possible imbeddings of $\Lambda = \mathsf{K}(\vartheta)$ in Σ are therefore determined by

$$\vartheta \to \vartheta_\nu \qquad (\nu = 1, \ldots, s).$$

This also gives all valuations: to obtain the value Φ of any element $\eta = g(\vartheta)$, we form the νth conjugate

$$\eta_\nu = g(\vartheta_\nu)$$

and compute this value according to Section 18.4:

$$\Phi(\eta) = \sqrt[n_\nu]{\varphi(N(\eta_\nu))}, \tag{18.18}$$

where n_ν is the degree of the polynomial F_ν and the norm is taken in the field $\Omega(\vartheta_\nu)$.

There exist precisely the same number of continuations of the valuation φ as there are irreducible factors of the polynomial $F(t)$ in $\Omega[t]$.

18.6 VALUATIONS OF ALGEBRAIC NUMBER FIELDS

The general theory of the preceding section is well illustrated by the example of an algebraic number field.

Let $\Lambda = \mathbb{Q}(\vartheta)$ be an algebraic number field, that is, a finite extension of the rational number field $\mathbb{Q}$ generated by adjunction of a primitive element ϑ. Let $F(x)$ be the normalized irreducible polynomial with root ϑ.

Up to equivalent valuations, the base field $\mathbb{Q}$ has a single Archimedean valuation $\varphi(a) = |a|$ and, for every prime number p, a non-Archimedean valuation, the p-adic valuation

$$\varphi_p(a) = p^{-m},$$

where m is the exponent of p in the factorization of the rational number a.

The field of real numbers $\mathbb{R}$ is the perfect extension field for the Archimedean valuation. If we adjoin i, the field becomes algebraically closed and $F(x)$ splits into linear factors:

$$F(x) = (x-\vartheta_1)(x-\vartheta_2)\cdots(x-\vartheta_n).$$

To obtain the real decomposition, we must combine every two conjugate complex factors to a real quadratic factor:

$$(x-a-bi)(x-a+bi) = (x-a)^2+b^2.$$

If r_1 is the number of real roots and r_2 the number of pairs of conjugate complex roots, then $F(x)$ splits into r_1+r_2 real irreducible factors.

To each such factor there corresponds a valuation of Λ which is obtained by imbedding Λ in the field of real or complex numbers with an isomorphism which takes ϑ into a real or complex root ϑ_ν; only one of two conjugate complex roots need hereby be used. The isomorphism takes every function of ϑ,

$$\eta = g(\vartheta) = c_0+c_1\vartheta+\cdots+c_{n-1}\vartheta^{n-1},$$

into the corresponding function of ϑ_ν:

$$\eta_\nu = g(\vartheta_\nu) = c_0+c_1\vartheta_\nu+\cdots+c_{n-1}\vartheta_\nu^{n-1}.$$

The associated Archimedean valuation of Λ is therefore

$$\Phi(\eta) = |\eta_\nu|.$$

The r_1+r_2 Archimedean valuations of η are therefore given by the absolute values of the real and complex numbers η_ν conjugate to η, whereby only one of every two conjugate complex roots is to be taken.

The r_1+r_2 Archimedean valuations of an algebraic number field are closely related to the units of the field. (See B. L. van der Waerden, *Abh. Math. Sem. Hamburg*, **6**, 259 (1928).)

The investigation in the p-adic case is altogether similar. The complete field associated with the valuation $\varphi = \varphi_p$ of $\mathbb{Q}$ is the field of p-adic numbers Ω_p. Let $F(x)$ be decomposed into irreducible factors in Ω_p:

$$F(x) = F_1(x)F_2(x)\cdots F_s(x). \tag{18.19}$$

We now adjoin a zero ϑ_ν of each of the irreducible polynomials F_ν to Ω_p and construct the isomorphisms which take $\eta = g(\vartheta)$ into $\eta_\nu = g(\vartheta_\nu)$ $(\nu = 1, \ldots, s)$. To these isomorphisms correspond the valuations

$$\Phi_\nu(\eta) = \Phi(\eta_\nu) = \sqrt[n_\nu]{\varphi(N_\nu(\eta_\nu))} \tag{18.20}$$

or, if we again take logarithms,

$$W_\nu(\eta) = \frac{1}{n_\nu}\, w_p(N_\nu(\eta_\nu)). \tag{18.21}$$

The norm $N_\nu(\eta_\nu)$ is here the product of all conjugates of η_ν which is obtained if ϑ_ν in $\eta_\nu = g(\vartheta_\nu)$ is replaced successively by all the roots of the polynomial $F(x)$. If $\vartheta_{\nu 1}, \vartheta_{\nu 2}, \ldots$ are these roots, then

$$N_\nu(\eta_\nu) = g(\vartheta_{\nu 1})\cdot g(\vartheta_{\nu 2})\cdots \tag{18.22}$$

is a symmetric function of the roots $\vartheta_{\nu 1}, \vartheta_{\nu 2}, \ldots$ which can therefore be expressed in terms of the coefficients of F_ν. We are therefore in a position to find all the values $W_\nu(\eta)$ with the help of (18.21) as soon as the factorization (18.19) is known.

Example: We wish to find all valuations of the quadratic number field $\Lambda = \mathbb{Q}(\sqrt{5})$.

The definining polynomial with root $\vartheta = \sqrt{5}$ is

$$F(x) = x^2-5.$$

In the field of real numbers $F(x)$ decomposes into two real linear factors:

$$F(x) = (x-\sqrt{5})(x+\sqrt{5}).$$

There are thus two imbeddings which are obtained by identifying ϑ with $-\sqrt{5}$ or $+\sqrt{5}$. If

$$\eta = a+b\vartheta$$

is an arbitrary field element, the associated valuations are

$$\varphi_0(\eta) = |a+b\sqrt{5}| \tag{18.23}$$

and

$$\varphi_1(\eta) = |a-b\sqrt{5}| \tag{18.24}$$

The two Archimedean valuations have thus been found. Now to the p-adic valuations!

The discriminant of $F(x)$ is 20. We first separate out the prime numbers 2 and 5 which divide the discriminant.

For all other prime numbers p $F(x)$ is free of multiple factors. There are thus only two possibilities: either $F(x)$ remains irreducible modulo p or $F(x)$ splits into two linear factors modulo p. If $x-c$ is one such factor, then $x+c$ is the other, for the sum of both zeros of x^2-5 is zero. Thus, in the second case

$$x^2-5 \equiv (x-c)(x+c) \pmod p \qquad (18.25)$$
$$5 \equiv c^2 \pmod p.$$

Hence there exists an integer c whose square modulo p is congruent to 5. We say also: 5 is a *quadratic residue* modulo p.

Conversely, if $c^2 \equiv 5(p)$, then the decomposition (18.25) holds. Thus *if* 5 *modulo p is not a quadratic residue, then* x^2-5 *is irreducible modulo p; if, however,* 5 *is a quadratic residue, then* x^2-5 *decomposes modulo p into two linear factors.*

In the first case $F(x)$ is also p-adic irreducible; in the second case it is decomposable into two linear factors in Ω_p by Hensel's lemma.

In the first case there is thus only one valuation belonging to the prime number p:

$$\Phi(\eta) = \sqrt{\varphi_p(N(\eta))}.$$

If we again put

$$\eta = a+b\vartheta = a+b\sqrt{5},$$

then

$$N(\eta) = (a+b\sqrt{5})(a-b\sqrt{5}) = a^2-5b^2,$$

and hence

$$\Phi(\eta) = \sqrt{\varphi_p(a^2-5b^2)} \qquad (18.26)$$

for all prime numbers p for which 5 is not a quadratic residue.

For prime numbers p for which 5 is a quadratic residue, there is a p-adic decomposition by Hensel's lemma:

$$x^2-5 = (x-\gamma)(x+\gamma). \qquad (18.27)$$

The p-adic number γ is found as follows. We solve the congruence

$$c^2 \equiv 5$$

first modulo p, then modulo p^2, and so on. Each time there are two solutions, c and $-c$. We thus obtain two sequences of nested residue classes modulo p, $p^2, \ldots$. One sequence defines the p-adic number γ, the other the p-adic number $-\gamma$.

The two continuations of the p-adic valuation φ_p of $\mathbb{Q}$ are finally obtained by identifying the field generator ϑ first with γ and then with $-\gamma$. If we again put

$$\eta = a+b\vartheta,$$

then the two valuations are

$$\Phi_1(\eta) = \varphi_p(a+b\gamma) \tag{18.28}$$

$$\Phi_2(\eta) = \varphi_p(a-b\gamma). \tag{18.29}$$

Since the p-adic valuation φ_p of Ω_p is known, Φ_1 and Φ_2 are also known.

We remark further that in specific cases the entire infinite sequence of residue classes modulo $p, p^2, \ldots$ is never needed; the procedure may be terminated after a finite number of steps. For in the case of the valuation $\varphi_p(a+b\gamma)$ it is only a question of which power of p divides the p-adic number $a+b\gamma$. If it is found after three steps, for example, that it is divisible by p^2 but not by p^3, then

$$\varphi_p(a+b\gamma) = p^{-2}.$$

It remains only to consider the two divisors of the discriminant $p = 2$ and $p = 5$.

In Ω_5 $F(x) = x^2-5$ is irreducible by the Eisenstein criterion (Exercise 18.11), since all coefficients after the first are divisible by 5 and the last coefficient is not divisible by 5^2. Thus (18.26) also holds for $p = 5$.

In Ω_2 the Eisenstein criterion is not applicable. If we put $x = 2y+1$, then

$$x^2-5 = (2y+1)^2-5 = 4(y^2+y-1),$$

and y^2+y-1 is irreducible modulo 2. Thus, x^2-5 is also 2-adic indecomposable, and (18.26) also holds for $p = 2$.

Exercises

18.15. The polynomial x^2+1 is real and 2-adic irreducible. It decomposes or fails to decompose modulo a prime number p depending on whether $p = 4k+1$ or $p = 4k-1$. (The multiplicative group of the residue class field GF(p) is cyclic of order $(p-1)$. It does or does not contain the fourth roots of unity, depending on whether $(p-1)$ is or is not divisible by 4.)

18.16. Find all valuations of the field of Gaussian numbers $a+bi$. How many Archimedean valuations are there? For which prime numbers p are there two valuations, and for which only one?

We have seen in Section 18.1 that there is a close relationship between classical ideal theory and valuation theory. We are now able to clarify this relationship.

Let $\mathbb{Z}$ again be the ring of integers in the rational number field $\mathbb{Q}$, and let $\mathfrak{o}$ be the ring of integral elements in the algebraic number field Λ. We thus have, as in Section 17.3, the inclusion relations

$$\begin{array}{ccc} \mathbb{Z} & \subseteq & \mathfrak{o} \\ \cap & & \cap \\ \mathbb{Q} & \subseteq & \Lambda. \end{array}$$

We again write exponential valuations. We thus consider valuations W of Λ which are continuations of the p-adic valuation w_p of $\mathbb{Q}$. The definition of w_p is as follows: if an integer m is exactly divisible by p^r and an integer n exactly divisible by p^s, then

$$w_p\left(\frac{m}{n}\right) = r-s.$$

We first prove the following statement.

For the elements a of $\mathfrak{o}$, $W(a)$ is nonnegative.

Suppose that $W(a)$ were negative. As an integral element, a satisfies an equation

$$a^n = c_1 a^{n-1} + \cdots + c_n, \tag{18.30}$$

where the c_i are elements of $\mathbb{Z}$. The left side of (18.30) would have a negative value

$$W(a^n) = nW(a),$$

whereas the right side would have a greater value. This is a contradiction.

The set of the a in $\mathfrak{o}$ such that $W(a) > 0$ is a prime ideal $\mathfrak{p}$ in $\mathfrak{o}$. Let π be an element of $\mathfrak{o}$ which is exactly divisible by the first power of $\mathfrak{p}$. If then a is exactly divisible by $\mathfrak{p}^r$, then, by Section 17.4,

$$a\mathfrak{o} = \mathfrak{p}^r\mathfrak{c}. \tag{18.31}$$

In $\mathfrak{c}$ there exists an element c which is not divisible by $\mathfrak{p}$. By (18.31) $\pi^r c$ is divisible by a:

$$\pi^r c = ab. \tag{18.32}$$

The left side is exactly divisible by $\mathfrak{p}^r$, and the factor a on the right-hand side is also; therefore b is not divisible by $\mathfrak{p}$ and hence $W(b) = 0$. Similarly, $W(c) = 0$, and it thus follows from (18.32) that

$$W(a) = W(\pi^r) = rW(\pi). \tag{18.33}$$

Since $W(\pi)$ is a positive constant, the valuation W is equivalent to the $\mathfrak{p}$-adic valuation

$$W_{\mathfrak{p}}(a) = r. \tag{18.34}$$

We have thus already obtained a principal result, as follows.

All non-Archimedean valuations of Λ are equivalent to the $\mathfrak{p}$-adic valuations defined by prime ideals $\mathfrak{p}$ of the ring $\mathfrak{o}$. To each prime ideal $\mathfrak{p}$ in $\mathfrak{o}$ distinct from the null ideal and the unit ideal there corresponds a class of equivalent, non-Archimedean valuations W, and conversely.

In the valuation W the prime number p has value 1, since W coincides on $\mathbb{Q}$ with the p-adic valuation w_p. We now apply (18.33) to $a = p$. The left side is equal to 1, so r on the right side cannot be zero. This means that the prime ideal $\mathfrak{p}$ must occur on the right-hand side in the factorization

$$(p) = p\mathfrak{o} = \mathfrak{p}_1^{e_1}\cdots\mathfrak{p}_s^{e_s}, \tag{18.35}$$

say, $\mathfrak{p} = \mathfrak{p}_\nu$. Then on the right in (18.33) we have to put $r = e_\nu$, and we obtain

$$1 = e_\nu W(\pi).$$

If we now multiply both sides of (18.33) by e_ν, we obtain from (18.34)

$$e_\nu W(a) = W_{\mathfrak{p}}(a), \tag{18.36}$$

or in words: *to obtain the normalized* $\mathfrak{p}$*-adic valuation* $W_{\mathfrak{p}}(a)$ *from the valuation* $W(a)$, *all values* $W(a)$ *must be multiplied by the exponent* e_ν *with which the prime ideal* $\mathfrak{p} = \mathfrak{p}_\nu$ *occurs in* (18.35).

The number s of distinct prime ideals occurring on the right-hand side of (18.35) is equal to the number of distinct continuations W of the p-adic valuation w_p of the field $\mathbb{Q}$. It is therefore equal to the number of prime factors on the right-hand side of (18.19), which was there also denoted by s.

Criterion for Integral Elements: *An element a of the field Λ belongs to the ring $\mathfrak{o}$ if and only if a has a nonnegative value in every* $\mathfrak{p}$*-adic valuation of the field* Λ.

We have already proved the "only if." Now let $a = b/c$ be an element of Λ, where b and c are elements of $\mathfrak{o}$. We decompose the principal ideals (b) and (c):

$$(b) = \mathfrak{p}_1^{r_1} \cdots \mathfrak{p}_m^{r_m} \tag{18.37}$$

$$(c) = \mathfrak{p}_1^{s_1} \cdots \mathfrak{p}_m^{s_m}. \tag{18.38}$$

By including factors $\mathfrak{p}^0$ if necessary, we may assume that the same prime ideals $\mathfrak{p}_\nu$ occur in (18.37) and (18.38). The value $W_\nu(a)$ in the $\mathfrak{p}$-adic valuation for the prime ideal $\mathfrak{p}_\nu$ is

$$W_\nu(a) = r_\nu - s_\nu.$$

If all these values are positive or zero, then the ideal (b) is divisible by (c). From this it follows that

$$b = cd,$$

and hence $a = b/c = d$ lies in $\mathfrak{o}$, which was to be proved.

The theorem just proved may also be formulated as follows.

Theorem: *The ring $\mathfrak{o}$ is the intersection of the valuation rings of all* $\mathfrak{p}$*-adic valuations of the quotient field Λ where $\mathfrak{p}$ runs through all prime ideals of the ring with the exceptions of* (0) *and* (1).

A similar theorem is true for arbitrary integral domains which are integrally closed in their quotient fields. (See W. Krull, "Idealtheorie," *Ergebn. Math.*, Vol. 4, Heft 3.

18.7 VALUATIONS OF A FIELD Δ(x) OF RATIONAL FUNCTIONS

Suppose that an indeterminate x is adjoined to an arbitrary field Δ, the "field of constants." We seek all valuations of the field $\Delta(x)$ of rational functions such that all constants of Δ have value 1.

In particular, all sums $1+1+\cdots+1$ then have value 1; the valuation is therefore non-Archimedean. If we write it in exponential form,

$$\varphi = e^{-\omega},$$

then, by assumption, $w(a) = 0$ for all constants a.

There are two possibilities:

1. $w(f) \geqq 0$ for all polynomials $f(x)$.
2. There exists a polynomial f with $w(f) < 0$.

It may happen that all $w(f) = 0$. In this case all quotients f/g also have value 0, and the valuation is trivial.

If we disregard this case, then in case 1 there exists a polynomial f with $w(f) > 0$. If we decompose f into prime factors, then at least one prime factor has a value > 1.

If $p(x)$ is this prime factor and $v = w(p)$ its value, then any polynomial not divisible by $p(x)$ has value 0. For suppose that $q(x)$ were a polynomial not divisible by $p(x)$ with value > 0; then since p and q are relatively prime, we would have

$$1 = Ap + Bq,$$

where A and B are again polynomials. It would follow that

$$w(Ap) = w(A) + w(p) > 0$$

$$w(Bq) = w(B) + w(q) > 0,$$

and hence from the basic property of non-Archimedean valuations that

$$w(1) = w(Ap + Bq) > 0,$$

which is impossible.

If now $f(x)$ is an arbitrary polynomial and we put

$$f(x) = p(x)^m q(x),$$

where $q(x)$ is not divisible by $p(x)$, then we can immediately find the value of $f(x)$:

$$w(f) = mw(p) + w(q) = mv.$$

For quotients of polynomials we have, as always,

$$w = \left(\frac{f}{g}\right) = w(f) - w(g).$$

Thus, *in case* 1 *the valuation is equivalent to a p-adic valuation defined by the prime polynomial* $p = p(x)$. These valuations are altogether analogous to the p-adic valuations of the rational number field $\mathbb{Q}$.

The case of an algebraically closed field of constants Δ is especially simple. In this case there are only linear prime polynomials:

$$p(x) = x - a.$$

To each a of Δ there belongs precisely one prime polynomial $p = x-a$ and therefore one p-adic valuation. It is called the valuation belonging to the *place* a if a is thought of as, say, a point in the complex plane. A polynomial has value m in this valuation if it is exactly divisible by $(x-a)^m$ or, as is also said, if a is a zero of mth order of the polynomial. The same holds for a rational function $\varphi = f/g$ if the numerator is divisible by $(x-a)^m$ and the denominator is not divisible by $x-a$. If this situation is reversed, then φ has a "pole of mth order" at the place a, and the value $w(\varphi)$ is $-m$.

Consideration of case 1 has now been completed. *We now show that in case 2, up to equivalent valuations, there is only one valuation, namely*

$$w\left(\frac{f}{g}\right) = -m+n,$$

where m is the degree of the numerator f and n is the degree of the denominator g.

Proof: Let $p(x)$ be a polynomial of lowest degree with value $w(p)<0$. The degree of $p(x)$ cannot be zero, since all constants have value zero by hypothesis. The degree can also not be greater than 1. For if

$$p(x) = a_0x^n+a_1x^{n-1}+\cdots+a_n, \qquad n>1, \qquad a_0 \neq 0,$$

then the polynomial x, as a polynomial of lower degree, would have a value $w(x) \geqq 0$, and therefore a_0x^n would have a value $\geqq 0$. The remaining terms, $a_1x^{n-1}+\cdots+a_n$, as a polynomial of lower degree, would also have a value $\geqq 0$. Therefore, the sum

$$p(x) = a_0x^n+(a_1x^{n-1}+\cdots+a_n)$$

would also have a value $\geqq 0$, contrary to hypothesis.

Therefore $p(x)$ is linear:

$$p(x) = x-c.$$

If now

$$q(x) = x-b = (x-c)+(c-b)$$

is another linear polynomial, then by a previous remark

$$w(q) = \min(w(x-c), w(c-b)) = w(p),$$

since $w(x-c)<w(c-b)$.

Thus all linear polynomials have the same negative value $w(p) = w(q) = -v$.

We may always go over to an equivalent valuation and choose $v = 1$. All linear polynomials then have the value -1.

The powers x^k now all have the value $-k$. This is not affected by a constant factor:

$$w(ax^k) = -k.$$

Finally, every polynomial $f(x)$ is a sum of terms ax^k. By the previous remark, the value $w(f)$ is equal to the minimum of the values of the terms:

$$w(f) = -n,$$

where f has degree n. This completes the proof.

In the case of a number field there is an essential difference between the one Archimedean and the infinitely many p-adic valuations. In the case of a function field, however, the valuation according to degree is of the same type as the p-adic valuations. This may be stated more strongly as follows: the valuation according to degree can be taken into any of the p-adic valuations by means of a very simple field isomorphism. Indeed, if we put

$$x = \frac{1}{y-c}, \tag{18.39}$$

then a quotient of polynomials of degrees m and n

$$\varphi(x) = \frac{f(x)}{g(x)} = \frac{ax^m + \cdots}{bx^n + \cdots}$$

goes over, by the substitution (18.39) and multiplication of numerator and denominator by $(y-c)^{m+n}$, into a quotient of polynomials in y whose numerator is exactly divisible by $(y-c)^n$ and whose denominator is exactly divisible by $(y-c)^m$. The value of the quotient $\psi(y)$ in the valuation belonging to the place c is therefore equal to the difference of degrees $n-m$. The isomorphism (18.39) thus transforms the valuation of the field $\Delta(x)$ according to degree into the valuation belonging to the place c of the isomorphic field $\Delta(y)$.

According to (18.39), to the "place" $y = c$ there corresponds the "place" $x = \infty$. The valuation according to degree is therefore called *the valuation of the function field* $\Delta(x)$ *belonging to the place* ∞. By including the place ∞, the complex plane becomes a sphere, and on the sphere all points are equivalent, since the linear fractional transformations

$$y = \frac{ax+b}{cx+d} \tag{18.40}$$

take any place into any other place. Clearly, (18.39) is only a special case of (18.40).

We now ask which complete extension fields belong to the different "places" of a field. We have seen earlier (Section 18.2) that the complete extension field belonging to $p = x-c$ is the field of all formal power series

$$\alpha = a_{-m}(x-c)^{-m} + \cdots + a_0 + a_1(x-c) + a_2(x-c)^2 + \cdots.$$

The coefficients of this power series are entirely arbitrary constants. The series always converges in the sense of the p-adic valuation however the coefficients are chosen. In the sense of function theory the series need not converge if the a_k are complex numbers: the radius of convergence may very well be zero.

The value $w(\alpha)$ of the power series above is $-m$ if a_{-m} is the first nonzero coefficient.

Similarly, to the place ∞ there belongs the complete field of all power series in x^{-1}:

$$\beta = b_{-m}x^m + \cdots + b_0 + b_1x^{-1} + b_2x^{-2} + \cdots.$$

18.8 THE APPROXIMATION THEOREM

As previously remarked, with each valuation φ of a field K there is associated a limit concept: $\lim a_\nu = a$ means $\lim \varphi(a_\nu - a) = 0$. We immediately verify that

$$\lim \frac{a^\nu}{1+a^\nu} \begin{cases} =0 & \text{if } \varphi(a)<1 \\ =1 & \text{if } \varphi(a)>1. \end{cases}$$

We recall that two valuations φ and ψ are called equivalent in $\lim \varphi(a_\nu) = 0$ implies $\lim \psi(a_\nu) = 0$ and conversely.

In Section 18.2 the following equivalence criterion was proved.

Lemma 1: *Two valuations φ and ψ are equivalent if $\varphi(a)<1$ implies $\psi(a)<1$.*

We next prove the second lemma.

Lemma 2: *Let $\varphi_1, \ldots, \varphi_n$ $(n>1)$ be a finite number of inequivalent valuations of the field K. Then there exists a field element a such that*

$$\varphi_1(a)>1 \quad \text{and} \quad \varphi_\nu(a)<1 \quad (\nu = 2, \ldots, n).$$

The proof is by induction on n. First let $n = 2$. Since the valuations φ_1 and φ_2 are not equivalent, by Lemma 1 there exists a b with the properties

$$\varphi_1(b)<1 \quad \text{and} \quad \varphi_2(b) \geqq 1$$

and a c with the properties

$$\varphi_1(c) \geqq 1 \quad \text{and} \quad \varphi_2(c)<1.$$

The element $a = b^{-1}c$ now has the desired properties:

$$\varphi_1(a)>1 \quad \text{and} \quad \varphi_2(a)<1.$$

Assuming that the assertion is true for $n-1$ valuations, there exists a b such that

$$\varphi_1(b)>1 \quad \text{and} \quad \varphi_\nu(b)<1 \quad (\nu = 2, \ldots, n-1).$$

By what was just proved for the case $n = 2$, there exists a c such that

$$\varphi_1(c)>1 \quad \text{and} \quad \varphi_n(c)<1.$$

We distinguish two cases.

Case 1: $\varphi_n(b) \leqq 1$. We form $a_r = cb^r$. Then

$$\varphi_1(a_r)>1$$
$$\varphi_n(a_r)<1,$$

and, for sufficiently large r,

$$\varphi_\nu(a_r)<1 \quad (\nu = 2, \ldots, n-1).$$

We may therefore put $a = a_r$.

Case 2: $\varphi_n(b) > 1$. We form

$$d_r = \frac{cb^r}{1+b^r}.$$

The sequence $\{d_r\}$ converges to c in the valuations φ_1 and φ_n and to 0 in the other valuations φ_ν. Hence

$$\lim \varphi_1(d_r) = \varphi_1(c) > 1$$
$$\lim \varphi_n(d_r) = \varphi_n(c) < 1$$
$$\lim \varphi_\nu(d_r) = 0 \qquad (\nu = 2, \ldots, n-1).$$

Thus, for sufficiently large r, $a = d_r$ has the desired properties:

$$\varphi_1(a) > 1 \tag{18.41}$$
$$\varphi_\nu(a) < 1 \qquad (\nu = 2, \ldots, n).$$

Lemma 3: *If $\varphi_1, \ldots, \varphi_n$ are inequivalent valuations, then there exists a field element b which is arbitrarily close to 1 in the valuation φ_1 and arbitrarily close to 0 in the valuations $\varphi_2, \ldots, \varphi_n$.*

Proof: The case $n = 1$ is trivial. In the case $n > 1$ we take an a with the properties (18.41) and form

$$b_r = \frac{a^r}{1+a^r}.$$

The sequence $\{b_r\}$ converges to 1 in the valuation φ_1 and to 0 in the valuations $\varphi_2, \ldots, \varphi_n$. This gives the assertion.

After these preparations we now prove the following.

Approximation Theorem: *Let $\varphi_1, \ldots, \varphi_n$ be inequivalent valuations. Given field elements $a_1, \ldots, a_n$, there exists a field element a which is arbitrarily close to a in the valuation φ_ν:*

$$\varphi_\nu(a_\nu - a) < \varepsilon \qquad (\nu = 1, \ldots, n). \tag{18.42}$$

Proof: By Lemma 3 there exist elements $b_\nu (\nu = 1, \ldots, n)$ close to 1 in the valuation φ_ν and close to 0 in all other valuations. The sum

$$a = a_1 b_1 + \cdots + a_n b_n$$

is then arbitrarily close to a_ν in the valuation φ_ν.

The proof of the approximation theorem given here was taken from a lecture by E. Artin.

Chapter 19

ALGEBRAIC FUNCTIONS OF ONE VARIABLE

The classical theory of algebraic functions over the field of complex numbers culminates in the Riemann-Roch theorem. There are function theory, geometric, and algebraic proofs of this theorem. A beautiful presentation of the function theory method of proof using geometric ideas may be found in C. Jordan, *Cours d'Analyse*, Chapter VIII. Among the geometric methods of proof the *metodo rapido* of Severi deserves special mention.[1] The purely algebraic proof of Dedekind and Weber (*J. Reine u. Angew. Math.*, Vol. 92, 1881) was simplified by Emmy Noether and generalized to perfect fields of constants. For arbitrary fields of constants the Riemann-Roch theorem was first proved by F. K. Schmidt (*Math. Z.*, Vol. 41, 1936; further literature cited there). A still simpler proof has been given by André Weil in *J. Reine u. Angew. Math.*, Vol. 179, 1938; we follow his method here.

19.1 SERIES EXPANSIONS IN THE UNIFORMIZING VARIABLE

Let K be an algebraic function field of one variable, that is, a finite extension of the rational function field $\Delta(x)$. The choice of the independent variable x is quite arbitrary: in place of x we may choose any transcendental element over Δ. We are interested only in the invariant properties of the function field, that is, those which are independent of the choice of x.

The elements of K which are algebraic over Δ are called *constants*. They form the *field of constants* Δ^*. The field Δ^* is *algebraically closed* in K; that is, all elements of K which are algebraic over Δ^* lie in Δ^*.

The starting point for the present theory of algebraic functions is the *valuation concept*. Just as in Section 18.7, only those valuations of the function field K will be considered in which all nonzero constants c^* of Δ^* have value $\varphi(c^*) = 1$.

[1]For the latest presentation of this method, see F. Severi, *Acta Pont. Accad. Sci.*, 1952. The *metodo rapido* has also influenced the proof of Weil, which will be presented here.

As in Section 18.7, we see immediately that all these valuations are non-Archimedean. We again write them exponentially:

$$\varphi(z) = e^{-w(z)}. \tag{19.1}$$

Thus $w(c^*) = 0$ for all $c^* \neq 0$ of Δ^*.

Exercise

19.1. If $w(c) = 0$ for all $c \neq 0$ of Δ, then $w(c^*) = 0$ for all $c^* \neq 0$ of Δ^*.

By a *place* of the field K we mean a class of equivalent valuations. The basis for this somewhat curious designation will be recognized if one thinks of the case of the rational function field $\Delta(x)$ treated in Section 18.7 with the complex numbers as the field of constants. If one imagines the complex plane transformed into a sphere by adjoining a point ∞ and if the points of this sphere are called places, then to each such place (c or ∞) there corresponds precisely one class of equivalent valuations. According to Section 18.7, all valuations of the field of rational functions $\Delta(x)$ are obtained in this way.

A similar approach can be taken for an algebraic function field over the field of complex numbers, by considering the *Riemann surface* of the function field.[2] It was shown in Section 18.1 that to each point P of this surface there belongs a class of equivalent valuations of the function field K. In this case also it can be shown[3] that all valuations in which all constants c have the value $w(c) = 0$ are obtained in this way.

In the following the theory of places and uniformizing variables will be developed in a purely algebraic manner without reference to the concept of a Riemann surface. The reader may, however, wish to think of a point on a Riemann surface whenever the discussion involves a "place."

To each place, that is, to each class of equivalent valuations of the function field K, there corresponds by Section 18.1 a valuation ring $\mathfrak{J}$ and a valuation ideal $\mathfrak{p}$ consisting of all field elements z with $w(z) \neq 0$. By Lemma 1 (Section 18.8), two valuations belonging to the same valuation ideal $\mathfrak{p}$ are equivalent. Hence, to each valuation ideal there corresponds a single place. We shall henceforth denote the place by the same letter $\mathfrak{p}$ that is used for the valuation ideal.

The field K is by hypothesis a finite extension of the field $\Delta(x)$ of rational functions. All valuations of K are thus obtained by first finding all valuations of $\Delta(x)$ according to Section 18.7 and then extending these valuations to K by imbedding K in all possible ways in a splitting field Λ of a polynomial $F(t)$ over a complete field Ω according to Section 18.5. The exponential valuation w of K can first be extended to the same type of valuation w of Ω; by Section 18.4 it

[2]See H. Weyl, *Die Idee der Riemannschen Flache*, 3. Aufl., Teubner, Stuttgart, 1955.
[3]For a proof see *Algebra*, Vol. I, 4. to 6. Aufl., pp. 280–282.

can then be extended uniquely to a valuation W of Λ so that, for each element z of Λ,

$$\Phi(z) = \sqrt[m]{\varphi(N(z))}$$

or, going back to the exponential valuations w and W,

$$W(z) = \frac{1}{m} w(N_\Lambda(z)),$$

where m is the field degree of Λ over Ω. For a given valuation w there are only a finite number of possibilities for the continuation W. In the classical theory this corresponds to the fact that over a point of the sphere there are only a finite number of points of the Riemann surface of the function field K.

By Section 18.7, the valuations w of $\Delta(x)$ are all *discrete*; that is, there exists a least positive value w_0 of which all values $w(z)$ are multiples. The valuations W of K are thus again discrete.

As before, we normalize the valuations $W(z)$ by the requirement that the smallest positive $W(z)$ be equal to 1. All the $W(z)$ then become integers. The normalized valuation depends only on the place $\mathfrak{p}$ and will be denoted by $W_\mathfrak{p}$ or simply by $\mathfrak{p}$. For each place there is a *uniformizing variable* π with $W_\mathfrak{p}(\pi) = 1$. The integer $W_\mathfrak{p}(z)$ is called the *order of the function z at the place* $\mathfrak{p}$. If it is positive and equal to k, then the place $\mathfrak{p}$ is a *zero of order k* or a *k-fold zero* of the function z. If the order is negative and equal to $-h$, then the place $\mathfrak{p}$ *is a pole of order* $-h$ or an *h-fold* pole of the function z.

The residue class ring $\overline{\mathfrak{J}} = \mathfrak{J}/\mathfrak{p}$ is by Section 18.1 always a field: the residue class field of the valuation. It contains the field $\overline{\Delta^*}$ of those residue classes which are represented by constants of Δ^*. Since $\overline{\Delta^*}$ is isomorphic to Δ^*, we may identify $\overline{\Delta^*}$ with Δ^* and interpret $\overline{\mathfrak{J}}$ as an extension field of Δ^*. The field of constants Δ^* is in turn an extension of the base field Δ.

We now prove that $\overline{\mathfrak{J}}$ *is a finite extension of* Δ.

Proof: Since π does not belong to Δ^*, π is transcendental over Δ, and hence K is algebraic over $\Delta(\pi)$. Here K arises from $\Delta(\pi)$ by adjunction of finitely many quantities; K is thus finite over $\Delta(\pi)$, say of degree m.

Suppose now that there were $m+1$ residue classes $\overline{\omega}_1, \ldots, \overline{\omega}_{m+1}$ in $\overline{\mathfrak{J}}$ which were linearly independent over Δ. We select representatives $\omega_1, \ldots, \omega_{m+1}$ of these residue classes in $\mathfrak{J}$. These $m+1$ quantities must be linearly dependent over $\Delta(\pi)$. There thus exists a relation

$$f_1(\pi)\omega_1 + \cdots + f_{m+1}(\pi)\omega_{m+1} = 0, \tag{19.2}$$

where $f_1(\pi), \ldots, f_{m+1}(\pi)$ are polynomials of $\Delta[\pi]$ which are not all zero. We may assume that these polynomials are not all divisible by π. Modulo $\mathfrak{p}$ they reduce to their constant terms $c_1, \ldots, c_{m+1}$; it thus follows from (19.2) that

$$c_1\omega_1 + \cdots + c_{m+1}\omega_{m+1} \equiv 0(\mathfrak{p})$$

or

$$c_1\overline{\omega}_1 + \cdots + c_{m+1}\overline{\omega}_{m+1} = 0,$$

contrary to the assumed linear independence of the $\bar{\omega}_i$. Therefore $\bar{\mathfrak{J}}$ has at most degree m over Δ.

It has thus been shown that $\bar{\mathfrak{J}}$ is finite over Δ. Since Δ^* is a subfield of $\mathfrak{J}$, it follows that Δ^* is likewise finite over Δ. If Δ is algebraically closed, then $\bar{\mathfrak{J}} = \Delta^* = \Delta$.

We shall henceforth consider Δ^* rather than Δ as base field and omit the asterisk. We thus assume that Δ is algebraically closed in K.

The degree of $\bar{\mathfrak{J}}$ over Δ will subsequently be denoted by $f_\mathfrak{p}$ or simply by f. In the classical case of an algebraically closed field of constants, $f = 1$.

We now wish to expand the elements z of the field K in power series in the uniformizing variable π. Let $(\bar{\omega}_1, \ldots, \bar{\omega}_f)$ be a basis for $\bar{\mathfrak{J}}$ over Δ, and let ω_i be an element of the residue class $\bar{\omega}_i$. If now z is an element of order b, then $z\pi^{-b}$ has order 0 and so belongs to $\mathfrak{J}$. The following congruence then holds modulo $\mathfrak{p}$:

$$z\pi^{-b} \equiv c_1\omega_1 + \cdots + c_f\omega_f (\mathfrak{p}); \tag{19.3}$$

the coefficients c_i are uniquely determined elements of Δ. The difference

$$z\pi^{-b} - (c_1\omega_1 + \cdots + c_f\omega_f) \tag{19.4}$$

is an element of $\mathfrak{p}$ and is thus a multiple of π:

$$z\pi^{-b} = c_1\omega_1 + \cdots + c_f\omega_f + z'\pi,$$
$$z = (c_1\omega_1 + \cdots + c_f\omega_f)\pi^b + z'\pi^{b+1}.$$

The remainder $z_1 = z'\pi^{b+1}$ has order $b+1$ at least, and the procedure can be repeated. After s steps we obtain

$$z = \sum_{k=b}^{b+s-1} (c_{k1}\omega_1 + \cdots + c_{kf}\omega_f)\pi^k + z_s,$$

where the remainder z_s has order $b+s$ at least.

For $s \to \infty$ the remainder z_s has limit zero, and we obtain

$$z = \sum_{k=b}^{\infty} (c_{k1}\omega_1 + \cdots + c_{kf}\omega_f)\pi^k \tag{19.5}$$

with uniquely determined coefficients c_{ki}. The initial exponent b may be negative, but in any case only finitely many terms with negative exponents occur in the series (19.5).

The procedure can be modified so that instead of π^b any element π_b of order b is chosen and a congruence of the type (19.3) is written for $z\pi_b{}^{-1}$. We then obtain instead of (19.5) a series expansion in the π_k:

$$z = \sum_{k=b}^{\infty} (c_{k1}\omega_1 + \cdots + c_{kf}\omega_f)\pi_k \tag{19.6}$$

In (19.6) the π_k are abitrary but fixed functions of order k. The coefficients c_{ki} are again uniquely determined elements of Δ.

The approximation theorem proved in Section 18.8 can now be formulated for function fields as follows.

Theorem I: *If for finitely many places finite segments of the series* (19.5) *are arbitrarily prescribed, then there always exists a function z in the field* K *whose series expansion at these places begins with just these segments.*

This theorem is called the *theorem of independence.*

We also have the following.

Theorem II: *A nonconstant function z has only finitely many zeros and poles.*

Proof: Every valuation W of the field K is a continuation of a valuation w of the field $\Delta(z)$. There are only two places of $\Delta(z)$ at which z has a positive or negative order, namely the places $z = 0$ and $z = \infty$. Only in the valuations w belonging to these places is $w(z) \neq 0$. Each of these valuations w can be continued in finitely many ways to valuations W of K. There are thus only finitely many places K with $W(z) \neq 0$.

Using the same method, we can show that every nonconstant function has at least one zero and at least one pole. Indeed, the valuation of $\Delta(z)$ belonging to the place $z = 0$ or ∞ can be continued in at least one way to a valuation of K. From this we obtain the following.

Theorem III: *A function z without a pole is a constant.*

The series expansions (19.5) and (19.6) hold not only for elements of the field K, but also for elements of the complete field Ω_K. If z is such an element and b is its order, then $z\pi^{-b}$ is an element of order zero. This element can be approximated arbitrarily closely (that is, with an error of arbitrarily high order) by an element y of $\mathfrak{J}$. In our case an approximation with an error of order 1 is already sufficient. For the element y we again have the congruence

$$y \equiv c_1\omega_1 + \cdots + c_f\omega_f \qquad (\mathfrak{p}).$$

The difference $y-(c_1\omega_1+\cdots+c_f\omega_f)$ is therefore divisible by π. Since the difference $z\pi^{-b}-y$ is likewise divisible by π, we obtain a representation as a multiple of π for the sum of these two differences, that is, for (19.4). The reasoning now continues as before.

19.2 DIVISORS AND MULTIPLES

Let K again be an algebraic function field in one variable over the constant field Δ. The functions of K will henceforth be denoted only by the letters, u, v, w, x, y, z, ϑ, and π.

Finitely many places $\mathfrak{p}$ with arbitrary integral exponents d define a divisor D of the field K. We write D symbolically as a product of finitely many factors

$$D = \Pi\, \mathfrak{p}^d. \tag{19.7}$$

The factors of the product may be interchanged in any manner. If an exponent d is zero, then the factor $\mathfrak{p}^d$ may be omitted in D. If all the d are zero, then $D = (1)$ is the *unit divisor.* If all $d \geqq 0$, then D is called an *integral divisor.*

Two divisors are multiplied by adding the exponents of equal factors $\mathfrak{p}$. To each divisor D with exponents d there is an inverse divisor D^{-1} with exponents $-d$, so that $D^{-1}D = (1)$. The divisors thus form an Abelian group, the *divisor group* of the field K. The individual places $\mathfrak{p}$ are called *prime divisors*. They generate the divisor group.

Each function z defines a divisor

$$(z) = \Pi\, \mathfrak{p}^d,$$

where the exponent d is equal to the order of z at the place $\mathfrak{p}$. To a constant z there corresponds the unit divisor. To a product yz there corresponds the product of the divisors (y) and (z):

$$(yz) = (y)\,(z).$$

The degree of a prime divisor $\mathfrak{p}$, that is, the degree of the residue class field $\bar{\mathfrak{J}} = \mathfrak{J}/\mathfrak{p}$ over Δ, will always be denoted by f as in Section 19.1. The sum of the degrees of the factors occurring in (19.7),

$$n(D) = \sum df,$$

is called the *degree of the divisor D*.

Instead of $(z)D$, we write simply zD. A function z is called a *multiple of the divisor D* if zD^{-1} is an integral divisor, that is, if, for all places $\mathfrak{p}$ of the field,

$$W_{\mathfrak{p}}(z) \geqq d. \tag{19.8}$$

The multiples of a divisor D are thus those functions z which have a zero of multiplicity at least h at all places with $d = h > 0$, which have a pole of at most multiplicity k at all places with $d = -k$, and which are *finite* at all other places, that is, have no other poles.

The multiples of a divisor A^{-1} form a Δ-module which will be denoted by $\mathfrak{M}(A)$. We shall now show that $\mathfrak{M}(A)$ has finite rank over Δ.

Let $A = \Pi\, \mathfrak{p}^a$. Since in the product there are only finitely many factors $\mathfrak{p}^a$ with $a > 0$, there are only a finite number of places $\mathfrak{p}$ which are admissible poles for the multiples z of A^{-1}. The series expansion of z at such a place can be written as follows:

$$z = (c_{-a,\,1}w_1 + \cdots + c_{-a,\,f}w_f)\pi^{-a} + \cdots;$$

here the ω_i previously used have been denoted by w_i.

The number of coefficients $c_{-i,\,j}$ belonging to the negative powers $\pi^{-a}, \ldots, \pi^{-1}$ is af for the single place $\mathfrak{p}$; the total number for all admissible poles is therefore

$$m = \sum af,$$

where the summation extends over all places $\mathfrak{p}$ with $a > 0$. We assert that there cannot be more than $m+1$ linearly independent multiples z of A^{-1}.

If there were $m+2$ such multiples $z_1, \ldots, z_{m+2}$, then we could form linear combinations

$$z = b_1 z_1 + \cdots + b_{m+2} z_{m+2} \tag{19.9}$$

with constant coefficients and impose the condition that all coefficients of negative powers in the expansion of z be zero. This would make m linear conditions for the $m+2$ coefficients $b_1, \ldots, b_{m+2}$. Each linear condition imposed on the coefficients b_i reduces the rank of the module of functions (19.9) by at most 1; the functions z which satisfy the linear conditions $c_{-i,j} = 0$ would therefore form a module of rank at least $(m+2)-m=2$. But these functions z have no poles and are therefore constants by Section 19.1, Theorem III. The constants form a module of rank 1 over Δ. Hence there can be only $m+1$ linearly independent multiples of A^{-1}; that is, the rank of $\mathfrak{M}(A)$ is at most $m+1$.

The object of the following investigation is the determination of the rank $l(A)$ of $\mathfrak{M}(A)$, that is, the number of linearly independent multiples of the divisor A^{-1}. Here $l(A)$ is also called the *dimension of A*. For integral divisors the proof just given affords the inequality

$$l(A) \leqq n(A)+1. \tag{19.10}$$

Now $A = \Pi\, \mathfrak{p}^a$ is said to be *divisible by* $B = \Pi\, \mathfrak{p}^b$ if AB^{-1} is integral; this means that $a \geqq b$ for all $\mathfrak{p}$. It is clear then that

$$n(A) \geqq n(B) \qquad \text{and} \qquad l(A) \geqq l(B).$$

We shall derive an inequality for the difference $n(A)-l(A)$. The method is the same as above. Let the multiples of A^{-1} be

$$z = b_1 z_1 + \cdots + b_l z_l \tag{19.11}$$

with constant coefficients b_i and $l = l(A)$. In order that the function z belong to $\mathfrak{M}(B)$ as well as $\mathfrak{M}(A)$, in the expansion

$$z = (c_{-a,1}w_1 + \cdots + c_{-a,f}w_f)\pi^{-a} + \cdots$$

the coefficients of the powers $\pi^{-a}, \pi^{-a+1}, \ldots, \pi^{-b-1}$ must all be zero. This gives $(a-b)f$ linear equations for each place and thus a total of

$$\sum (a-b)f = \sum af - \sum bf = n(A)-n(B)$$

linear equations for the coefficients $b_1, \ldots, b_l$ in (19.11). Each linear equation reduces the rank by at most 1; therefore

$$l(B) \geqq l(A) - [n(A)-n(B)]$$

or

$$n(A)-l(A) \geqq n(B)-l(B). \tag{19.12}$$

Equation (19.12) holds if A is divisible by B. In particular, taking for A an integral divisor and $B = (1)$, the right side of (19.12) becomes

$$0-1 = -1,$$

and we again obtain inequality (19.10).

The following theorem is almost obvious.

Theorem: *If $z \neq 0$, then $\mathfrak{M}(A)$ and $\mathfrak{M}(zA)$ have the same rank:*

$$l(zA) = l(A).$$

Proof: If $y_1, \ldots, y_l$ are linearly independent multiples of $(zA)^{-1} = z^{-1}A^{-1}$, then

$$y_1 z, \ldots, y_l z$$

are linearly independent multiples of A^{-1}, and conversely.

Two divisors A and zA which differ only by a factor (z) are said to be *equivalent*. We have thus proved that *equivalent divisors have the same dimension.*

Exercises

19.2. In the field $\mathsf{K} = \Delta(x)$ of rational functions let $A = \Pi\, \mathfrak{p}^a$ be a divisor. Show that the multiples of A^{-1} are given by

$$z = f(x)\, \Pi\, p(x)^{-a},$$

where $p(x)$ are the prime polynomials which by Section 18.7 belong to the prime divisors $\mathfrak{p}$ distinct from $\mathfrak{p}_\infty$ occurring in A.

19.3. Using Exercise 19.2, show that

$$l(A) = n(A)+1 \qquad \text{if } n(A) \geqq 0$$
$$l(A) = 0 \qquad \text{if } n(A) < 0.$$

19.3 THE GENUS g

Let z be a nonconstant function of the field K. The divisor (z) can be represented as the quotient of two integral divisors without common prime factor $\mathfrak{p}$:

$$(z) = CD^{-1}. \tag{19.13}$$

Now C is called the *divisor of the numerator* and D the *divisor of denominator* of z. Let the degree of K over $\Delta(z)$ be n. The degree of $C = \Pi\, \mathfrak{p}^c$ is

$$n(C) = \sum cf$$

and correspondingly for D.

We now prove the important equality

$$n(C) = n(D) = n. \tag{19.14}$$

The prime factors of $C = \Pi\, \mathfrak{p}^c$ we denote by $\mathfrak{p}, \mathfrak{p}', \ldots$, and their exponents by $c, c', \ldots$. An integral function u for $\mathfrak{p}$ of the field K has a series expansion at the place $\mathfrak{p}$

$$u = \sum_0^\infty (a_{k1} w_1 + \cdots + a_{kf} w_f)\pi^k. \tag{19.15}$$

We break the series off after the term π^{c-1} and thus write

$$u \equiv \sum_{k=0}^{c-1} \sum_{i=1}^{f} a_{ki} w_i \pi^k \qquad (\text{mod } \pi^c); \tag{19.16}$$

we do the same for the places $\mathfrak{p}'$, and so on.

By the theorem of independence (Theorem I, Section 19.1), there exist cf functions u_{ki}, each of whose initial segments (19.16) for the place $\mathfrak{p}$ consists of the single term $w_i \pi^k$ and whose initial segments for all other places $\mathfrak{p}', \ldots$ are zero. Similarly, there exist $c'f'$ functions u'_{ki} each of whose initial segments for the place $\mathfrak{p}'$ consists of a single $w'_i \pi'^k$, and so on. We now assert the following.

The $cf + c'f' + \cdots = n(C)$ *functions* $u_{ki}, u'_{ki}, \ldots$ *are linearly independent over* $\Delta(z)$.

Suppose that there were a linear dependence

$$\sum f_{ki}(z) u_{ki} + \sum f'_{ki}(z) u'_{ki} + \cdots = 0, \tag{19.17}$$

where $f_{ki}, f'_{ki}, \ldots$ are polynomials in z. We may assume that the constant terms $c_{ki}, c'_{ki}, \ldots$ of these polynomials are not all zero. If we now substitute the series expansions (19.15) for the place $\mathfrak{p}$ in (19.17) for $u_{ki}, u'_{ki}, \ldots$ and z and compute modulo π^c as in (19.16), then the polynomials $f_{ki}(z)$ reduce to their constant terms c_{ki}, the u_{ki} to $w_i \pi^k$, and the other u'_{ki} to zero. From (19.17) we thus obtain

$$\sum_{k=0}^{c-1} \sum_{i=1}^{f} c_{ki} w_i \pi^k \equiv 0 \qquad (\pi^c).$$

Because of the uniqueness of the series expansion (19.15), this is only possible if all the $c_{ki} = 0$. Similarly, all $c'_{ki} = 0$, and so on. We have thus reached a contradiction.

From the linear independence just proved it follows that

$$n \geqq n(C).$$

By replacing z everywhere by z^{-1}, it can be shown in the same way that

$$n \geqq n(D).$$

Now let $(u_1, \ldots, u_n)$ be a basis for K over $\Delta(z)$. We may assume that the u_j remain finite at all places where z is finite. Indeed, if u_j has a pole $\mathfrak{p}$ where z is finite, then to this pole there corresponds a valuation $W_{\mathfrak{p}}$ which induces a valuation of the field $\Delta(z)$, and this is not the valuation w_∞ belonging to the place $z = \infty$. By Section 18.7, the valuations of the field $\Delta(z)$ distinct from w_∞ are all p-adic; that is, they belong to prime polynomials $p = p(z)$, where p has positive order at the place in question. For sufficiently large d the product $p^d u_j$ therefore no longer has a pole at $\mathfrak{p}$. Thus all the poles of the u_j where z is finite can be successively removed by multiplying the basis elements u_j with suitable polynomials in z.

The poles of z are all contained in the divisor of the denominator D. For sufficiently large m_i, u_i is therefore a multiple of D^{-m_i-1}. We now choose m greater than all the m_i:

$$m \geqq m_i+1 \qquad (i = 1, \ldots, n).$$

The $\sum (m-m_i)$ field elements

$$z^\mu u_i \qquad (0 \leqq \mu < m-m_i)$$

are linearly independent over Δ and are multiples of D^{-m}; they are thus contained in $\mathfrak{M}(D^m)$. From this it follows that

$$\sum (m-m_i) \leqq l(D^m) \leqq n(D^m)+1$$

or

$$nm - \sum m_i \leqq l(D^m) \leqq m\cdot n(D)+1. \tag{19.18}$$

Letting m go to infinity, from (19.18) we obtain

$$n \leqq n(D),$$

and hence, since it has already been shown that $n \geqq n(D)$,

$$n = n(D). \tag{19.19}$$

Similarly,

$$n = n(C). \tag{19.20}$$

Now (19.19) and (19.20) imply

$$n((z)) = n(CD^{-1}) = 0. \tag{19.21}$$

From (19.21) it further follows that

$$n(zA) = n(A), \tag{19.22}$$

that is, *equivalent divisors have not only the same dimension $l(A)$ but also the same degree $n(A)$.*

Substituting (19.19) into (19.18), we obtain

$$n(D)\cdot m - \sum m_i \leqq l(D^m)$$

or

$$n(D^m) - l(D^m) \leqq \sum m_i. \tag{19.23}$$

If B divides D^m, then, by (19.12),

$$n(B)-l(B) \leqq n(D^m)-l(D^m),$$

and hence, by (19.23),

$$n(B)-l(B) \leqq \sum m_i. \tag{19.24}$$

Now let A be an arbitrary divisor. We wish to show that (19.24) also holds for A. For this it is sufficient to show that there exists a divisor $uA = B$ equivalent to A which divides a power D^m.

Let $\mathfrak{p}$ be a prime factor which occurs with positive exponent in $A = \prod \mathfrak{p}^d$. If all these $\mathfrak{p}$ are poles of z, then A itself divides D^m and we are done. If $\mathfrak{p}$ is not a pole of z, then as before we can find a polynomial $p = p(z)$ which has positive order at the place $\mathfrak{p}$. We now multiply A by $\mathfrak{p}^{-d}$ and hereby remove the factor $\mathfrak{p}^d$ in A. By repeating this procedure, we can remove all factors $\mathfrak{p}^d$ with $d>0$ which do not belong to poles of z. We thus finally find a divisor $B = uA$ equivalent of A which divides D^m and for which (19.24) holds. But then (19.24) also holds for A:

$$n(A)-l(A) \leqq \sum m_i, \tag{19.25}$$

that is, *the difference $n(A)-l(A)$ is bounded for all A.*

The least upper bound of $n(A)-l(A)+1$ for all divisors A is called the *genus* g of the field K.

For $A = (1)$, $n(A)-l(A) = 0-1$, and hence $g \geqq 0$. The genus is thus a non-negative integer which is a numerical invariant of the function field K.

By the definition of genus we have, for all A,

$$n(A)-l(A)+1 \leqq g$$

or

$$l(A) \geqq n(A)-g+1, \tag{19.26}$$

where equality holds for at least one divisor A. Inequality (19.26) might well be called the Riemann part of the Riemann-Roch theorem.

We put

$$l(A) = n(A)-g+1+i(A) \tag{19.27}$$

and call $i(A)$ the *speciality index* of the divisor A. The divisor A is called *special* if $i(A)>0$. If A is not special, then $n(A)-l(A)$ has the greatest possible value $g-1$. There exist divisors A which are not special. Our task will be to determine the speciality index $i(A)$ and so to prove the complete Riemann-Roch theorem.

Exercises

19.4. The field $\mathsf{K} = \Delta(z)$ of rational functions has genus zero and prime divisors of degree 1.

19.5. If K has genus zero and a prime divisor of degree 1, then K is a field of rational functions $\Delta(z)$. (Apply formula (19.26) to $A = \mathfrak{p}$.)

19.4 VECTORS AND COVECTORS

In the series expansion of the functions of a field K at a place $\mathfrak{p}$, expressions such as

$$v = c_1 w_1 + \cdots + c_f w_f \tag{19.28}$$

appear as coefficients of the powers of π. These expressions form (for each place $\mathfrak{p}$) and f-dimensional vector space L_f over Δ.

The power series for the place $\mathfrak{p}$ may be written more simply as

$$V_{\mathfrak{p}} = \sum_{a}^{\infty} v_k \pi^k \tag{19.29}$$

or

$$V_{\mathfrak{p}} = \sum_{a}^{\infty} v_{\mathfrak{p}k} \pi^k \tag{19.30}$$

if the dependence of the coefficients v_k on the place $\mathfrak{p}$ is made explicit.

If to each place $\mathfrak{p}$ a power series (19.30) with arbitrary coefficients $v_{\mathfrak{p}k}$ of L_f is assigned such that in the totality of all these power series there are only finitely many terms with negative exponents, then the system of power series is called a *vector* V. The power series $V_{\mathfrak{p}}$ are called the *components* of the vector V. They may also be defined, independently of the special choice of uniformizing variable π and basic vectors w_i in (19.28), as elements of the complete extension field $\Omega_{\mathsf{K}}(\mathfrak{p})$ belonging to the place $\mathfrak{p}$. Only finitely many of these elements $V_{\mathfrak{p}}$ may have negative order $W_{\mathfrak{p}}(V_{\mathfrak{p}})$; otherwise they may be chosen quite arbitrarily.

A vector V is said to be *divisible by a divisor* $D = \Pi\, \mathfrak{p}^d$ if the series (19.30) at each place $\mathfrak{p}$ begins with π^d:

$$w_{\mathfrak{p}}(V_{\mathfrak{p}}) \geqq d \qquad \text{for all } \mathfrak{p}.$$

In particular, the functions u of the field K are vectors, since each function u can be expanded at each place in a power series (19.30), and in all these power series there is altogether only a finite number of terms with negative exponents.

Corresponding to the vector space L_f there is a dual space D_f according to Section 4.3. The elements of D_f are linear forms on L_f.

From each $v = \sum c_i w_i$ of L_f and each α of D_f we may form a scalar product

$$v \cdot \alpha = c_1 \alpha_1 + \cdots + c_f \alpha_f.$$

In a similar way, we shall now construct the dual space of covectors corresponding to the infinite-dimensional space $\mathfrak{V}$ of vectors V.

If to each place $\mathfrak{p}$ a sequence $\{\alpha_{\mathfrak{p}k}\}$ $(k = b, b+1, \ldots)$ of elements of D_f is assigned so that in all these sequences there is altogether only a finite number of negative indices k, then the system of these sequences is called a *covector* λ. The *scalar product* of a vector V and a covector λ is defined as follows:

$$V \cdot \lambda = \sum_{\mathfrak{p}} \sum_{j+k=-1} v_{\mathfrak{p}j} \cdot \alpha_{\mathfrak{p}k}. \tag{19.31}$$

Since there are only finitely many $v_{\mathfrak{p}j}$ with negative j and only finitely many $\alpha_{\mathfrak{p}k}$ with negative k, there are only finitely many terms in the sum (19.31). The individual terms are scalar products $v \cdot \alpha$ and are thus elements of Δ.

The operator $\cdot\lambda$ is a mapping of the space $\mathfrak{V}$ of vectors V into the field of constants which has the following properties:

(*a*) $(V+W) \cdot \lambda = V \cdot \lambda + W \cdot \lambda$

(*b*) $(cV) \cdot \lambda = c(V \cdot \lambda)$

(*c*) $V \cdot \lambda = 0$ if V is divisible by a divisor D depending only on λ.

Statements (*a*) and (*b*) are clear. To prove (*c*), we note that there are only finitely many $\mathfrak{p}$ for which the sequence $\{\alpha_{\mathfrak{p}k}\}$ begins with a negative index $k = -d$. If from these places $\mathfrak{p}$ with exponents d we form the divisor

$$D = \Pi\, \mathfrak{p}^d,$$

then (*c*) is satisfied.

The set of all vectors V which are divisible by a divisor D is called a *neighborhood of zero* in the vector space $\mathfrak{V}$. Property (*c*) states that the linear functional λ maps a certain neighborhood of zero onto zero. Property (*c*) is therefore a type of continuity property.

We now prove the following.

Every mapping $\cdot\lambda$ *of* $\mathfrak{V}$ *into* Δ *with properties* (*a*), (*b*), *and* (*c*) *can be defined in terms of sequences* $\{\alpha_{\mathfrak{p}k}\}$ *in the manner indicated.*

Proof: Every vector V can be represented as a sum of a vector divisible by D and finitely many vectors $V_{\mathfrak{p}j}$ which in their expansion at the place $\mathfrak{p}$ have only a single term $v\pi^j$ and whose other components are zero:

$$(V_{\mathfrak{p}j})_{\mathfrak{p}} = v\pi^j$$

$$(V_{\mathfrak{p}j'})_{\mathfrak{p}'} = 0 \qquad \text{for } \mathfrak{p}' \neq \mathfrak{p} \quad \text{or} \quad j' \neq j.$$

Here, as always, $v = \sum c_i w_i$ is an element of the vector space L_f. If we apply the mapping $\cdot\lambda$ to the vector $V_{\mathfrak{p}j}$ just defined, then we obtain an element $V_{\mathfrak{p}j}\cdot\lambda$ of Δ which depends linearly on v and can therefore be written as $v\cdot\alpha$, where α is an element of D_f. This element α we call $\alpha_{\mathfrak{p}k}$ where k is determined from

$$j+k = -1.$$

Since $V_{\mathfrak{p}j}$ is not divisible by D, it follows that $j<d$ and hence $k>-d$; therefore, in the sequences $\{\alpha_{\mathfrak{p}k}\}$ there is altogether only a finite number of negative indices. It follows further from (*a*) and (*c*) that

$$V\cdot\Omega = \sum_{\mathfrak{p}}\sum_{j} V_{\mathfrak{p}j}\cdot\Omega = \sum_{\mathfrak{p}}\sum_{j+k=-1} v_{\mathfrak{p}j}\cdot\alpha_{\mathfrak{p}k},$$

and this completes the proof.

On the basis of this theorem, the covectors λ can also be defined as mappings of $\mathfrak{V}$ into Δ with properties (*a*), (*b*) and (*c*). This definition is invariant; that is, it does not depend on the choice of the w_i and π.

19.5 DIFFERENTIALS. THE THEOREM ON THE SPECIALITY INDEX

The speciality index $i(B)$ will now be determined with the help of the covectors. We begin with two lemmas.

Lemma: *If the divisor D is not special and if A is a multiple of D, then A is likewise not special.*

Proof: By (19.12),

$$n(A)-l(A) \geqq n(D)-l(D).$$

Thus, if $n(D)-l(D)$ already has the maximum value $g-1$, then $n(A)-l(A)$ must also have the maximum value $g-1$.

Corollary: *Every divisor B has a multiple A which is not special.*

Proof: Suppose that D is not special. Choose A to be a common multiple of B and D. The assertion now follows immediately from the lemma.

We now put $A = \Pi\, \mathfrak{p}^a$ and $B = \Pi\, \mathfrak{p}^b$. Let A be a multiple of B, so that $b \leqq a$ and $\mathfrak{M}(B) \subseteqq \mathfrak{M}(A)$. We assume that B is special and A is not. Then

$$l(A) = n(A)-g+1 \tag{19.32}$$

$$l(B) = n(B)-g+1+i(B). \tag{19.33}$$

As in Section 19.2, we write $\sum (a-b)f$ linear equations which an element of $\mathfrak{M}(A)$,

$$u = b_1 u_1 + \cdots + b_l u_l, \tag{19.34}$$

must satisfy in order that it belong to $\mathfrak{M}(B)$. If the series expansion for u at the place $\mathfrak{p}$ begins:

$$u = (c_{-a,1} w_1 + \cdots + c_{-a,f} w_f)\pi^{-a} + \cdots, \tag{19.35}$$

then the $(a-b)f$ conditions for the place $\mathfrak{p}$ are

$$c_{j\nu} = 0 \qquad (-a \leqq j < -b, \quad 1 \leqq \nu \leqq f). \tag{19.36}$$

The $c_{j\nu}$ of course depend on the place $\mathfrak{p}$. We should really write $c_{j\nu}(\mathfrak{p})$, but we shall neglect to do so for ease of notation.

If the $\sum (a-b)f = n(A)-n(B)$ equations (19.36) were independent, then we would have

$$l(A)-l(B) = n(A)-n(B).$$

However, by (19.32) and (19.33) the difference $l(A)-l(B)$ is smaller by a term $i(B)$ than $n(A)-n(B)$, and hence there are $i(B)$ linear dependencies between the left sides of equations (19.36); that is, there are $i(B)$ linearly independent relations

$$R\{c_{j\nu}\} = \sum_{\mathfrak{p}} \sum_{j=-a}^{-b-1} \sum_{\nu=1}^{f} c_{j\nu}\gamma_{j\nu} = 0, \tag{19.37}$$

which must be satisfied for each element u of $\mathfrak{M}(A)$.

Equations (19.37) can be written somewhat more simply if the sum over f is interpreted as a scalar product:

$$\sum_1^f c_{j\nu}\gamma_{j\nu} = v_j \cdot \beta_j.$$

Here, as always, $v_j = \sum c_{j\nu} w_\nu$, and $\beta_j = \beta_j(\mathfrak{p})$ is the sequence $(\gamma_{j1}, \ldots, \gamma_{jf})$. To make the connection with previous notation, we put $v_j = v_{\mathfrak{p}j}$ and

$$\beta_j(\mathfrak{p}) = \alpha_{\mathfrak{p}k} \qquad (j+k = -1).$$

Then (19.37) becomes

$$R\{c_{j\nu}\} = \sum_{\mathfrak{p}} \sum_{j+k=-1} v_{\mathfrak{p}j} \cdot \alpha_{\mathfrak{p}k} = 0 \tag{19.38}$$

with

$$b \leqq k \leqq a-1.$$

We now replace A by a multiple

$$A' = \Pi\, \mathfrak{p}^{a'} \qquad (a' \geqq a).$$

Then

$$\mathfrak{M}(B) \subseteq \mathfrak{M}(A) \subseteq \mathfrak{M}(A').$$

Since A', as a multiple of A, is not special, there again exist $i(B)$ linearly independent relations

$$R'\{c_{j\nu}\} = \sum_{\mathfrak{p}} \sum_{j+k=-1} v_{\mathfrak{p}j} \cdot \alpha'_{\mathfrak{p}k} = 0 \tag{19.39}$$

with $b \leqq k \leqq a'-1$, valid for all u of $\mathfrak{M}(A')$.

The relations R, or more precisely their coefficient systems $\{\alpha_{\mathfrak{p}k}\}$, form a Δ-module of rank $i(B)$. Similarly, the R' form a Δ-module of rank $i(B)$.

If terms with $k > a-1$ are omitted in a relation R', then a relation R is obtained which is valid for all u of $\mathfrak{M}(A)$. By means of this "projection" every R' gives rise to an R, and the mapping $R' \to R$ is linear. If $R' \neq 0$ were to go into $R = 0$ under the projection, then R' would contain only terms with $k > a-1$, and hence

$$-a' \leqq j < -a.$$

Such a relation R' would hold for all u of $\mathfrak{M}(A')$. If we again write down the conditions which an element of $\mathfrak{M}(A')$ must satisfy in order to belong to $\mathfrak{M}(A)$, then the relation R' would state that there was a dependency between these $n(A')-n(A)$ conditions. This would imply that

$$l(A')-l(A) < n(A')-n(A),$$

which is impossible, since (19.32) is valid for both A' and A.

The mapping $R' \to R$ is therefore injective. It maps the module of the R' isomorphically onto a module of the same rank $i(B)$ in the module of the R; it is therefore also surjective. This means that *each relation R can be continued in a unique manner to a relation R'*.

If now, beginning with a, we let an exponent a' tend to infinity while always continuing the relation R, we obtain a uniquely determined infinite sequence

$$\{\alpha_{\mathfrak{p}k}\} \qquad (k = b, b+1, \ldots). \tag{19.40}$$

The same can be done for each place $\mathfrak{p}$. We thus obtain a system of sequences (19.40) for all places $\mathfrak{p}$, that is, a covector λ. The relations (19.39) may now be written as follows:

$$u \cdot \lambda = 0. \tag{19.41}$$

The relation (19.41) holds for all u of $\mathfrak{M}(A')$. But now for any function u of the field a divisor A' can be found which is divisible both by B and by (u^{-1}). Then uA' is integral; that is, u belongs to $\mathfrak{M}(A')$ and therefore (19.41) holds. *Hence* (19.41) *holds for all functions u of the field* K.

Since there are $i(B)$ linearly independent relations R, there are also $i(B)$ linearly independent covectors defined by (19.40) with the property (19.41). We now make a definition (following A. Weil).

Definition 1: A covector λ with the property (19.41) for all z of K is called a *differential* of the field K.

The relation of the Weil differentials to the differentials of classical function theory will be established in Section 19.8.

Definition 2: A covector λ is called a *multiple of* $B = \Pi\, \mathfrak{p}^b$ if only $\alpha_{\mathfrak{p}k}$ with $k \geqq b$ occur in the definition of the covector.

From the definition of a covector it follows immediately that *for each covector* λ *there exists a divisor B such that* λ *is a multiple of B.*

Using Definitions 1 and 2, we may now combine what we have proved in this section in the following theorem.

Theorem on the Speciality Index: *The speciality index* $i(B)$ *is equal to the number of linearly independent differentials* λ *which are multiples of B.*

Definition 3: A differential is said to be *everywhere finite* or a *differential of first kind* if it is a multiple of the unit divisor (1), that is, if all the $\alpha_{\mathfrak{p}k}$ with negative index k are zero.

To find the number of linearly independent differentials of first kind, we have only to apply the theorem on the speciality index to the divisor $B = (1)$. Formula (19.27) gives

$$\begin{aligned} i(1) &= l(1)-n(1)+g-1 \\ &= 1-0+g-1 = g, \end{aligned}$$

and it thus follows that *the number of linearly independent differentials of first kind is equal to the genus g.*

We obtain another application of the theorem on the speciality index if we put $B = C^{-1}$, where C is an integral divisor different from (1). In this case $l(B) = 0$, since the only function which is a multiple of the integral divisor $B^{-1} = C$ is the zero function. Furthermore, $n(B) = -n(C)$, and hence

$$i(C^{-1}) = n(C)+g-1. \tag{19.42}$$

In particular, if we choose $C = \mathfrak{p}^n$, so that $B = \mathfrak{p}^{-n}$, then $n(C) = nf$ and we obtain

$$i(\mathfrak{p}^{-n}) = nf+g-1. \tag{19.43}$$

We thus have the following theorem.

Theorem: *If f is the degree of the prime divisor* $\mathfrak{p}$, *then there are* $nf+g-1$ *linearly independent differentials which are multiples of* $\mathfrak{p}^{-n}$.

Exercises

19.6. Let the base field Δ be algebraically closed. Then except for the differentials of first kind there is no differential which is a multiple of $\mathfrak{p}^{-1}$; that is, there is no differential with only one simple pole $\mathfrak{p}$.

19.7. Under the same hypotheses there is for each $n > 1$ an *elementary differential of second kind* $\omega(\mathfrak{p}^n)$ which has a pole of order n at $\mathfrak{p}$. Any differential which is a multiple of $\mathfrak{p}^{-n}$ can be written as a linear combination of $\omega(\mathfrak{p}^2)$, $\omega(\mathfrak{p}^3)$, ..., $\omega(\mathfrak{p}^n)$ and the g linearly independent differentials of first kind.

19.8. Under the same hypotheses there is for any two places $\mathfrak{p}_1$ and $\mathfrak{p}_2$ an *elementary differential of third kind* $\omega(\mathfrak{p}_1, \mathfrak{p}_2)$ which has a simple pole at each of $\mathfrak{p}_1$ and $\mathfrak{p}_2$. Any differential can be written as a linear combination of the elementary differentials of second and third kind and the differentials of first kind.

19.6 THE RIEMANN-ROCH THEOREM

We have now nearly reached our objective. We first define the *product* $u\lambda$ of a function u and a covector λ. The product is defined as a linear mapping of $\mathfrak{V}$ into Δ:

$$V \cdot u\lambda = Vu \cdot \lambda. \tag{19.44}$$

The operation $\cdot u\lambda$ obviously has properties (*a*), (*b*) and (*c*) of Section 19.4; (19.44) therefore defines a covector.

If λ is a differential, then $u\lambda$ is also a differential:

$$v \cdot u\lambda = vu \cdot \lambda = 0 \qquad \text{for all} \quad v.$$

The following lemmas are almost obvious.

Lemma 1: *If λ is a multiple of $D = \Pi\, \mathfrak{p}^d$, then $V \cdot \lambda = 0$ for all vectors V divisible by D^{-1}, and conversely.*

Proof: Let the covector λ be given by the sequences $\{\alpha_{\mathfrak{p}k}\}$. If λ is a multiple of D, then only indices with $k \geqq d$ occur in these sequences. Further, if V is given by the power series

$$V_{\mathfrak{p}} = \sum v_{\mathfrak{p}j}\pi^j \tag{19.45}$$

and if V is divisible by D^{-1}, then only terms with $j \geqq -d$ occur in the power series (19.45). The scalar product

$$V \cdot \lambda = \sum_{j+k=-1} v_{\mathfrak{p}j}\alpha_{\mathfrak{p}k} \tag{19.46}$$

is zero, since the sum $j+k$ can never equal -1. Conversely, if $V \cdot \lambda = 0$ for all V divisible by D^{-1}, then only terms with $k \geqq d$ can occur in the sequences $\{\alpha_{\mathfrak{p}k}\}$ and thus λ is a multiple of D.

Lemma 2: *If λ is a multiple of D, then $u\lambda$ is a multiple of uD.*
Proof: If V is divisible by D^{-1}, then $V\cdot\lambda = 0$ by Lemma 1; hence $Vu\cdot\lambda = 0$ if Vu is divisible by D^{-1}, that is, $V\cdot u\lambda = 0$ if V is divisible by $(uD)^{-1}$.

Now let λ be a differential. By Section 19.5, there exists a divisor D of which λ is a multiple. Let $B = \mathfrak{p}^{-n}$, where $\mathfrak{p}$ is a prime divisor of degree f. The divisor $B^{-1}D = \mathfrak{p}^n D$ has degree

$$n(B^{-1}D) = nf+n(D).$$

Therefore, by the Riemann part of the Riemann-Roch theorem, the number of linearly independent multiples u of BD^{-1} is

$$l(B^{-1}D) \geqq nf+n(D)-g+1. \tag{19.47}$$

If u is a multiple of BD^{-1}, then uD is a multiple of B. By Lemma 2, $u\lambda$ is a multiple of uD, and hence $u\lambda$ is a multiple of B. The total number of linearly independent differentials which are multiples of B is $i(B)$. Hence it follows from (19.47) that

$$nf+n(D)-g+1 \leqq i(B). \tag{19.48}$$

For $n>0$ it follows (19.43) that

$$i(B) = nf+g-1. \tag{19.49}$$

Substituting this into (19.48), we obtain

$$n(D) \leqq 2g-2. \tag{19.50}$$

The degree of the divisor D is therefore bounded above. For given λ there is thus a maximal divisor D_λ such that λ is a multiple of D_λ but it is not a multiple of $D_\lambda\mathfrak{p}'$ however $\mathfrak{p}'$ is chosen. The uniquely determined maximal divisor D_λ of which λ is a multiple is called *the divisor of the differential* λ.

We now prove the following.

All differentials ω are equal to $u\lambda$, where λ is an arbitrary fixed differential.
Proof: Suppose that there is a differential ω which is not equal to $u\lambda$. Then

$$u\lambda \neq v\omega \qquad \text{for all } u \text{ and } v \neq 0. \tag{19.51}$$

We have seen in the paragraph following (19.47) above that there are at least

$$nf+n(D_\lambda)-g+1$$

linearly independent differentials $u\lambda$ which are multiples of $B = \mathfrak{p}^{-n}$. Similarly, there are at least

$$nf+n(D_\omega)-g+1$$

linearly independent differentials $v\omega$ which are multiples of B. All these differentials are independent, since no linear combination of the $u\lambda$ is equal to a linear combination of the $v\omega$. There is thus a total of

$$2nf+\text{const}$$

linearly independent differentials which are multiples of B. But by (19.49) there are only $nf+g-1$ such differentials. For large n this gives a contradiction. All differentials are therefore equal to $u\lambda$ as asserted.

We now replace B by an arbitrary divisor A and again ask how many linearly independent differentials $\omega = u\lambda$ there are which are multiples of A. If $u\lambda$ is a multiple of A, then λ is a multiple of $u^{-1}A$. The maximal divisor D_λ is therefore divisible by $u^{-1}A$, so that uD_λ is divisible by A, and hence u is a multiple of $AD_\lambda{}^{-1}$. Conversely, if u is a multiple of $AD_\lambda{}^{-1}$, then $u\lambda$ is a multiple of A, since the argument is reversible. It therefore follows that

$$i(A) = l(A^{-1}D_\lambda). \tag{19.52}$$

Substituting this into (19.27), we obtain the following complete theorem.

Riemann-Roch Theorem: *If A is an arbitrary divisor of the field K and λ is an arbitrary nonzero differential, then*

$$l(A) = n(A)-g+1+l(A^{-1}D_\lambda). \tag{19.53}$$

We mention some further corollaries.

1. Putting $A = (1)$, we obtain, from either (19.52 or (19.53),

$$l(D_\lambda) = g. \tag{19.54}$$

2. Putting $A = D_\lambda$, we obtain, from (19.53),

$$n(D_\lambda) = 2g-2. \tag{19.55}$$

3. If λ is a multiple of D, then $u\lambda$ is a multiple of uD and conversely. Hence if D_λ is the divisor of the differential λ, then uD_λ is the divisor of the differential $u\lambda$. The divisors $D_\omega = uD_\lambda$ of the differentials $\omega = u\lambda$ are thus all equivalent. The class of these divisors D_ω is called the *differential class* or the *canonical class*.

4. In general, a *divisor class* consists of all divisors uA which are equivalent to a divisor A. All the divisors uA of the class have the same dimension $l(A)$ and the same degree $n(A)$; $l(A)$ is therefore called the *dimension of the class* and $n(A)$ the *degree of the class*.

The dimension of the class $\{A\}$ may also be interpreted as follows. If u is divisible by A^{-1}, then uA is an integral divisor. To the elements u of the module $\mathfrak{M}(A)$ there thus correspond the integral divisors uA of the class $\{A\}$. If $u_1, \ldots, u_r$ are linearly independent, then the divisors $u_1A, \ldots, u_rA$ are also called linearly independent. The rank $l(A)$ of the module $\mathfrak{M}(A)$ is thus the maximum number of linearly independent integral divisors of the class $\{A\}$.

5. If $n(A)<0$, then there is no integral divisor equivalent to A and hence $l(A) = 0$.

6. If $n(A)>2g-2$, then $n(A^{-1}D_\lambda)<0$ and hence $l(A^{-1}D_\lambda) = 0$ by 5. From this it follows by (19.52) that $i(A) = 0$, and thus *a divisor A with $n(A)>2g-2$ is not special.*

Exercises

19.9. There is only one class $\{A\}$ with $l(A) \geqq g$ and $n(A) = 2g-2$, namely the canonical class.

19.10. An integral divisor B with $l(B) > g$ is not special.

This completes the development of the general theory for arbitrary base fields Δ. We now wish to establish the connection with the classical theory in which Δ is the field of complex numbers. For this purpose, we must first briefly consider some separability questions.

The general Riemann-Roch theorem can be extended also to skew fields which are finite extensions of a field of rational functions $\Delta(z)$. See E. Witt, "Riemann-Rochscher Satz und ζ-Funktion im Hyperkomplexen," *Math. Ann.*, **110**, 12 (1934).

19.7 SEPARABLE GENERATION OF FUNCTION FIELDS

A *field* K *of algebraic functions in r variables* is a finite extension of the field $\Delta(x_1, \ldots, x_r)$ of the rational functions in r algebraically independent quantities $x_1, \ldots, x_r$.

If the field K is generated from $\Delta(x_1, \ldots, x_r)$ by adjunction of $x_{r+1}, \ldots, x_n$, then

$$\mathsf{K} = \Delta(x_1, \ldots, x_r, x_{r+1}, \ldots, x_n),$$

where all the x_i are algebraic functions of the independent $x_1, \ldots, x_r$.

For such function fields we have the following theorem.

Theorem on Separable Generation: *If the field of constants is perfect, then $x_1, \ldots, x_n$ can be enumerated in such a manner that all the x_i are separable algebraic functions of the independent $x_1, \ldots, x_r$.*

Proof: We proceed by induction on n for given r. The case $n = r$ is trivial. Then let $n > r$, and suppose that the assertion is true for $\Delta(x_1, \ldots, x_{n-1})$. We may then assume that $x_1, \ldots, x_{n-1}$ are separable functions of $x_1, \ldots, x_r$.

Here x_n is an algebraic function of $x_1, \ldots, x_r$ and so satisfies an equation

$$f(x_1, \ldots, x_r, x_n) = 0, \tag{19.56}$$

which may be assumed to be integrally rational in all the x_i. If the field elements $x_1, \ldots, x_r$ and x_n are replaced by indeterminates $X_1, \ldots, X_r$ and X_n, then $f(X_1, \ldots, X_n)$ is irreducible as a polynomial in X_n. If f is decomposable as a polynomial in $X_1, \ldots, X_n$, then one of the factors contains only $X_1, \ldots, X_r$. Such a factor may be omitted from (19.56). We may therefore assume that f is irreducible as a polynomial in the X_i.

If x_n is separable over $\Delta(x_1, \ldots, x_r)$, then there is nothing more to prove. If

x_n is inseparable, then the characteristic of the field is a prime number p, and the polynomial f contains only powers of X_n which can be written as powers of $X_n{}^p$. If this were also the case for the powers of $X_1, \ldots, X_r$ occurring in f, then

$$f = \sum a_s X_1^{ps_1} \cdots X_r^{ps_r} X_n^{ps_n}. \tag{19.57}$$

Now in a perfect field Δ every a_s is a pth power:

$$a_s = b_s{}^p.$$

It would then follow that

$$f = (\sum b_s X_1{}^{s_1} \cdots X_r{}^{s_r} X_n{}^{s_n})^p.$$

However, this is impossible, since f is irreducible. Therefore, one of the variables $X_1, \ldots, X_r$, say X_1, must occur in f with an exponent which is not divisible by p.

From (19.56) it now follows that x_1 is a separable algebraic function of $x_2, \ldots, x_r$ and x_n. All the x_i are dependent on $x_1, \ldots, x_r$ and hence also on x_n, $x_2, \ldots, x_r$. Since the transcendency degree of $\Delta(x_1, \ldots, x_n)$ is equal to r, it follows that $x_n, x_2, \ldots, x_r$ are independent. The field $\Delta(x_1, \ldots, x_{n-1})$ is separable over the field $\Delta(x_1, \ldots, x_r)$, and this field is separable over $\Delta(x_n, x_2, \ldots, x_r)$; thus all the x_i are separable over $\Delta(x_n, x_2, \ldots, x_r)$. If the x_i are now renumbered so that the numbers 1 and n are interchanged, then the assertion follows.

A. Weil has given necessary and sufficient conditions for separable generation in the case of imperfect fields. See my paper: "Über Weil's Neubegründung der Algebraischen Geometrie," *Abh. Math. Sem. Hamburg*, **22**, 158 (1958).

19.8 DIFFERENTIALS AND INTEGRALS IN THE CLASSICAL CASE

Classical function theory deals with Abelian integrals

$$\int w\,dz,$$

where z is an independent variable, that is, a nonconstant function, and w is an arbitrary function from the field K. Change to another variable t is accomplished by the formula

$$\int w\,dz = \int w\frac{dz}{dt}\,dt.$$

In the algebraic theory we may omit the integral sign and consider *Abelian differentials* $w\,dz$. Change to another variable t again takes place by the formula

$$w\,dz = w\frac{dz}{dt}\,dt.$$

In order that dz/dt be meaningful, we must hereby assume that z is separable over $\Delta(t)$ (see Section 10.5). Consideration is therefore restricted to t for which the field K is separable over $\Delta(t)$. Such t exist if the field K is separably generated and thus, in particular, when Δ is perfect.

We shall assume for simplicity that the field of constants Δ is algebraically closed. Translation of the theory to the case of arbitrary complete perfect fields of constants is left to the reader.

Let the variable z be chosen once and for all so that K is separable over $\Delta(z)$. In order to investigate the behavior of a differential wdz at a place $\mathfrak{p}$, we choose a uniformizing variable π for this place and expand z in a power series:

$$z = P(\pi) = \sum c_k \pi^k. \tag{19.58}$$

The irreducible equation $F(z, \pi) = 0$ relating z to π is satisfied if the power series $P(\pi)$ is substituted for z:

$$F(P(\pi), \pi) = 0. \tag{19.59}$$

The left-hand side is a power series in π, all of whose coefficients are zero. They remain zero if the power series is formally differentiated, where formal differentiation of a power series $P(\pi)$ is defined by

$$P'(\pi) = \sum kc_k \pi^{k-1}.$$

We thus obtain, from (19.59),

$$F_z'(z, \pi) \cdot P'(\pi) + F_\pi'(z, \pi) = 0, \tag{19.60}$$

where z has again been substituted for $P(\pi)$ and the partial derivatives of F with respect to z and π have been denoted by F_z' and F_π', respectively.

Now $F_\pi'(z, \pi) \neq 0$, since π is separable over $\Delta(z)$. By (19.60) $F_z'(z, \pi)$ cannot be zero, and hence z is separable over $\Delta(\pi)$. The differential quotient $dz/d\pi$ is therefore defined and satisfies the equation

$$F_z'(z, \pi) \cdot \frac{dz}{d\pi} + F_\pi'(z, \pi) = 0. \tag{19.61}$$

Comparison of (19.60) and (19.61) gives

$$\frac{dz}{d\pi} = P'(\pi) = \sum kc_k \pi^{k-1}. \tag{19.62}$$

The variable z is thus differentiable with respect to any uniformizing variable, and the power series for the differential quotient is found by termwise differentiation of the power series for z.

The differential wdz can now likewise be expressed in terms of the uniformizing variable π:

$$wdz = w \frac{dz}{d\pi} d\pi. \tag{19.63}$$

The power series for $w(dz/d\pi)$ is of course found by multiplying the power series for w by the power series (19.62). Let the result be

$$w \frac{dz}{d\pi} = \sum \alpha_{\mathfrak{p}k} \pi^k. \tag{19.64}$$

If no negative exponents occur in the series (19.64), then the differential wdz is said to *finite* at the place $\mathfrak{p}$. If only exponents from a onward occur with nonzero coefficients, then $\mathfrak{p}$ is a *zero* of order a of the differential. If negative exponents occur, then $\mathfrak{p}$ is a *pole* of the differential. The *order* of a differential at the place $\mathfrak{p}$ is the smallest exponent k to which there belongs a nonzero coefficient $\alpha_{\mathfrak{p}k}$. All these concepts are clearly independent of the choice of the uniformizing variable.

The poles of a differential wdz are to be found among the poles of w and z, for wdz cannot have a pole where w and z remain finite. Hence, *every differential wdz has only finitely many poles.*

The *residue* of the differential wdz at the place $\mathfrak{p}$ is the coefficient of π^{-1} in the expansion (19.63). In the classical theory the residue can be obtained by integrating the differential wdz over a small circle about the point $\mathfrak{p}$ of the Riemann surface and dividing by $2\pi i$. We now prove quite generally that the residue is independent of the choice of the uniformizing variable.

The power series (19.63) may be interpreted as the sum of three types of terms: terms with $k < -1$, a term with $k = -1$, and a power series without negative exponents. This power series of course has residue zero and may be disregarded. The term $\alpha_{-1}\pi^{-1}$ gives the residue α_{-1}, and it is easily seen that the differential

$$\alpha_{-1}\pi^{-1}d\pi$$

likewise gives the residue α_{-1} when expressed in terms of a new uniformizing variable τ. It therefore suffices to consider the terms

$$\pi^{-n}d\pi \qquad (n>1) \tag{19.65}$$

and show that they again give zero residue after a transformation

$$\pi = \tau + a_2\tau^2 + \cdots$$
$$d\pi = (1 + 2a_2\tau + \cdots)d\tau. \tag{19.66}$$

The transformation (19.66) can be carried out quite formally in the domain of power series in τ with coefficients from the integral polynomial domain of the indeterminates $a_2, a_3, \ldots$. The integral polynomial domain can be imbedded in the rational polynomial domain. The rational numbers form a field of characteristic zero even when the original coefficient field Δ has characteristic p.

Now the proof is easy. The differential (19.65) is the differential of the function

$$(-n+1)^{-1}\pi^{-n+1}.$$

If this function is expanded in terms of τ, then a rational power series

$$\varrho_{-n+1}\tau^{-n+1} + \cdots + \varrho_{-1}\tau^{-1} + \varrho_0 + \varrho_1\tau + \cdots$$

is obtained.

The differential of this power series is a power series, in which the term τ^{-1}. does not occur, times $d\tau$. The residue after the transformation is therefore zero, as was to be proved.

All these considerations continue to hold if w is not a function from the field, but rather any power series in π which has only finitely many terms with negative exponents.

Now let V be a vector in the sense of Section 19.4, that is, a system of power series $V_{\mathfrak{p}}$ for the individual places $\mathfrak{p}$. At any place $\mathfrak{p}$ we can expand the product

$$Vwdz$$

in a power series and determine the residue. If

$$V_{\mathfrak{p}} = \sum v_{\mathfrak{p}j}\pi^j \tag{19.67}$$

is the $\mathfrak{p}$-component of the vector V and if

$$w\frac{dz}{d\pi}d\pi = (\sum \alpha_{\mathfrak{p}k}\pi^k)d\pi \tag{19.68}$$

is the expansion of the differential, then the residue is

$$r_{\mathfrak{p}} = \sum_{j+k=-1} v_{\mathfrak{p}j}\alpha_{\mathfrak{p}k}. \tag{19.69}$$

Since the vector V as well as the differential wdz have only a finite number of poles, there is altogether only a finite number of nonzero residues $r_{\mathfrak{p}}$. We may therefore form the sum

$$\sum r_{\mathfrak{p}} = \sum_{\mathfrak{p}} \sum_{j+k=-1} v_{\mathfrak{p}j}\alpha_{\mathfrak{p}}.$$

This sum is precisely the scalar product of the vector V with the covector

$$\lambda = \{a_{\mathfrak{p}j}\} \tag{19.70}$$

in the sense of Section 19.4. We thus have the following result.

Each differential wdz uniquely defines a covector λ such that the scalar product $V\cdot\lambda$ is precisely the sum of the residues of the product $Vwdz$:

$$V\cdot\lambda = \sum r_{\mathfrak{p}} = \sum_{\mathfrak{p}} \sum_{j+k=-1} v_{\mathfrak{p}j}\alpha_{\mathfrak{p}k}. \tag{19.71}$$

We now ask what becomes of the scalar product if the vector V is replaced by a function v from the field K. The scalar product $V\cdot\lambda$ then becomes equal to the sum of the residues of the differential

$$vwdz = udz,$$

where u is again a function from the field.

Residue Theorem: *The sum of the residues of a differential udz is always zero.*

In classical function theory this theorem follows immediately from Cauchy's integral theorem. Hasse[4] has given a general proof which is valid in perfect fields of constants. A simplified version of Hasse's proof, due to P. Roquette, will be presented in Section 19.9.

[4] H. Hasse, "Theorie der Differentiale in Algebraischen Funktionenkörpern," *J. Reine u. Angew. Math.*, **172**, 55 (1934).

It follows from the residue theorem that the covector λ defined by a differential wdz is a differential in the sense of Weil.

In particular, dz defines a differential in the sense of Weil, which we shall likewise call dz. This differential is nonzero, since it is easy to find a vector V such that Vdz has a nonzero residue sum. If dz has order m at a place $\mathfrak{p}$, it is sufficient to choose the vector V so that its component $V_{\mathfrak{p}}$ is equal to π^{-m-1} and all other components are zero.

From the fact that the differential defined by dz is nonzero it follows by Section 19.6 that all differentials ω are obtained from this differential by multiplication with functions u. In other words, *all Weil differentials are classical differentials.*

19.9 PROOF OF THE RESIDUE THEOREM

I am indebted to a personal communication from P. Roquette for the following proof. The proof goes through for arbitrary perfect base fields, but it will here be presented only for base fields which are algebraically closed.

Let z again be chosen so that K is separable over $\Delta(z)$. We put $L = \Delta(z)$; K is then a finite separable extension of L, and we may put $\mathsf{K} = L(\vartheta)$.

Equating the coefficients of t^{n-1} and of $t^{\mathfrak{p}}$ on the right and left of (18.17), we obtain

$$N(\vartheta) = \Pi\, N(\vartheta_\nu) \tag{19.72}$$

$$S(\vartheta) = \sum S(\vartheta_\nu). \tag{19.73}$$

The same formulas hold not only for the generator ϑ, but also for any arbitrary element u of the field K. To see this, we first form the norm and trace of u in the field $L(u)$. Let us denote this norm and trace by $n(u)$ and $s(u)$; then what we have previously proved for ϑ is now true for u:

$$n(u) = \Pi\, n(u_\nu) \tag{19.74}$$

$$s(u) = \sum s(u_\nu). \tag{19.75}$$

We now apply formulas (6.21) and (6.22):

$$N(u) = n(u)^g \tag{19.76}$$

$$S(u) = g \cdot s(u), \tag{19.77}$$

where g is the degree of K over $L(u)$. We thus obtain quite generally

$$N(u) = \Pi\, N(u_\nu) \tag{19.78}$$

$$S(u) = \sum S(u_\nu). \tag{19.79}$$

Let us now recall how ϑ_ν and u_ν are defined. By Section 18.5, all valuations Φ_ν of K which are continuations of a given valuation φ of L are defined by imbeddings $\vartheta \to \vartheta_\nu$. Each such imbedding maps the field $\mathsf{K} = L(\vartheta)$ isomorphically into a complete field $\Omega_\nu = \Omega(\vartheta_\nu)$. This field Ω_ν is the complete extension field of K for the valuation Φ_ν.

Instead of valuations, we shall speak of places. Let us denote the places of the field K by $\mathfrak{p}$ and those of the field L by $\mathfrak{q}$. If a valuation of K belonging to the place $\mathfrak{p}$ is a continuation of a valuation of L belonging to the place $\mathfrak{q}$, then we call $\mathfrak{p}$ a *divisor* of $\mathfrak{q}$ and write $\mathfrak{p}/\mathfrak{q}$. Each $\mathfrak{q}$ has only finitely many divisors $\mathfrak{p}_\nu$, which correspond to the factors $F_\nu(t)$ in (18.17). To each $\mathfrak{p}_\nu$ there belongs a complete field Ω_ν which consists of power series with respect to a uniformizing variable Π. If we assign to each function u its power series u_ν, then we obtain the isomorphism $\vartheta \to \vartheta_\nu$ $u \to u_\nu$ mentioned above.

The norm $N(u_\nu)$ formed in Ω_ν over Ω is also called the *local norm* of u for the place $\mathfrak{p}$ and is denoted by $N_\mathfrak{p}(u)$; similarly for the trace. Formulas (19.78) and (19.79) may now be written as follows:

$$N(u) = \prod_{\mathfrak{p}/\mathfrak{q}} N_\mathfrak{p}(u) \tag{19.80}$$

$$S(u) = \sum_{\mathfrak{p}/\mathfrak{q}} S_\mathfrak{p}(u). \tag{19.81}$$

A vector V over K was defined as a system of components $V_\mathfrak{p}$, one for each place $\mathfrak{p}$. We can now define the trace SV of a vector V as a vector over L, by the formula

$$(SV)_\mathfrak{q} = \sum_{\mathfrak{p}/\mathfrak{q}} S_\mathfrak{p}(V_\mathfrak{p}). \tag{19.82}$$

The traces on the right-hand side are again to be formed in the complete fields $\Omega_\mathfrak{p} = \Omega_\nu$. In particular, if the vector belonging to a function u is taken for V, then SV is equal to $S(u)$ by (19.81).

The trace mapping $V \to SV$ is a linear mapping of the module $\mathfrak{Y}(\mathsf{K})$ of all vectors over K into the module $\mathfrak{Y}(L)$ of vectors over L. There is thus a dual mapping S^* of the module $\mathfrak{Y}^*(L)$ of covectors over L into the module $\mathfrak{Y}^*(\mathsf{K})$ of covectors over K which is defined as follows:

$$V \cdot S^*\varrho = SV \cdot \varrho \qquad \text{for all } V. \tag{19.83}$$

In particular, if ϱ is a Weil differential, that is, $v \cdot \varrho = 0$ for every v in L, then $S^*\varrho$ is also a Weil differential:

$$u \cdot S^*\varrho = Su \cdot \varrho = 0 \qquad \text{for all } u.$$

We prove the residue theorem first for the field $L = \Delta(z)$ of rational functions. Let vdz be a classical differential in L. The rational function

$$v = \frac{f(z)}{g(z)}$$

can be split into a polynomial and a fractional remainder in which the numerator has lower degree than the denominator:

$$\frac{f(z)}{g(z)} = q(z) + \frac{r(z)}{g(z)}.$$

The differential $q(z)dz$ has no residues. A uniformizing variable of the pole ∞ is $y = z^{-1}$, and we have

$$\begin{aligned} q(z)dz &= (\sum c_k z^k)dz \\ &= \sum (-c_k) y^{-k-2} dy, \end{aligned}$$

in which no term containing y^{-1} occurs.

According to Section 5.10, the remainder can be decomposed into partial fractions

$$\frac{r(z)}{g(z)} = \sum_a \{c_1(z-a)^{-1} + \cdots + c_s(z-a)^{-s}\}.$$

It therefore suffices to prove the residue theorem for a single partial fraction $c(z-a)^{-k}$. For $k > 1$ there are no residues. It therefore suffices to consider the differential

$$c(z-a)^{-1}dz.$$

This differential has a residue c at the place a and a residue $-c$ at the place ∞. The sum of the residues is therefore zero, and we are through.

The general case of the residue theorem will now be reduced to the case $L = \Delta(z)$ just completed by means of the dual trace mapping.

Let us denote the residue of the differential udz at the place $\mathfrak{p}$ by $\text{res}_\mathfrak{p}(udz)$ If V is a vector, we likewise denote the residue of the product Vdz at the place $\mathfrak{p}$ by $\text{res}_\mathfrak{p}(Vdz)$.

According to formula (19.71), the differential dz defines a covector which we denote by λ_{dz}. Thus, for each vector V,

$$V \cdot \lambda_{dz} = \sum_\mathfrak{p} \text{res}_\mathfrak{p} Vdz. \tag{19.84}$$

We call two covectors λ and μ *almost equal* if in the products $V \cdot \lambda$ and $V \cdot \mu$ defined by (19.31) the contributions of the individual places $\mathfrak{p}$ are always equal (for all V) except for a finite number of places $\mathfrak{p}'$.

Theorem 1: *There exists a Weil differential* μ_{dz} *which is almost equal to* λ_{dz}. *This* μ_{dz} *is uniquely determined by this property.*

Proof: The differential dz also defines a covector λ_0 in the field $L = \Delta(z)$ of rational functions. Since the residue theorem holds in L, λ_0 is a Weil differential. The dual trace $S^*(\lambda_0)$ is thus also a Weil differential which we denote by μ_{dz}:

$$\mu_{dz} = S^*(\lambda_0).$$

With each place $\mathfrak{p}$ of K there is associated a place $\mathfrak{q}$ of L. If the uniformizing variable $z-a$ or z^{-1} at the place $\mathfrak{q}$ is also a uniformizing variable for $\mathfrak{p}$, then the place $\mathfrak{p}$ is said to be *unramified* over L. We may then put $\Pi = z-a$ (or $\Pi = z^{-1}$). The complete field $\Omega_\mathfrak{p}$ belonging to the place $\mathfrak{p}$ is in this case simply equal to the field Ω of power series in $z-a$, and the residue of a power series at the place $\mathfrak{p}$ is equal to the residue at the place $\mathfrak{q}$.

Almost all places, that is, all but a finite number, are unramified over L.

Indeed, if $\mathsf{K} = L(\vartheta)$ and $F(z, t)$ is the irreducible polynomial in t with zero ϑ, then $F(z, t)$ may be assumed to be a polynomial in z and t. The discriminant of F is a polynomial in z which has only finitely many zeros. For all other values $z = a$, $F(a, t)$ decomposes into distinct prime factors:

$$F(a, t) = c(t-b_1)\cdots(t-b_n).$$

From this it follows by Hensel's lemma (Section 18.4) that $F(z, t)$ splits completely into linear factors in the complete field of power series in $z-a$. In the factorization (18.17) all the factors $F_\nu(t)$ are therefore linear, and all the fields $\Omega_\nu = \Omega(\vartheta_\nu)$ are equal to Ω. But then $z-a$ is a uniformizing variable for all the places belonging to these fields. All these places are therefore unramified.

If $\mathfrak{p}$ is unramified, then the place $\mathfrak{p}$ makes the same contribution to the covectors μ_{dz} and λ_{dz}. Indeed, if V is a vector which is different from zero only at this single place p, then it may be assumed that V is a power series in $z-a$ or z^{-1}. The local trace of V is then equal to V itself, and

$$\begin{aligned} V\cdot\mu_{dz} &= V\cdot S^*\lambda_0 = SV\cdot\lambda_0 = V\cdot\lambda_0 \\ &= \operatorname{res}_{\mathfrak{q}} Vdz = \operatorname{res}_{\mathfrak{p}} Vdz = V\cdot\lambda_{dz}. \end{aligned}$$

From this it follows that μ_{dz} is almost equal to λ_{dz}.

It remains to demonstrate the uniqueness of μ_{dz}. We actually shall prove something more general: *if two Weil differentials λ and μ are almost equal, then they are equal.*

We put $\varrho = \lambda-\mu$ and wish to show that $V\cdot\varrho$ is zero for any arbitrary vector V. By (19.31), the scalar product $V\cdot\varrho$ is a sum of contributions from the places $\mathfrak{p}$. We may hereby restrict our consideration to contributions from those $\mathfrak{p}$ belonging to a finite set M, since the contributions of the other $\mathfrak{p}$ to the covector ϱ are zero. For the $\mathfrak{p}$ in the set M we can approximate V by a function u of K in such a manner that the contributions of these $\mathfrak{p}$ to $(u-V)\cdot\varrho$ are zero (Section 19.1, Theorem I). It then follows that

$$(u-V)\cdot\varrho = 0,$$

and hence that $V\cdot\varrho = u\cdot\varrho = 0$, since ϱ is a Weil differential. This completes the proof of Theorem 1.

Now let y be another element such that K is separable over $\Delta(y)$. We wish to show that

$$\mu_{dz} = \frac{dz}{dy}\mu_{dy}. \tag{19.85}$$

Since both sides are Weil differentials, it suffices to show that the two sides are almost equal. Now μ_{dy} is almost equal to λ_{dy}, and μ_{dz} is almost equal to λ_{dz}. It is therefore sufficient to show that

$$\lambda_{dz} = \frac{dz}{dy}\lambda_{dy}. \tag{19.86}$$

This follows immediately from definition (19.84):

$$V\cdot\lambda_{dz} = \sum_{\mathfrak{p}} \mathrm{res}_{\mathfrak{p}}\, V dz = \sum_{\mathfrak{p}} \mathrm{res}_{\mathfrak{p}}\, V\frac{dz}{dy}\, dy$$

$$= V\frac{dz}{dy}\cdot\lambda_{dy} = V\cdot\frac{dz}{dy}\lambda_{dy}.$$

Finally, we shall show that

$$\lambda_{dz} = \mu_{dz}. \tag{19.87}$$

Let $\mathfrak{p}$ be a place, and let y be a uniformizing variable. In Section 19.8 it was shown that z is separable over $\Delta(y)$. Since K is separable over $\Delta(z)$ and $\Delta(z)$ is separable over $\Delta(y)$, it follows that K is separable over $\Delta(y)$. Furthermore, $\mathfrak{p}$ is unramified over $\Delta(y)$; the $\mathfrak{p}$-components of λ_{dy} and μ_{dy} are therefore equal:

$$(\lambda_{dy})_{\mathfrak{p}} = (\mu_{dy})_{\mathfrak{p}}.$$

From this it follows that

$$(\lambda_{dz})_{\mathfrak{p}} = \left(\frac{dz}{dy}\lambda_{dy}\right)_{\mathfrak{p}} = \left(\frac{dz}{dy}\mu_{dy}\right)_{\mathfrak{p}} = (\mu_{dz})_{\mathfrak{p}}.$$

Since this is true for any $\mathfrak{p}$, the assertion (19.87) follows.

We thus need no longer distinguish between λ_{dz} and μ_{dz}. Since μ_{dz} was a Weil divisor, λ_{dz} is also a Weil divisor, that is, the residue theorem holds.

Chapter 20

TOPOLOGICAL ALGEBRA

Topological algebra is the study of groups, rings, and skew fields which are also topological spaces and in which the algebraic operations are continuous in the sense of the topology. They are called topological groups, rings, and fields or briefly T-groups, T-rings, and skew T-fields.

20.1 THE CONCEPT OF A TOPOLOGICAL SPACE

A topological space is a set T in which certain subsets are distinguished as *open sets*. The open sets must have the following properties.

I. *The intersection of finitely many open sets is again open.*
II. *The union of any set of open sets is again open.*

Example 1: Let T be any ordered set which contains more than one element. An open interval in T is defined by $a<x<b$, $a<x$, or $x<b$. An open set is a set which with each element y contains an open interval which contains y.
Example 2: Let T be the field of complex numbers. A disk about a is defined by $|z-a|<\varepsilon$. An open set is a set which with each element a contains a disk about a.
Example 3: The same definition holds for any field with a valuation, but we must now write $\varphi(z-a)$ in place of $|z-a|$. Any field with a valuation is therefore a topological space.

It follows from I that the entire space T is open, since it is the intersection of any empty set of open sets. Similarly, it follows from II that the empty set is open, since it is the union of an empty set of open sets.

A subset M is said to be *closed* in T if its complement in T is open. The following rules, which are equivalent to I and II, hold for closed sets.

I′. *The union of a finite number of closed sets is again closed.*
II′. *The intersection of a set of closed sets is again closed.*

The elements of the set T are called *points* of the space T. An open set which contains the point p is called an *open neighborhood* of p. Any set which contains an open neighborhood of p is called a *neighborhood* of p and is denoted by $U(p)$.

A subset T' of a topological space T is again a topological space if the intersections of T' with the open sets of T are designated as *open sets in* T'. Properties I and II are obviously satisfied.

The *closure* $\bar{M}$ of a subset M of T is the intersection of all closed sets containing M.

Exercises

20.1. A point p belongs to the closure $\bar{M}$ if and only if every neighborhood of p contains a point of M.

20.2. Kuratowski defines a topological space as a set T in which to each subset M there corresponds a closure $\bar{M}$ with the following properties.

(*a*) The closure of $M \cup N$ is $\bar{M} \cup \bar{N}$.
(*b*) $\bar{M}$ contains M.
(*c*) The closure of $\bar{M}$ is $\bar{M}$.
(*d*) The closure of the empty set is empty.

He further makes the following definition: if $\bar{M} = M$, then M is said to be closed; if the complement of M in T is closed, then M is called open. Show that Kuratowski's definition of a topological space is equivalent to the definition given above.

Hint. From (*a*) it follows first of all that $M \subseteq N$ implies $\bar{M} \subseteq \bar{N}$. It then follows from (*a*), (*b*), and (*c*) that $\bar{M}$ is the intersection of all closed sets $\bar{N} = N$ which contain M. Rules I'. and II'. now follow. Conversely, (*a*), (*b*), (*c*), and (*d*) follow from I'. and II'.

A set M is called *dense in* T if the closure of M is equal to T or, what is the same thing, if every neighborhood of any point of T contains a point of M.

20.2 NEIGHBORHOOD BASES

A system of neighborhoods $U(p)$ of a point p form a *basis for the neighborhoods of* p if every neighborhood of p contains a neighborhood $U(p)$ of the system. For this it is sufficeint that every open neighborhood of p contain a neighborhood $U(p)$ of the system. For example, the open neighborhoods of p form a basis for the neighborhoods of p. In our Example 1 the open intervals containing p form a basis for the neighborhoods of p. In Example 2 the disks about a form a basis for the neighborhoods of a.

Topological spaces are frequently defined by first giving a basis for the neighborhoods of any point and then defining the open sets in terms of this basis just as was done in our examples. To each point p we thus first assign certain *basis sets* $U(p)$ which satisfy the following conditions.

U_1: *For each p there are basis sets U(p), and each such set contains p.*

U_2: *For any two basis sets U(p) and V(p) there is a set W(p) which is contained in both of them.*

The *open sets M* are defined in terms of these basis sets as those sets which with each of their points p contain an entire basis set $U(p)$. The open sets so defined clearly have properties I and II; we thus have a topological space. However, in order that the basis sets $U(p)$ be neighborhoods in the sense of this topology, they must satisfy a further condition. A sufficient condition is that the $U(p)$ themselves be open sets.

U_3: *If q lies in U(p), then U(p) contains a basis set V(q).*

The following weaker condition is necessary and sufficient.

U_3': *Each basis set U(p) contains a basis set V(p) such that for each point q of V(p) there is a basis set W(q) contained in U(p).*

If U_3' is satisfied, then a set U' can be defined in $U(p)$ which consists of all points q such that a basis set $W(q)$ is contained in $U(p)$. This set is open and contains p. Thus $U(p)$ contains an open neighborhood of p, that is, $U(p)$ is a neighborhood of p.

We no longer need the term "basis set": we shall subsequently always refer to the basis sets $U(p)$ as *basis neighborhoods*. The set of all basis neighborhoods of all points p is called a *neighborhood basis* or a *neighborhood system* of the topological space T.

The concept of a neighborhood system is due to Hausdorff. He used only open neighborhoods. Conditions U_1, U_2, and U_3 are the first three neighborhood axioms of Hausdorff. The fourth axiom is the Hausdorff separation axiom which we shall formulate in Section 20.4.

Example 4: In an n-dimensional vector space over the field of real numbers let a *cube* of edge length 2ε about the vector $(b_1, \ldots, b_n)$ be defined as the set of vectors $(a_1, \ldots, a_n)$ with the property

$$|a_i - b_i| < \varepsilon.$$

The cubes satisfy conditions U_1, U_2, and U_3. The vector space is therefore a topological space with the cubes as neighborhood basis.

A topological space is called *discrete* if all sets are open. The individual points then form a neighborhood system.

Exercises

20.3. In order that two systems of sets $U(p)$ and $V(p)$ define the same topological space, it is necessary and sufficient that each set $U(p)$ contain a $V(p)$ and each $V(p)$ contain a $U(p)$.

20.4. The topology of the vector space defined by the cubes is independent of the choice of basis of the vector space.

20.3 CONTINUITY. LIMITS

A function $p' = f(p)$ which maps a topological space T into a topological space T' is said to be *continuous at the point* p_0 if for every neighborhood U' of $f(p_0)$ in T' there is a neighborhood U of p_0 in T whose image is entirely contained in U'.

Similarly, a function $f(p, q)$ with arguments p and q in T_1 and T_2 and values in T_3, is called continuous at the point (p_0, q_0) if for every neighborhood W of $f(p_0, q_0)$ there are neighborhoods U and V of p_0 and q_0 such that $f(p, q)$ lies in W if p lies in U and q in V.

If a function is continuous at every point, then we speak of a *continuous function* or a *continuous mapping*. A mapping $p' = f(p)$ is continuous if and only if the pre-image of any open set U' in T' (that is, the set of elements of T whose images lie in U') is open in T.

A one-to-one continuous mapping of T onto T' which is continuous in both directions is called *topological*. A topological mapping takes open sets into open sets and closed sets into closed sets.

A sequence of points $\{p_\nu\}$ in a topological space T is called *convergent* with *limit* p if each neighborhood $U(p)$ contains all the points of the sequence after a certain index:

$$p_\nu \in U(p) \quad \text{for} \quad \nu \geqq k.$$

Only neighborhoods $U(p)$ of a neighborhood basis of p need hereby be considered, since each neighborhood contains such a basis neighborhood.

Exercises

20.5. A continuous mapping preserves the limit relation.

20.6. A continuous function of a continuous function is continuous.

20.4 SEPARATION AND COUNTABILITY AXIOMS

In addition to axioms I and II, the most important topological spaces satisfy the following *first separation axiom*.

T_1: *If $p \neq q$, then there exists a neighborhood of p not containing q.*

A space with property T_1 is called a T_1-space. An equivalent formulation is the following.

The closure of a single point consists only of the point itself.

The *second* or *Hausdorff separation axiom* is stronger than the T_1 property.

T_2: *If $p \neq q$, then there exist neighborhoods $U(p)$ and $U(q)$ which are disjoint.*

If T_2 is satisfied, the space is called a *Hausdorff space*.

The *first countability axiom* states the following.

A_1: *Every point p has a countable neighborhood basis.*

We shall not need the stronger second countability axiom.

The topological spaces which are important for our purposes all satisfy the first separation and countability axioms. In the case of topological groups, and hence also in the case of topological rings and skew fields (which are additive groups, among other things), the second separation axiom follows as a consequence of the first.

In the introduction to topology given here only the necessary basic concepts have been mentioned. The reader who wishes to learn more about topology might first study the excellent textbook by Alexandroff and Hopf, *Topologie I* (Springer, Grundlehren, Band XLV, 1935) and then consult the more recent literature.

Exercises

20.7. In a Hausdorff space a sequence of points $\{p_\nu\}$ can have only one limit.

20.8. If A_1 is satisfied, then the closure of a set M consists of all limits of convergent sequences $\{p_\nu\}$ in M. The set M is closed if all these limits lie in M.

20.5 TOPOLOGICAL GROUPS

A *topological group* (or briefly a T-group) is a topological space that is at the same time a group in which xy is a continuous function of x and y and x^{-1} is a continuous function of x. Thus, in addition to the four group axioms and the two basic properties of open sets, the following two conditions are also required.

TG_1: *For every neighborhood $U(ab)$ of a product ab there exist neighborhoods $V(a)$ and $W(b)$ such that the product $V(a)W(b)$ is contained in $U(ab)$.*

TG_2: *For every neighborhood $U(a^{-1})$ there exists a neighborhood $V(a)$ such that $V(a)^{-1}$ is contained in $U(a^{-1})$.*

Here M^{-1} denotes the set of inverses x^{-1} of the elements x of M.

It clearly suffices to require TG_1 and TG_2 for the neighborhoods U of a neighborhood basis; also $V(a)$ and $W(b)$ may always be taken to be basis neighborhoods.

Examples of topological groups are the following:

(*a*) The additive group of real or complex numbers;
(*b*) The real n-dimensional vector space (Section 20.2, Example 4);
(*c*) The multiplicative group of real or complex nonzero numbers.

Every group G becomes a *discrete topological group* if the discrete topology is adopted, that is, if all sets in G are open.

For further examples see Exercise 20.10 and Section 20.7, Example 5.

Now TG_1 and TG_2 imply the following.

TG': *For every neighborhood* $U(a^{-1}b)$ *there exist neighborhoods* $V(a)$ *and* $W(b)$ *such that* $V(a)^{-1}W(b)$ *is contained in* $U(a^{-1}b)$.

TG'': *For every neighborhood* $U(ab^{-1})$ *there exist neighborhoods* $V'(a)$ *and* $W'(b)$ *such that* $V'(a)W'(b)^{-1}$ *is contained in* $U(ab^{-1})$.

Exercise

20.9. Show that either of conditions TG' or TG'' alone can replace the two conditions TG_1 and TG_2.

We now prove the following.

A T_1*-group is a* T_2*-group.*

Proof: Let $a \neq b$; then $a^{-1}b \neq e$. By T_1 there exists a neighborhood $U(a^{-1}b)$ which does not contain e. By TG' there exists a $V(a)$ and $W(b)$ such that $V(a)^{-1}W(b)$ lies in $U(a^{-1}b)$ and thus does not contain e. Then $V(a)$ and $W(b)$ are disjoint. This proves T_2.

By the same method we can prove the following.

If in a T-group there exists a neighborhood of p which does not contain q, then there exist two disjoint neighborhoods $U(p)$ *and* $U(q)$. *Thus there also exists a neighborhood* $U(q)$ *which does not contain p.* In this case p and q are said to be *separable.* The points q which are not separable from p form the closure of the set $\{p\}$.

Two T-groups G and H are called *topologically isomorphic* if there exists an isomorphism which is also a topological mapping of G onto H.

20.6 NEIGHBORHOODS OF THE IDENTITY

If a neighborhood basis for e is given, then all the neighborhoods of e are known: they are the sets $U(e)$ which contain at least one basis neighborhood. Furthermore, the neighborhoods of the other points are also known, for if $U(e)$ is a neighborhood of e, then $aU(e)$ is a neighborhood of a, and all neighborhoods of a can be obtained in this manner. We may call $aU(e)$ a neighborhood of e translated to a.

We thus see that the topology of a T-group is completely determined if a basis for the neighborhoods of e is known. We denote the neighborhoods of such a basis by U (or also $V, W, \ldots$).

What properties must the sets U have in order that G with the translated neighborhoods $U(a) = aU(e)$ become a topological group?

The following properties are in any case necessary.

E_1: *Each* U *contains* e (this follows from U_1, Section 20.2).

E_2: *For each* U *there exists a* V *such that* $V \cdot V$ *is contained in* U.

E_3: *For each* U *there exists a* V *such that* V^{-1} *is contained in* U (this follows from TG_2, Section 20.5).

E_4: *Every transformed set aUa^{-1} contains a V.*

E_5: *Every intersection $U \cap V$ contains a W* (this follows from U_2, Section 20.2).

Proof of E_2: For every U there exist by TG_1 a V' and W' such that $V'W'$ is contained in U. By U_2 a V is contained in the intersection $V' \cap W'$.

Proof of E_4: Since $a^{-1}xa$ is a continuous function of x, there exists for U a V such that $a^{-1}Va$ is contained in U; hence V is contained in aUa^{-1}.

Now suppose that in a group G a system of sets U is given which has properties E_1 through E_5. We form the translated sets aU and take these as basis neighborhoods for the point a. These basis neighborhoods clearly have properties U_1 and U_2 (Section 20.2). We shall show that they also have the property U_3'.

Let $U(a) = aU$. By E_2 there exists a V such that $V \cdot V$ is contained in U. If now x is a point of aV, then xV is contained in aVV and therefore also in aU. This proves U_3'.

We must now prove TG_1 and TG_2 (Section 20.5).

Let a neighborhood abU be given. By E_2 there exists a V such that $V \cdot V$ is contained in U. By E_4 there is a W in bVb^{-1}. Now

$$aW \cdot bV \subseteqq a \cdot bVb^{-1} \cdot bV = abVV \subseteqq abU,$$

wherewith TG_1 is proved.

Let a neighborhood $a^{-1}U$ be given. By E_3 there exists a V such that V^{-1} is contained in U. There is a W in $a^{-1}Va$ by E_4.

Now $aW \subseteqq Va$, and hence

$$(aW)^{-1} \subseteqq (Va)^{-1} = a^{-1}V^{-1} \subseteqq a^{-1}U,$$

which proves TG_2.

In order to make a group a T-group, we must therefore only prescribe a neighborhood basis of the identity element and prove E_1 through E_5.

Conditions E_2 and E_3 may be combined to a single condition.

E_{2+3}: *For every U there exists a V such that $V^{-1}V \subseteqq U$.*

Now E_4 may be omitted in the case of Abelian groups. If they are written in additive notation, then the neighborhoods of zero have only three conditions to satisfy.

1: *Every U contains zero.*

2: *For every U there exists a V such that $V - V \subseteqq U$.*

3: *Every intersection $U \cap V$ contains a W.*

In order that a T-group defined by neighborhoods of the identity be a T_1-group, the following separation axiom must be satisfied.

E_6: *For every $a \neq e$ there exists a U which does not contain a.*

Now E_1 and E_6 may be combined to a single condition, as follows.

E_{1+6}: *The intersection of all U consists of the identity element alone.*

The corresponding condition for additive groups is the following.

The intersection of all U consists of zero alone.

If G is not a T_1-group, then there are other elements p in addition to e in all the neighborhoods of e; these elements are thus not separable from e. They clearly form a normal subgroup N of G. By Section 20.5, N is the closure of the set $\{e\}$ and hence N is closed. The factor group G/N is a T_1-group.

Exercise

20.10. In a group G let a sequence of nested normal subgroups

$$H_1 \supset H_2 \supset \cdots$$

be given. If these normal subgroups are taken as basis neighborhoods of the identity, then properties E_1–E_5 are satisfied and G becomes a T-group. But E_6 is satisfied only if the intersection of all the H_i consists of the identity alone.

20.7 SUBGROUPS AND FACTOR GROUPS

Every subgroup of a T-group is again a T-group. The closed subgroups are especially important. We first prove the following.

An open subgroup is also closed.

Proof: Let the subgroup H be open in G. The cosets aH are then likewise open in G. The union of all the cosets excepting H is thus again open. This union is the complement of H in G; hence H is closed.

Example 5: Let R be the ring of all matrices with n rows and n columns over the field of real numbers. The units in R are those matrices A which have inverses A^{-1}. These units form a group G. If a cubic neighborhood of a matrix A is defined as the set of matrices B such that

$$|b_{ik} - a_{ik}| < \varepsilon$$

(cf. Section 20.2, Example 4), then R becomes an additive and G a multiplicative topological group. Let us consider the subgroup of matrices A in G whose determinants are positive. This subgroup is open in G; it is therefore also closed.

Now let H be a normal subgroup of G. It will at first not be required that H be closed. We form the factor group

$$G/H = \bar{G}.$$

Under the homomorphism $a \to \bar{a}$ of G onto $\bar{G}$, the basis neighborhoods U of e go into certain subsets $\bar{U}$ of $\bar{G}$ which trivially satisfy conditions E_1–E_5. The sets $\bar{U}$ therefore define a topology in $\bar{G}$. The mapping $a \to \bar{a}$ is continuous in this topology; this follows directly from the definition of continuity. We thus have the following.

Every factor group G/H of a T-group is a T-group, and the mapping $a \to \bar{a}$ is continuous.

We now ask under what condition the factor group satisfies the separation axiom T_1. The answer is as follows.

If the normal subgroup H is closed in G, then G/H is a T_1-group and conversely.
Proof: Let H be closed in G. Every coset aH is then also closed in G. If $\bar{a} \neq \bar{e}$, then e does not lie in aH; that is, e is contained in the open complement of aH. There thus exists a neighborhood U of e which does not intersect aH. The image $\bar{U}$ in $\bar{G}$ then does not contain $\bar{a}$. Thus $\bar{G}$ satisfies condition E_6; therefore $\bar{G}$ is a T_1-group.

Suppose now that $\bar{G}$ is a T_1-group. The set of $\bar{a} \neq \bar{e}$ is then an open set in $\bar{G}$. Since the mapping $a \to \bar{a}$ is continuous, the pre-image of this open set is again open. But this pre-image is precisely the complement of H. Hence H is closed in G.

Exercise

20.11. Let H be a subgroup and N be a normal subgroup of G. If N is closed in G, then the intersection $D = N \cap H$ is closed in H, and the natural isomorphism of H/D to NH/N is continuous.

20.8 T-RINGS AND SKEW T-FIELDS

A *topological ring* (or briefly a T-ring) is a topological space which is also a ring and in which $x+y$, $-x$, and xy are continuous functions. Instead of this, we may require that $x-y$ and xy be continuous functions of x and y. Thus we have the following.

***TR*$_1$:** *For every neighborhood $U(a-b)$ there exist $V(a)$ and $W(b)$ such that all differences of elements of $V(a)$ and $W(b)$ lie in $U(a-b)$.*

***TR*$_2$:** *For every neighborhood $U(ab)$ there exist $V(a)$ and $W(b)$ such that all products of elements of $V(a)$ and $W(b)$ lie in $U(ab)$.*

In the case of a *skew T-field* it is required in addition that x^{-1} be a continuous function of x, as follows.

***TS*:** *For every neighborhood $U(a^{-1})$ there exists a $V(a)$ whose inverse is contained in $U(a^{-1})$.*

If *TS* is satisfied, the ring topology is said to be a *field topology*.

Commutative skew T-fields are naturally called T-fields.

A ring is an Abelian group with respect to addition. In order to define a topology in this group, it is sufficient by Section 20.5 to define basis neighborhoods $U, V, \ldots$ of zero which satisfy condition 1, 2, and 3 (Section 20.6). In order that multiplication also be continuous, the following condition must be satisfied.

4: *For a, b, and U there exist V and W such that*

$$(a+V)(b+W) \subseteq ab+U.$$

A topological skew field must, moreover, satisfy the following condition which is equivalent to *TS*.

For $a \neq 0$ and U there exists a V such that

$$(a+V)^{-1} \subseteqq a^{-1}+U. \tag{20.1}$$

Putting $aU = U'$ and $Va^{-1} = V'$, so that $U = a^{-1}U'$ and $V = V'a$, it follows from (20.1) that

$$a^{-1}(1+V')^{-1} \subseteqq a^{-1}(1+U')$$

or

$$(1+V')^{-1} \subseteqq 1+U'. \tag{20.2}$$

It therefore suffices to require (20.1) for $a = 1$. The axiom *TS* is thus equivalent to the following condition.

5: *For every neighborhood U of zero there exists a neighborhood V of zero such that*

$$(1+V)^{-1} \subseteqq 1+U. \tag{20.3}$$

Examples of *T*-fields are all fields with valuations, in particular the fields of real, complex, and *p*-adic numbers as well as their subfields.

The ring of real n by n matrices is a *T*-ring. A basis neighborhood of zero consists in this case of the matrices whose elements are less than ε in absolute value.

Further examples may be obtained by considering a sequence of nested two-sided ideals.

$$\mathfrak{g}_1 \supseteqq \mathfrak{g}_2 \supseteqq \cdots$$

in a ring $\mathfrak{o}$ and taking these ideals as basis neighborhoods of zero. Conditions 1 through 4 are then satisfied. A T_1-ring is obtained if the intersection of all the $\mathfrak{g}_\nu$ consists of zero alone.

The ring topology defined by the sequence $\{\mathfrak{g}_\nu\}$ is called the *$\{\mathfrak{g}_\nu\}$-adic topology*. In particular, if the $\mathfrak{g}_\nu$ are powers of a prime ideal $\mathfrak{p}$ in a commutative ring $\mathfrak{o}$,

$$\mathfrak{p} \supseteqq \mathfrak{p}^2 \supseteqq \mathfrak{p}^3 \cdots,$$

then we speak of a *$\mathfrak{p}$-adic topology*. We shall later see that in many important cases the intersection of all the powers of $\mathfrak{p}$ is the null ideal. In all these cases, then, the separation axiom T_1 is satisfied.

In Section 18.1 the sequence of powers $\mathfrak{p}^\nu$ of a prime ideal $\mathfrak{p}$ was employed under very restrictive conditions to construct a valuation of the ring $\mathfrak{o}$. If only a ring topology rather than a valuation is required, then these restrictive conditions are unnecessary.

Exercises

20.12. Condition 4 can be replaced by three subconditions:

(*a*) For a and U there exists a V such that $aV \subseteqq U$.

(*b*) For b and U there exists a V such that $Vb \subseteqq U$.
(*c*) For U there exists a V such that $VV \subseteqq U$.

20.13. In the skew field of quaternions over the field of real numbers (Section 13.2, Example 2) the neighborhoods of zero can be defined as follows: U_ε consists of the quaternions $a+bj+ck+dl$ whose norm

$$(a-bj-ck-dl)(a+bj+ck+dl) = a^2+b^2+c^2+d^2$$

is less than ε. Show that the quaternion field with this topology is a skew T_1-field.

20.9 GROUP COMPLETION BY MEANS OF FUNDAMENTAL SEQUENCES

In Section 18.2, for a field with a valuation we constructed an extension field in which the Cauchy convergence theorem holds. Fundamental sequences $\{a_\nu\}$ provided the means of doing this; they were characterized by the fact that for sufficiently large μ and ν, $a_\mu - a_\nu$ belongs to any neighborhood of zero. An analogous construction will now be carried out for T-groups following D. van Dantzig.[1]

A sequence $\{x_\nu\}$ in a T-group is called a *fundamental* or *Cauchy sequence* if the quotients $x_\mu^{-1}x_\nu$ for $\mu \geqq m$ and $\nu \geqq n$ lie in any given neighborhood of the identity element.

A T-group is called *weakly complete* if every fundamental sequence has a limit in the group.

We set ourselves the task of extending any T-group satisfying axioms T_1 and A_1 to a weakly complete T-group.

I am indebted to H. R. Fischer for the proof of the following lemma.

Lemma: *Let $\{x_\nu\}$ be a fundamental sequence. Then for any U there exist a V and an m such that*

$$x_\mu^{-1}Vx_\mu \subseteqq U \qquad \text{for } \mu \geqq m. \tag{20.4}$$

Proof: Choose W so that $WWW \subseteqq U$ and m so that

$$x_\mu^{-1}x_\nu \in W \qquad \text{for } \mu \geqq m,\ \nu \geqq m.$$

Then, in particular, for $\mu \geqq m$, $x_\mu^{-1}x_m$ and $x_m^{-1}x_\mu$ are contained in W. By E_4 we can choose a V in $x_m W x_m^{-1}$. Then

$$x_\mu^{-1}Vx_\mu \subseteqq x_\mu^{-1}x_m W x_m^{-1}x_\mu \subseteqq WWW \subseteqq U \qquad \text{for } \mu \geqq m.$$

From this lemma we obtain the following corollary.

Corollary I: *If $\{x_\mu\}$ and $\{y_\mu\}$ are fundamental sequences, then $\{x_\mu y_\mu\}$ is also a fundamental sequence.*

[1]D. van Dantzig, "Zur Topologischen Algebra I: Komplettierungstheorie," *Math. Ann.* **107**, 587 (1933).

Proof: We have

$$(x_\mu y_\mu)^{-1}x_\nu y_\nu = y_\mu^{-1}(x_\mu^{-1}x_\nu)y_\mu \cdot y_\mu^{-1}y_\nu.$$

In the product on the right both factors are contained in arbitrarily small neighborhoods of e: the first factor by the lemma and the second factor by the definition of a fundamental sequence. The product is therefore also contained in any given neighborhood U of e. Then $\{x_\mu y_\mu\}$ is called the *product* of the fundamental sequences $\{x_\mu\}$ and $\{y_\mu\}$.

Another corollary of the lemma is the following.

Corollary II: *If $\{x_\mu\}$ is a fundamental sequence and $\{y_\mu\}$ converges to e, then*

$$\{x_\mu^{-1}y_\mu x_\mu\}$$

also converges to e.

Proof: By the lemma, $x_\mu^{-1}Vx_\mu \subseteqq U$ for sufficiently large μ, and y_μ lies in V for sufficiently large μ; $x_\mu^{-1}y_\mu x_\mu$ therefore lies in U for sufficiently large μ.

The following *completion axiom* is necessary in order that G be extendable to a complete topological group.

TG_3: *If $\{x_\mu\}$ is a fundamental sequence, then $\{x_\mu^{-1}\}$ is also a fundamental sequence.*

In an Abelian group TG_3 is automatically satisfied, for if $x_\mu^{-1}x_\nu$ lies in U, then

$$x_\nu x_\mu^{-1} = (x_\nu^{-1})^{-1}x_\mu^{-1}$$

also lies in U. In the general case, however, TG_3 is not a consequence of the other axioms.

From Corollary I and TG_3 it follows immediately that the fundamental sequences form a group F. The identity element of the group F is the sequence $\{e\}$.

We now make the group F into a topological group by defining the neighborhoods $\bar{U}$ of the identity element $\{e\}$ as follows: U consists of all fundamental sequences $\{x_\nu\}$ whose elements x_ν lie in $\bar{U}$ for sufficiently large ν.

These neighborhoods $\bar{U}$ satisfy conditions E_1–E_5 (Section 20.6). This is obvious in the case of E_1–E_3 and E_5, whereas E_4 is precisely the lemma above: if $\{x_\mu\}$ is a fundamental sequence, then there exists a V such that

$$x_\mu^{-1}Vx_\mu \subseteqq U \quad \text{or} \quad V \subseteqq x_\mu U x_\mu^{-1}$$

for sufficiently large μ.

Thus F is a topological group. In this group the sequences which converge to e form a subgroup, which by Corollary II is a normal subgroup N. We now prove that N is closed in F.

If a fundamental sequence $\{x_\mu\}$ does not belong to N, and thus does not converge to e, then there exists a neighborhood U which does not contain almost all elements of the sequence. By E_2 and E_3 there exists a V such that

$$VV^{-1} \subseteqq U.$$

This V defines a neighborhood $\bar{V}$ in F consisting of all fundamental sequences

$\{y_\mu\}$ almost all of whose elements y_μ lie in V. We now assert that the neighborhood $\{x_\mu\}\bar{V}$ of $\{x_\mu\}$ in F is contained in the complement of N in F.

Suppose that $\{x_\mu\}\bar{V}$ had a fundamental sequence

$$\{x_\mu\}\,\{y_\mu\} = \{x_\mu y_\mu\} = \{z_\mu\}$$

in common with N, where the y_μ almost all lie in V and $\{z_\mu\}$ converges to e. Almost all the z_μ then also lie in V; hence the

$$x_\mu = z_\mu y_\mu^{-1}$$

almost all lie in VV^{-1} and therefore in U, contrary to the definition of U. Hence, $\{x_\mu\}\bar{V}$ has no element in common with N.

The complement of N in F is thus an open set; that is, N is closed in F. From this it follows by Section 20.7 that F/N is a T_1-group.

In F the constant fundamental sequences $\{a\}$ form a subgroup G' which is topologically isomorphic to the given group G. Because of the separation axiom T_1, this subgroup has only the element $\{e\}$ in common with N. We may identify the constant sequences $\{a\}$ with the elements a and hence G with G'. If we now form cosets modulo N, then G' goes into a factor group G'' which is a subgroup of F/N and as such is again a T-group. This T-group is topologically isomorphic to G' and hence also to G; it may therefore be identified with G.

We now put $F/N = \tilde{G}$. Therefore G is imbedded in a T_1-group $\tilde{G}$. We now prove the following.

Corollary III: *If the fundamental sequence $\{x_\mu\}$ defines the element $\tilde{x}$ of $\tilde{G}$ then*

$$\lim x_\mu = \tilde{x}. \tag{20.5}$$

Proof: Let us denote the fundamental sequence $\{x_\mu\}$ considered as an element of F by $\bar{x}$. Under the homomorphism which maps F onto $F/N = \tilde{G}$, $\bar{x}$ goes into $\tilde{x}$. The mapping is continuous; (20.5) will therefore be proved as soon as the corresponding relation is demonstrated in F:

$$\lim x_\mu = \bar{x} \text{ in } F. \tag{20.6}$$

Relation (20.6) means

$$\bar{x}^{-1}x_\mu \text{ is in } \bar{U} \text{ for sufficiently large } \mu$$

or, by the definition of $\bar{U}$,

$$x_\nu^{-1}x_\mu \text{ is in } U \text{ for sufficiently large } \mu \text{ and } \nu.$$

But this is clear, since $\{x_\mu\}$ is a fundamental sequence.

We are now ready to prove the main theorem.

Corollary IV: *$\tilde{G}$ is weakly complete.*

The proof is entirely similar to the proof given in Section 11.2 for the real numbers. Let $\{\tilde{x}_1, \tilde{x}_2, \ldots\}$ be a sequence of elements of $\tilde{G}$ which satisfy the Cauchy convergence criterion

$$\tilde{x}_\mu^{-1}\tilde{x}_\nu \in \tilde{V} \quad \text{for} \quad \mu \geqq m \quad \text{and} \quad \nu \geqq m.$$

We choose a countable basis $\{U_1, U_2, \ldots\}$ for the neighborhoods of e in G. For each U_λ we choose a V_λ such that

$$V_\lambda^{-1} V_\lambda V_\lambda \subseteqq U_\lambda.$$

We may, moreover, assume that

$$V_1 \supseteqq V_2 \supseteqq V_3 \cdots.$$

The neighborhoods V_λ define neighborhoods $\bar{V}_\lambda$ in F, and these in turn define neighborhoods $\tilde{V}_\lambda$ in $\tilde{G}$. Each $\tilde{x}_\mu$ is by Corollary III a limit of a sequence of elements of G; corresponding to $\tilde{x}_\mu$ we can therefore choose a y_μ in G such that

$$\tilde{x}_\mu^{-1} y_\mu \in \tilde{V}_\mu.$$

We now show that the y_μ form a fundamental sequence. We have

$$y_\mu^{-1} y_\nu = (y_\mu^{-1} \tilde{x}_\mu)(\tilde{x}_\mu^{-1} \tilde{x}_\nu)(\tilde{x}_\nu^{-1} y_\nu) \in \tilde{V}_\mu^{-1}(\tilde{x}_\mu^{-1} \tilde{x}_\nu)\tilde{V}_\nu. \tag{20.7}$$

For each λ there exists an $m \geqq \lambda$ such that

$$\tilde{x}_\mu^{-1} \tilde{x}_\nu \in \tilde{V}_\lambda \qquad \text{for} \quad \mu \geqq m,\ \nu \geqq m.$$

From (20.7) it now follows that, for $\mu \geqq m \geqq \lambda$ and $\nu \geqq m \geqq \lambda$,

$$y_\mu^{-1} y_\nu \in \tilde{V}_\mu^{-1} \tilde{V}_\lambda \tilde{V}_\nu \subseteqq \tilde{V}_\lambda^{-1} \tilde{V}_\lambda \tilde{V}_\lambda \subseteqq \tilde{U}_\lambda,$$

that is, $y_\mu^{-1} y_\nu \in U_\lambda$. The y_μ therefore form a fundamental sequence in G. This fundamental sequence defines an element y of $\tilde{G}$ and has the limit $\tilde{y}$ by Corollary III. The $\tilde{x}_\mu$ have precisely the same limit, for

$$\tilde{y}^{-1} \tilde{x}_\mu = (\tilde{y}^{-1} y_\mu)(y_\mu^{-1} \tilde{x}_\mu),$$

and for sufficiently large μ both factors lie in arbitrarily small neighborhoods of e. The sequence $\{\tilde{x}_\mu\}$ therefore has a limit in $\tilde{G}$, and the group is weakly complete.

The T_1-groups which do not satisfy the countability axiom A_1 can likewise be completed under suitable conditions. In this case, however, following Bourbaki (*Eléments de Mathématique*, Book III, Chapter III; Actualités scient.), Cauchy filters must be used in place of fundamental sequences both to define the concept of "completeness" and to construct the complete extension. This will now be examined in more detail.

Exercise

20.14. If G satisfies axioms T_1 and A_1, then every weakly complete subgroup H is closed in G. (Use Exercise 20.8.)

20.10 FILTERS

Let M be a fixed set. Subsets of M will be denoted by $A, B, \ldots$. Collections of such sets will be denoted by capital German letters $\mathfrak{F}, \mathfrak{G}, \ldots$.

A collection of sets $\mathfrak{F}$ is called a *filter* if it has the following properties.

F_1: *Every set A which contains a set of $\mathfrak{F}$ belongs to $\mathfrak{F}$.*

F_2: *Any intersection of finitely many sets of $\mathfrak{F}$ belongs to $\mathfrak{F}$.*

F_3: *The empty set does not belong to $\mathfrak{F}$.*

From F_2 it follows that M itself, as the intersection of an empty collection of subsets of M, also belongs to $\mathfrak{F}$. Instead of F_2 we may require the following.

F_2': *The intersection of two sets of $\mathfrak{F}$ belongs to $\mathfrak{F}$.*

F_2'': *M belongs to $\mathfrak{F}$.*

Example 1: The neighborhoods of a point p in a topological space M form a filter, the *neighborhood filter* of the point p.

A nonempty collection of sets $\mathfrak{B}$ is called a *filter basis* if it has the following properties.

B_1: *The intersection of two sets of $\mathfrak{B}$ contains a set of $\mathfrak{B}$.*

B_2: *The empty set does not belong to $\mathfrak{B}$.*

If these properties are satisfied, then we can form a filter $\mathfrak{F}$ which consists of the subsets of M which contain at least one set of $\mathfrak{B}$. This filter is called the *filter generated by* $\mathfrak{B}$, and $\mathfrak{B}$ is called a *basis* of the filter $\mathfrak{F}$.

Example 2: The basis neighborhoods of a point p in a topological space M form a basis for the neighborhood filter of p.

Example 3: Let a sequence of elements of M be given:

$$a_1 a_2 a_3 \cdots.$$

Omitting finitely many terms of the sequence, we form a set A from the remaining terms. These sets A form a filter basis $\mathfrak{B}$. The filter generated by $\mathfrak{B}$ consists of the subsets of M which contain almost all terms of the sequence.

Henceforth M shall be a topological group G. Let V be a neighborhood of the identity element e. A set A is called *small of order* V if all quotients $x^{-1}y$ of elements in A lie in V:

$$x^{-1}y \in V, \text{ and hence } y \in xV, \text{ for } x \text{ and } y \text{ in } A.$$

We say that a set $\mathfrak{B}$ *contains arbitrarily small sets* if for every neighborhood V of e there is a set A in $\mathfrak{B}$ which is small of order V.

A *Cauchy filter* is a filter which contains arbitrarily small sets.

A Cauchy filter basis $\mathfrak{B}$ in G is a basis containing arbitrarily small sets. The filter generated by a Cauchy filter basis is a Cauchy filter.

A filter basis $\mathfrak{B}$ *converges* to a if in every neighborhood of a there is a set A of $\mathfrak{B}$. We then write

$$\lim \mathfrak{B} = a.$$

In a T_1-group the limit a is uniquely determined.

In Section 20.9 a T_1-group was called weakly complete if every Cauchy sequence had a limit in the group. However, this concept is really useful only if the group satisfies the first countability axiom. In the general case we need a stronger concept. We define: G is *strongly complete* if every Cauchy filter in G converges.

The present concept of completeness is indeed stronger than that used previously: *every strongly complete T-group is weakly complete.*

Proof: Let G be strongly complete, and let $\{x_\nu\}$ be a fundamental sequence in G. The sets A obtained by omitting finitely many terms of the sequence are arbitrarily small by the definition of a Cauchy sequence. These sets form a Cauchy filter basis $\mathfrak{B}$ which generates a Cauchy filter $\mathfrak{F}$. This filter has a limit a in G. Every neighborhood of a contains almost all terms x_ν of the sequence; the sequence therefore has limit a in G.

Following Bourbaki, we now prove the next statement.

If a set D is dense in a T-group G and if every Cauchy filter basis in D converges to a limit in G, the G is strongly complete.

Proof: Let $\mathfrak{F}$ be a Cauchy filter in G. We must show that $\mathfrak{F}$ converges.

For every neighborhood V of e and every set A of the filter $\mathfrak{F}$ we form a product set AV. These sets form a filter basis $\mathfrak{B}$, for if AV and $A'V'$ are two such sets, then the set

$$(A \cap A')(V \cap V')$$

is contained in the intersection of AV and $A'V'$. We now show that $\mathfrak{B}$ is a Cauchy filter basis.

Let U be a neighborhood of e, and let V be a neighborhood such that $V^{-1}VV$ is contained in U. We choose A small of order V. For any two elements av and $a'v'$ of AV we then have

$$(av)^{-1}a'v' = v^{-1}(a^{-1}a')v' \in V^{-1}VV \subseteq U,$$

and hence AV is small of order U. Therefore $\mathfrak{B}$ is a Cauchy filter basis.

The intersections of the product sets AV with D are nonempty, since A contains at least one element a, and in every neighborhood aV of a there is at least one point of D. The intersections $AV \cap D$ therefore form a Cauchy filter basis on D. This has a limit b in G by hypothesis. In every neighborhood of b there is a set AV and hence also a subset $Ae = A$. Therefore $\mathfrak{F}$ converges to b, and this completes the proof.

Exercises

20.15. If a filter $\mathfrak{F}$ converges to a, then $\mathfrak{F}$ is a Cauchy filter.

20.16. If a filter basis $\mathfrak{B}$ converges to a, then the filter $\mathfrak{F}$ generated by $\mathfrak{B}$ likewise converges to a, and conversely.

20.17. A T-group which is weakly complete and satisfies the first countability axiom is also strongly complete.

Hint. Let $V_1, V_2, \ldots$ be a countable neighborhood basis of e, and let $\mathfrak{F}$ be a Cauchy filter. For each n there is a set A_n in the filter which is small of order V_n. Form the intersections

$$D_n = A_1 \cap A_2 \cap \cdots \cap A_n$$

and choose x_n in D_n. Then $\{x_n\}$ is a fundamental sequence whose limit is also the limit of the filter $\mathfrak{F}$.

20.11 GROUP COMPLETION BY MEANS OF CAUCHY FILTERS

To prepare the way for strong completion, we first prove a *lemma* which is the analogue of the lemma of Section 20.9 and is proved in a similar manner.

Let $\mathfrak{F}$ be a Cauchy filter. Then for every neighborhood U of e a neighborhood V and an A in F exist such that

$$x^{-1}Vx \subseteq U \qquad \text{for all } x \text{ in } A.$$

Proof: Choose W so that $WWW \subseteq U$. Choose A so that

$$x^{-1}y \in W \qquad \text{for } x \text{ and } y \text{ in } A.$$

Choose a fixed y in A. Then $x^{-1}y$ and $y^{-1}x$ are in W if x is in A. By E_4 (Section 20.6) we may choose a V in yWy^{-1}. Then $x^{-1}Vx \subseteq (x^{-1}y)W(y^{-1}x) \subseteq WWW \subseteq U$ for all x in A.

By the *product* of two filters $\mathfrak{F}$ and $\mathfrak{G}$ we mean the filter generated by the products AB (A in $\mathfrak{F}$, B in $\mathfrak{G}$). The product is associative:

$$\mathfrak{F}\cdot\mathfrak{G}\mathfrak{H} = \mathfrak{F}\mathfrak{G}\cdot\mathfrak{H}. \tag{20.8}$$

Indeed, both sides of (20.8) are equal to the filter generated by the products ABC (A in $\mathfrak{F}$, B in $\mathfrak{G}$, and C in $\mathfrak{H}$).

I: *If $\mathfrak{F}$ and $\mathfrak{G}$ are Cauchy filters, then $\mathfrak{F}\mathfrak{G}$ is also a Cauchy filter.*

Proof: We have

$$(xy)^{-1}x'y' = y^{-1}(x^{-1}x')y\cdot(y^{-1}y'). \tag{20.9}$$

If x and x' are contained in an appropriately chosen set A of $\mathfrak{F}$ and similarly y and y' are in an appropriately chosen set B of $\mathfrak{G}$, then $x^{-1}x'$ and $y^{-1}y'$ lie in arbitrarily small neighborhoods of e. By the lemma, $y^{-1}(x^{-1}x')y$ then lies in an arbitrarily small neighborhood U. The product (20.9) therefore lies in an arbitrarily small neighborhood of e, and this completes the proof.

II: *If $\mathfrak{F}$ is a Cauchy filter and $\mathfrak{G}$ converges to e, then $\mathfrak{F}^{-1}\mathfrak{G}\mathfrak{F}$ converges to e.*

Proof: If x and x' are contained in a set A of the filter $\mathfrak{F}$ and if y is contained in a set $\mathfrak{B}$ of the filter $\mathfrak{G}$, and thus by appropriate choice of B in an arbitrarily small neighborhood V of e, then

$$\begin{aligned} x^{-1}yx' &= x^{-1}yx\cdot x^{-1}x' \\ &\subseteq x^{-1}Vx\cdot U. \end{aligned} \tag{20.10}$$

By appropriate choice of V and A, $x^{-1}VX$ is contained in an arbitrarily small neighborhood U of e by the lemma. The product (20.10) therefore lies in $U \cdot U$ and thus in an arbitrarily small neighborhood of e.

Exercise

20.18. The sets A which contain the element e form a Cauchy filter $\mathfrak{E}$. This filter is the identity element of filter multiplication:

$$\mathfrak{E}\mathfrak{F} = \mathfrak{F}\mathfrak{E} = \mathfrak{F} \qquad \text{for all } \mathfrak{F}.$$

As in Section 20.9, we must now introduce a *group-completion axiom*, which is a stronger version of TG_3.

GC: *If $\mathfrak{F}$ is a Cauchy filter, then $\mathfrak{F}^{-1}$ is also a Cauchy filter.*

This means that if the products $x^{-1}y$ (x and y in $A \in \mathfrak{F}$) are contained in arbitrarily small neighborhoods of e, then the products yx^{-1} are also contained in arbitrarily small neighborhoods of e. In Abelian groups this is trivial.

The Cauchy filters form a *semigroup* under multiplication in the sense that the first three group axioms of Section 2.1 hold. Axiom 4 does not hold in general. It is true that for every Cauchy filter there is an inverse Cauchy filter $\mathfrak{F}^{-1}$, but the product $\mathfrak{F}^{-1}\mathfrak{F}$ in most cases is not equal to $\mathfrak{E}$.

Let $\hat{G}$ denote the semigroup of Cauchy filters in G. We make $\hat{G}$ into a topological space by defining the *basis neighborhoods* $\hat{U}$ of the identity element $\mathfrak{E}$; to each neighborhood U of e in G there corresponds a basis neighborhood $\hat{U}$ defined as follows: $\hat{U}$ consists of all filters $\mathfrak{F}$ which contain at least one set $A \subseteq U$.

The basis neighborhoods $\hat{U}$ so defined satisfy conditions E_1 through E_5 (Section 20.6). For $E_1 - E_3$ and E_5 this is trivial; to prove E_4 the lemma must be used.

Exercises

20.19. Prove E_4.

20.20. The filters converging to e are precisely those which lie in all neighborhoods $\hat{U}$.

Using the neighborhoods $\hat{U}$, we can form translated neighborhoods $\mathfrak{F}\hat{U}$ as in Section 20.6. Thus $\hat{G}$ becomes a topological space. Product formation $\mathfrak{F}\mathfrak{G}$ and formation of inverses $\mathfrak{F}^{-1}$ are continuous in the sense of this topology; $\hat{G}$ may therefore be designated a *topological semigroup*. The separation axiom T_1 is not satisfied in general (see Exercise 20.20).

The filters converging to e form a subsemigroup $\hat{N}$ in $\hat{G}$. Because of II, $\hat{N}$ is a normal subsemigroup in the sense that

$$\mathfrak{F}^{-1}\hat{N}\mathfrak{F} \subseteq \hat{N}$$

for all $\mathfrak{F}$.

These properties of $\hat{G}$ and $\hat{N}$, together with the obvious property

$$\mathfrak{F}^{-1}\mathfrak{F} \in \hat{N},$$

suffice for the formation of the factor group

$$\hat{G}/\hat{N} = \tilde{G}.$$

We need only examine the construction of the factor group in Section 2.5 to see that the property $a^{-1}a = e$ (in our case $\mathfrak{F}^{-1}\mathfrak{F} = \mathfrak{E}$) was not used, but rather only $\mathfrak{F}^{-1}\mathfrak{F} \in \hat{N}$. The factor group is a genuine group: each element has a genuine inverse. As in Section 20.7, we see that the factor group $\hat{G}/\hat{N}$ is a T-group. There is a continuous homomorphism of $\hat{G}$ onto $\hat{G}/\hat{N} = \tilde{G}$.

According to Exercise 20.20, $\hat{N}$ consists precisely of those filters $\mathfrak{F}$ which are not separable from the identity element $\mathfrak{E}$ of the group $\hat{G}$. Here $\hat{N}$ is closed by Section 20.6, and hence $\tilde{G} = \hat{G}/\hat{N}$ is a T_1-group.

Every element x of G defines a filter $\mathfrak{F}_x$ consisting of the sets A which contain x.

The filter contains the set $\{x\}$ and is therefore a Cauchy filter. Thus, to each x in G there corresponds an $\hat{x} = \mathfrak{F}_x$ in $\hat{G}$. The correspondence $x \to \hat{x}$ is continuous, and products go into products. The homomorphism $\hat{G} \to \tilde{G}$ takes the element $\hat{x}$ into an image $\tilde{x}$. There is thus a chain of continuous homomorphisms:

$$x \to \hat{x} \to \tilde{x}. \tag{20.11}$$

If two elements x and y are not separable in G, then they have the same image $\tilde{x}$ in $\tilde{G}$, and conversely.

Henceforth G shall be a T_1-group. Any two distinct elements x and y can then be separated, and the mapping $x \to \tilde{x}$ is therefore one-to-one. *Thus G is imbedded in $\tilde{G}$.*

Now let $\mathfrak{B}$ be a Cauchy filter basis in G. Since G is imbedded in $\tilde{G}$, we may interpret $\mathfrak{B}$ as a filter basis in $\tilde{G}$. On the other hand, $\mathfrak{B}$ generates a Cauchy filter $\mathfrak{F}$ in G, and to this filter there corresponds an element $\tilde{a}$ of $\tilde{G}$ under the homomorphism $\hat{G} \to \tilde{G}$. We now assert the following.

III: *The filter basis $\mathfrak{B}$ converges to $\tilde{a}$.*

Proof: By the definition of a Cauchy filter basis, for every neighborhood U of e there exists a set A in $\mathfrak{B}$ such that

$$y^{-1}x \in U \qquad \text{for all } x \text{ and } y \text{ in } A.$$

This may also be written

$$A^{-1}x \subseteq U \qquad \text{for all} \quad x \in A.$$

The set A^{-1} belongs to the filter $\mathfrak{F}^{-1}$ and the set $\{x\}$ to the filter $\hat{x}$; the product $\mathfrak{F}^{-1}\hat{x}$ therefore contains the set $A^{-1}\{x\} \subseteq U$. This means, according to the definition of neighborhoods $\hat{U}$ in $\hat{G}$, that

$$\mathfrak{F}^{-1}\hat{x} \in \hat{U} \qquad \text{for all} \quad x \in A.$$

Passing by means of the continuous homomorphism from $\hat{G}$ to $\tilde{G}$, we find

$$\tilde{a}^{-1}\tilde{x} \in \tilde{U},$$

and hence

$$\tilde{x} \in \tilde{a}\tilde{U}.$$

We have identified $\tilde{x}$ with x; it follows therefore that

$$x \in \tilde{a}\tilde{U} \quad \text{for all} \quad x \in A,$$

that is,

$$A \subseteq \tilde{a}\tilde{U}.$$

Thus, in the filter basis $\mathfrak{B}$ there exist sets A which are contained in arbitrarily small neighborhoods $\tilde{a}\tilde{U}$ of $\tilde{a}$; this means that $\mathfrak{B}$ converges to $\tilde{a}$ and the proof of III is herewith complete.

Since there is a nonempty set A in every neighborhood of $\tilde{a}$, points of G lie in every neighborhood of $\tilde{a}$. This means: *G is dense in $\tilde{G}$.*

From this, III, and the last theorem of Section 20.10, we obtain the following.

IV: *$\tilde{G}$ is strongly complete.*

Exercise

20.21. If the first countability axiom holds in G, then it also holds in $\tilde{G}$. Each element of $\tilde{G}$ is then the limit of a sequence $\{x_\nu\}$ in G, and the weak completion of G according to Section 20.9 produces the same result as the strong completion according to Section 20.11.

20.12 TOPOLOGICAL VECTOR SPACES

A *T-module M* is an additive Abelian T-group. By Section 20.6 the topology in M is defined by a system of neighborhoods U of zero which satisfy conditions 1, 2, and 3 (end of Section 20.6).

The concepts of Sections 20.9 and 20.11 carry over to additive T-groups. A sequence $\{x_\nu\}$ is called a fundamental sequence if the differences $x_\mu - x_\nu$ belong to any neighborhood V of zero for sufficiently large μ and ν. A set A is called *small of order V* if the differences $y - x$ ($x \in A$, $y \in A$) all lie in V. A filter which contains arbitrarily small sets is called a *Cauchy filter*. The module M is said to be *strongly complete* or simply *complete* if every Cauchy filter in M converges.

Since by Section 20.11 no completion axiom is required for commutative groups, any T_1-module M can be imbedded in a complete T_1-module $\tilde{M}$.

Now let there be an *operator domain* Ω for M with the property

$$\gamma(a+b) = \gamma a + \gamma b \tag{20.12}$$

for every operator γ. We assume that γx is a continuous function of x. For this it is necessary and sufficient that for every U there exist a V with the property

$$\gamma V \subseteq U.$$

If a filter $\mathfrak{F}$ contains arbitrarily small sets, then $\gamma\mathfrak{F}$ contains arbitrarily small sets γA; that is, $\gamma\mathfrak{F}$ is again a Cauchy filter. The completion theory of Section 20.11 may therefore be extended immediately to T_1-modules with operators; the complete module $\tilde{M}$ again has the same operator domain Ω.

It is sometimes expedient to write $a\gamma$ instead of γa. Then Ω is called a *right operator domain* and M a *right Ω-module.* In place of (20.12) we then have

$$(a+b)\gamma = a\gamma + b\gamma. \tag{20.13}$$

If Ω is a ring, then we require, in addition to (20.13), that

$$a(\beta+\gamma) = a\beta + a\gamma \tag{20.14}$$

$$a(\beta\gamma) = (a\beta)\gamma. \tag{20.15}$$

These relations also are preserved in going over to the complete module $\tilde{M}$.

If Ω is a T-ring, then it is required that the product $x\gamma$ be a continuous function of x and γ. This property is also transferred to $\tilde{M}$ so that $\tilde{M}$ becomes a complete right Ω-module.

If Ω is a skew field and if

$$a \cdot 1 = a \tag{20.16}$$

holds in addition to the composition rules already assumed (1 is here the identity element of Ω), then M is called a *vector space* over Ω. If Ω is a skew T-field, then continuity of $x\gamma$ as a function of x and γ is also required.

A simple example of a topological vector space over a skew T-field Ω is the *canonical n-dimensional vector space* Ω^n, which is defined as the set of all ordered sequences of n elements $(\beta_1, \ldots, \beta_n)$ of Ω. The multiplication of vectors by elements of Ω is defined by

$$(\beta_1, \ldots, \beta_n)\gamma = (\beta_1\gamma, \ldots, \beta_n\gamma).$$

A basis neighborhood U' of the zero vector consists of all vectors whose individual coordinates $\beta_1, \ldots, \beta_n$ all belong to a basis neighborhood U of zero in Ω. The neighborhood axioms are satisfied, and addition and multiplication are continuous.

If Ω is complete, then Ω^n is also complete.

Proof: A set A of vectors $(\beta_1, \ldots, \beta_n)$ is small of order U' if and only if the set of β_i for each i is small of order U. We call the set of the β_i the i-component of the set A and denote it by A_i. If now a Cauchy filter $\mathfrak{F}$ of sets A is given, then, for each i, A_i forms a Cauchy filter in Ω. If Ω is complete, all these Cauchy filters have limits γ_i in Ω. There is then for each U a set $A^{(1)}$ whose 1-component lies in $\gamma_1 + U$, a set $A^{(2)}$ whose 2-component lies in $\gamma_2 + U$, and so on up to $A^{(n)}$. The intersection

$$A = A^{(1)} \cap A^{(2)} \cap \cdots \cap A^{(n)}$$

then lies in $(\gamma_1, \ldots, \gamma_n) + U'$. The filter $\mathfrak{F}$ thus converges to the limit $(\gamma_1, \ldots, \gamma_n)$.

20.13 RING COMPLETION

A T_1-ring R is an additive T_1-group and can therefore be extended to a strongly complete group

$$\tilde{R} = \hat{R}/\hat{N}.$$

Here $\hat{R}$ is the additive group of Cauchy filters, and $\hat{N}$ is the normal subgroup which consists of filters having limit zero.

We wish to define a multiplication in $\hat{R}$ which makes $\hat{R}$ a ring and $\hat{N}$ a two-sided ideal in this ring and such that $\tilde{R} = \hat{R}/\hat{N}$ becomes a complete T-ring.

The neighborhoods of zero will again be denoted by $U, V, W, \ldots$. We first prove the following.

Lemma: *If $\mathfrak{F}$ is a Cauchy filter, then for every U there is a W and a set A in $\mathfrak{F}$ such that*

$$AW \subseteqq U \quad \text{and} \quad WA \subseteqq U.$$

Proof: There exists a U' such that

$$U' + U' \subseteqq U.$$

There exists a V such that

$$V \cdot V \subseteqq U'.$$

There is an A in $\mathfrak{F}$ with the property

$$x - y \in V \qquad \text{for all } x \text{ and } y \text{ in } A.$$

If y in A is fixed, then there is a $W \subseteqq V$ such that

$$yW \subseteqq U' \quad \text{and} \quad Wy \subseteqq U'.$$

Then for every x in A and z in W,

$$xz = (x-y)z + yz \in VV + yW \subseteqq U' + U' \subseteqq U,$$

and hence $AW \subseteqq U$. Then $WA \subseteqq U$ is proved in precisely the same way.

From the lemma we obtain the following corollaries.

I: *If $\mathfrak{F}$ and $\mathfrak{G}$ are Cauchy filters, then $\mathfrak{F}\mathfrak{G}$ is also a Cauchy filter.*

Proof: We have

$$xy - x'y' = x(y-y') + (x-x')y'. \tag{20.17}$$

Given U, we choose V so that

$$V + V \subseteqq U.$$

By the lemma, there is an A in $\mathfrak{F}$, a B in $\mathfrak{G}$, and a W such that

$$WB \subseteqq V \quad \text{and} \quad AW \subseteqq V.$$

If now xy and $x'y'$ are any two elements of AB (x and x' in A, y and y' in B), then it follows from (20.17) that

$$xy - x'y' \in V + V \subseteq U.$$

Therefore $\mathfrak{F}\mathfrak{G}$ is a Cauchy filter.

II: *If $\mathfrak{F}$ is a Cauchy filter and $\mathfrak{G}$ converges to zero, then $\mathfrak{F}\mathfrak{G}$ and $\mathfrak{G}\mathfrak{F}$ converge to zero.*

The proof of II follows immediately from the lemma.

By I the Cauchy filters form a ring $\hat{R}$. By II the filters converging to zero form a two-sided ideal $\hat{N}$ in this ring. The factor module

$$\tilde{R} = \hat{R}/\hat{N}$$

is therefore not only a complete T-module, but also a ring.

We now prove the continuity of multiplication in $\hat{R}$.

III: *If $\mathfrak{F}$ and $\mathfrak{G}$ are Cauchy filters and if $\hat{U}$ is a basis neighborhood of zero in $\hat{R}$ (as defined in Section* 20.11), *then there exist basis neighborhoods $\hat{V}$ and $\hat{W}$ such that*

$$(\mathfrak{F} + \hat{V})(\mathfrak{G} + \hat{W}) \subseteq \mathfrak{F}\mathfrak{G} + \hat{U}. \tag{20.18}$$

Proof: For any x, y, v, w in R,

$$(x+v)(y+w) = xy + xw + vy + vw. \tag{20.19}$$

Now let a neighborhood U of zero be given in R. We choose U' so that $U' + U' + U' \subseteq U$; we then choose, in accordance with the lemma, an A in $\mathfrak{F}$, a B in $\mathfrak{G}$, and neighborhoods V' and W' so that

$$AV' \subseteq U' \quad \text{and} \quad W'B \subseteq U'.$$

Finally, we choose $V \subseteq V'$ and $W \subseteq W'$ so that $VW \subseteq U'$. It then follows from (20.19) that for $x \in A$, $y \in B$, $v \in V$, and $w \in W$,

$$(x+v)(y+w) \in xy + U' + U' + U' \subseteq xy + U,$$

and hence

$$(A+V)(B+W) \subseteq AB + U.$$

Then III is herewith proved.

Thus $\hat{R}$ is a T-ring. Therefore $\tilde{R}$ is also a T-ring; it is even a T_1-ring, since the first separation axiom T_1 is satisfied in $\tilde{R}$.

Now $\tilde{R}$ is complete by Section 20.11. Hence *every T_1-ring can be imbedded in a complete T_1-ring.*

20.14 COMPLETION OF SKEW FIELDS

Let S be a skew T-field which satisfies the first separation axiom. By Section 20.13, S can be imbedded in a complete T-ring $\tilde{S} = \hat{S}/\hat{N}$. However $\tilde{S}$ is not necessarily a skew T-field, since the inverse of an element $w \neq 0$ of $\tilde{S}$ need not exist; even if it exists it need not depend continuously on w.

The following *completion axiom for skew fields* is necessary and sufficient in order that S admit an imbedding in a complete skew T-field.

SF: *If $\mathfrak{F}$ is a Cauchy filter in S which does not converge to zero, then $\mathfrak{F}^{-1}$ is a basis for a Cauchy filter.*

We shall show first of all that SF is necessary when S can be imbedded in a complete skew T-field. A Cauchy filter $\mathfrak{F}$ in S under the imbedding gives rise to a basis for a Cauchy filter which has a limit $a \neq 0$ in $\tilde{S}$. The inverse filter basis $\mathfrak{F}^{-1}$ then converges to a^{-1}, since the mapping $x \to x^{-1}$ is continuous. Therefore $\mathfrak{F}^{-1}$ is a basis for a Cauchy filter.

Now suppose that SF is satisfied. We wish to show that S is a complete skew T-field.

We first show that the previous axiom TS (Section 20.8) follows from SF.

Let U be a neighborhood of zero in S. We must show that there exists a neighborhood V such that

$$(1+V)^{-1} \subseteq 1+U.$$

The neighborhoods $1+V$ of the identity form a Cauchy filter $\mathfrak{F}$ which converges to one and hence does not converge to zero. By SF, $\mathfrak{F}^{-1} = \mathfrak{B}$ is then a basis for a Cauchy filter. The sets of $\mathfrak{B}$ are

$$A = (1+V)^{-1},$$

where zero is of course to be omitted from $1+V$. For every $y \neq 0$ in $1+V$ we have

$$1-y^{-1} = y^{-1}(y-1) \in AV. \tag{20.20}$$

By the lemma of Section 20.13, for every U there exists a W and an A' in $\mathfrak{B}$ such that

$$A'W \subseteq -U.$$

This A' has the form $(1+V')^{-1}$. We now choose V in the intersection $V' \cap W$. Then $A \subseteq A'$ and $V \subseteq W$; hence

$$AV \subseteq A'W \subseteq -U$$

$$1-y^{-1} \in -U$$

$$y^{-1}-1 \in U$$

$$y^{-1} \in 1+U.$$

This holds for all $y \neq 0$ in $1+V$, and we thus have

$$(1+V)^{-1} \subseteq 1+U, \tag{20.21}$$

as asserted.

We can now show that each element $a \neq 0$ of $\tilde{S}$ has an inverse. The element a is the limit of a Cauchy filter $\mathfrak{F}$ in S. By SF, $\mathfrak{F}^{-1}$ is then a basis for a Cauchy filter which in $\tilde{S}$ thus has a limit b. The product $\mathfrak{F}^{-1}\mathfrak{F}$ has the limit ba on the one hand and the limit 1 on the other; hence $ba = 1$.

To show that $\tilde{S}$ is a skew field, by Section 20.8 we need only show that for every basis neighborhood $\tilde{U}$ of zero there exists a basis neighborhood $\tilde{V}$ of zero such that

$$(1+\tilde{V})^{-1} \subseteqq 1+\tilde{U}.$$

The basis neighborhoods $\tilde{U}$ and $\tilde{V}$ arise from basis neighborhoods $\hat{U}$ and $\hat{V}$ in $\hat{S}$ under the homomorphism $\hat{S}\rightarrow\tilde{S}$. It therefore suffices to show that

$$(1+\hat{V})^{-1} \subseteqq 1+\hat{U}.$$

But this follows immediately from (20.21) if it is recalled how $\hat{U}$ and $\hat{V}$ are related to U and V.

We now combine these results.

If SF is satisfied, then $\tilde{S}$ is a skew T-field. Now SF is necessary and sufficient for the imbedding of S in a complete skew T-field.[2]

[2]For further studies of topological skew fields see:

I. Kaplanski, "Topological Methods in Valuation Theory," *Duke Math. J.*, **14**, 527 (1947).

H. J. Kowalsky and H. Dürbaum, "Arithmetische Kennzeichnung von Körpertopologien," *J. Reine u. Angew. Math.*, **191**, 135 (1953).

H. J. Kowalsky, "Zur Topologischen Kennzeichnung von Körpern," *Math. Nachr.*, **9**, 261 (1953).

L. S. Pontrjagin, *Topologische Gruppen*, Teubner, Leipzig, 1957.

INDEX